FOOD QUALITY AND NUTRITION
Research Priorities for Thermal Processing

Proceedings of COST Seminar held in Dublin, 22–25 November, 1977. The Seminar was held under the auspices of COST (European Cooperation in Scientific and Technical Research).

COST Member States are: Austria, Belgium, Denmark, Finland, France, Greece, Italy, Ireland, Luxembourg, Netherlands, Norway, Portugal, Spain, Sweden, Switzerland, Turkey, United Kingdom, Federal Republic of Germany and Yugoslavia.

FOOD QUALITY AND NUTRITION

Research Priorities for Thermal Processing

Edited by

W. K. DOWNEY
D.Sc., Ph.D.

*National Science Council,
Dublin, Ireland*

APPLIED SCIENCE PUBLISHERS LTD
LONDON

APPLIED SCIENCE PUBLISHERS LTD
RIPPLE ROAD, BARKING, ESSEX, ENGLAND

British Library Cataloguing in Publication Data

Downey, William Kevin
    Food quality and nutrition.
    1. Food—Preservation—Research—Congresses
    2. Food—Thermal properties—Research—Congresses
    I. Title
    664′.02842   TP371.2

ISBN 0-85334-803-0

WITH 80 TABLES AND 72 ILLUSTRATIONS

© COMMISSION OF THE EUROPEAN COMMUNITIES

Publication arrangements by: Commission of the European Communities, Directorate-General for Scientific and Technical Information and Information Management, Luxembourg

All rights reserved. No part of this publication may be reproduced, stored in a retrieval system, or transmitted in any form or by any means, electronic, mechanical, photocopying, recording, or otherwise, without the prior written permission of the publishers, Applied Science Publishers Ltd, Ripple Road, Barking, Essex, England

Printed in Great Britain by Galliard (Printers) Ltd, Great Yarmouth

CONTENTS

	Page No.
Foreword	xi
Introduction	xvii

PLENARY SESSIONS

OPENING SESSION

OPENING ADDRESS
Desmond O'Malley, TD 1

KEYNOTE ADDRESS
E. von Sydow 5

SESSION 1 CURRENT FOCUS OF R & D ON COMMODITY AREAS

QUALITY AND NUTRITIVE ASPECTS OF MEAT PROCESSING
AND DISTRIBUTION
M. Jul 17

MILK PROCESSING - R & D FOCUS AND NEEDS
W.K. Downey and P.F. Fox 41

R & D NEEDS FOR FISH AND FISH PRODUCTS
J.J. Connell 107

R & D REQUIREMENTS ON SOME ASPECTS OF THE
QUALITY AND NUTRITIVE VALUE OF THERMALLY
PROCESSED CEREALS
Christiane Mercier and J. Delort-Laval .. 119

NUTRITIONAL AND QUALITY ATTRIBUTES INVOLVED IN
THERMAL PROCESSING OF FRUIT AND VEGETABLES
C. Cantarelli 137

SESSION 2 PASTEURISATION/BLANCHING

PASTEURISATION OF MEAT, FISH AND CONVENIENCE
FOOD PRODUCTS
T. Ohlsson 163

PASTEURISATION AND THERMISATION OF MILK AND
BLANCHING OF FRUIT AND VEGETABLES
J. Foley and J. Buckley 191

SESSION 3 STERILISATION

MODERN HEAT PRESERVATION OF CANNED MEAT AND
MEAT PRODUCTS
F. Wirth 219

QUALITY ASPECTS OF THERMAL STERILISATION
PROCESSES
H. Burton 239

SESSION 4   DEHYDRATION

   THE STATE OF THE ART OF FOOD DEHYDRATION
   H.A. Leniger and S. Bruin 265

   QUALITY AND NUTRITIONAL ASPECTS OF FOOD
   DEHYDRATION
   F. Escher and B. Blanc 297

SESSION 5   CHILLING/FREEZING/THAWING

   INFLUENCE OF CHILLING, FREEZING AND THAWING
   ON FISH QUALITY - RECENT ASPECTS
   W. Vyncke 325

   EFFECTS OF FREEZING, STORAGE AND DISTRIBUTION
   ON QUALITY AND NUTRITIVE ATTRIBUTES OF FOODS,
   IN PARTICULAR OF FRUIT AND VEGETABLES
   J. Munoz-Delgado 353

SESSION 6   COOKING (DOMESTIC AND INSTITUTIONAL)

   EFFECTS OF DOMESTIC AND LARGE SCALE COOKING ON
   THE QUALITY AND NUTRITIVE VALUE OF VEGETABLES
   AND FRUITS
   Rosmarie Zacharias 387

   THE EFFECT OF HEAT ON PROTEIN RICH FOODS
   A.E. Bender 411

SEMINAR DINNER ADDRESS

   NUTRITIONAL NEEDS AND RESEARCH PRIORITIES:
   REPORT OF THE U.S. NATIONAL ACADEMY OF SCIENCES
   J. Bernstein 429

COMMODITY STUDY PANELS

NUTRITION COORDINATION GROUP

   NUTRITIONAL PRIORITIES, WITH PARTICULAR
   REFERENCE TO FROZEN FOODS
   D.H. Buss 457

   IMPORTANCE OF MEAT PRODUCTS FOR THE SUPPLY
   OF MICRONUTRIENTS IN HUMAN NUTRITION
   G. Brubacher 463

   THE NUTRITIONAL VALUE OF PROTEIN IN
   BREAKFAST CEREALS
   A. Dahlqvist, N.G. Asp and G. Jonsson 469

   METHODS FOR DETECTION OF PROTEIN QUALITY
   IN TRADITIONAL AND NEW PROTEIN FOODS
   G. Tomassi 471

EFFECTS OF THERMAL PROCESSING ON THE
NUTRITIVE VALUE OF MILK AND MILK PRODUCTS
J.W.G. Porter 477

THE ENZYMATIC ULTRAFILTRATE DIGEST (EUD) AMINO
ACID INDEX FOR PROTEIN QUALITY EVALUATION
F. Fidanza 481

BAKING CONDITIONS AND NUTRITIVE QUALITY OF
DIFFERENT BREAD VARIETIES
W. Seibel 487

DAIRY PANEL

MILK TECHNOLOGY WITH REFERENCE TO HUMAN
NUTRITION
B. Blanc 493

FACTORS AFFECTING THE HEAT STABILITY OF MILK AND
OF CONCENTRATED MILK AT STERILISATION TEMPERATURES
J.A.F. Rook 501

SOME AREAS FOR FUTURE RESEARCH INTO HEAT TREATMENT
AND LONG LIFE PROPERTIES OF MILK, MILK POWDER AND
YOGHURT
M. Naudts 505

SOME ASPECTS OF THE EFFECT OF HEAT-TREATMENT ON
INDUSTRIALLY MANUFACTURED CULTURED MILKS
R. Negri 511

SOME ASPECTS OF DAIRY CHEMISTRY REQUIRING
FURTHER R & D IN RELATION TO THERMAL PROCESSING
G.C. Cheeseman 517

OPTIMISATION OF QUALITY AND NUTRITIVE PROPERTIES
OF STERILISED DAIRY DRINKS FOR USE IN INFANT
NUTRITION AND INVALID DIETS
M. Caric 521

ENERGY SAVINGS IN EVAPORATION AND DRYING OF
MILK AND MILK PRODUCTS
H.G. Kessler 525

SOME ASPECTS ON THE EFFECTS OF THERMAL
PROCESSING ON QUALITY OF DAIRY PRODUCTS
M. Heikonen and P. Linko 531

MEAT PANEL

DSC STUDIES ON THE EFFECT OF THERMAL TREATMENT
ON MEAT PROTEIN QUALITY
O. Kvaale and H. Martens 537

FOOD DETERIORATING ENZYMES SURVIVING
HTST-STERILISATION
K. Östlund 543

REFRIGERATION STUDIES AT THE UK AGRICULTURAL
RESEARCH COUNCIL MEAT RESEARCH INSTITUTE
C. Bailey 547

HURDLE EFFECT AND ENERGY SAVING
L. Leistner 553

THE FUTURE USE OF VEGETABLE PROTEINS IN
MEAT PRODUCTS
A.W. Holmes 559

CEREALS PANEL

NUTRITIONAL ASPECTS OF PROCESSING WITH
SPECIAL REFERENCE TO CEREALS
D.A.T. Southgate 565

THERMIC PROCESSES AND NUTRITIVE VALUE
OF CEREALS
G. Fabriani 569

EFFECT OF THERMAL PROCESSING ON CEREAL
BASED FOOD SYSTEMS
J. Olkku and P. Linko 575

FREEZING PROBLEMS WITH GERMAN ROLLS,
BREAD VARIETIES AND OTHER BAKED GOODS
W. Seibel 581

INVESTIGATIONS INTO THE INFLUENCE OF THERMAL
PROCESSING ON CEREALS
Daniela Schlettwein-Gsell 585

POSSIBLE APPROACHES FOR REMOVING THE TOXICITY
OF CEREAL GLUTEN TO PATIENTS WITH COELIAC
DISEASE (GLUTEN ENTEROPATHY)
P.F. Fottrell, J. Phelan, F. Stevens,
B. Nicholl and C.F. McCarthy 589

FRUIT AND VEGETABLES PANEL

INDUSTRIAL R & D PRIORITIES IN FRUIT AND
VEGETABLES
M. Woods 597

SOME EFFECTS OF THERMAL PROCESS CONDITIONS
ON FOOD PROTEINS FROM VEGETABLE ORIGIN
V. Wenner 603

RESEARCH AND DEVELOPMENT PROBLEMS FOR THE
FRUIT AND VEGETABLE PRESERVATION INDUSTRY
S.D. Holdsworth 609

COOPERATIVE RESEARCH REQUIREMENTS ON FRUIT
AND VEGETABLES
J. Solms and F. Escher 615

RETENTION OF ASCORBIC ACID, THIAMINE AND
FOLIC ACID IN VEGETABLES IN MASS FEEDING
- THE ROLE OF SYSTEMS AND EQUIPMENT DESIGN
J. Ryley and G. Glew 617

THE QUALITY OF FRUIT AND VEGETABLE JUICES,
INFLUENCE OF PROCESSING METHODS ON FINAL
PRODUCT QUALITY
K. Gierschner 621

FISH PANEL

   SOME PROBLEMS IN FREEZING AND FROZEN
   STORAGE OF FISH
   G. Löndahl 625

   TWO FROZEN FISH PRODUCTS REQUIRING SPECIAL CARE
   F. Bramsnaes 629

   PROCESSING OF UNDER- AND NON-UTILISED FISH
   SPECIES FOR HUMAN CONSUMPTION - NEEDS FOR R & D
   T. Strøm 635

   NUTRITIONAL ASPECTS OF FROZEN FISH
   G. Varela, Olga Moreiras-Varela and
   M. de la Higuera 639

   PROBLEMS ASSOCIATED WITH THERMAL PROCESSING
   OF FISHERY PRODUCTS
   J.C. O'Connor 647

   RESEARCH PROPOSALS ON FISH TECHNOLOGY
   (TECHNOLOGICAL, QUALITATIVE AND NUTRITIONAL ASPECTS)
   Report prepared by the Western European Fish
   Technologists Association (WEFTA) Presented by:
   W. Vyncke 651

FINAL SESSION - SEMINAR CONCLUDING REPORTS

MEAT STUDY PANEL

   MEAT PANEL FINAL REPORT
   B. Krol and J.V. McLoughlin 667

DAIRY STUDY PANEL

   DAIRY PANEL FINAL REPORT
   J. Moore and P.F. Fox 673

FISH STUDY PANEL

   FISH PANEL FINAL REPORT
   A. Hansen and J. Somers 677

CEREALS STUDY PANEL

   CEREALS PANEL FINAL REPORT
   P. Linko and E. Markham 683

FRUIT AND VEGETABLES STUDY PANEL

   FRUIT AND VEGETABLES PANEL FINAL REPORT
   W. Spiess and A. Hunter 689

NUTRITION COORDINATION GROUP

   NUTRITION COORDINATION GROUP FINAL REPORT
   G. Varela, J.P. Kevany and F. Cremin . . 695

LIST OF PARTICIPANTS 699

FOREWORD

COST is the inter-Governmental framework proposed by the European Community and adopted by 19 European States in 1971 to promote co-operation in the field of science and technology. COST projects have been developed in the fields of informatics, telecommunications, transport, oceanography, environment, metallurgy, and meteorology.

In December 1974, Sweden (Professor E. von Sydow) submitted outline proposals for co-operation in the field of industrial food technology and, in April 1975, the COST Committee of Senior Officials established an ad hoc Food Technology Committee to follow up these proposals. This committee first developed a project concerned with the 'Physical Aspects of Food' and it is hoped that the project will commence during 1978.

With regard to projects concerned with the 'Quality and Nutritive Value of Food', the Committee accepted the recommendations of a Working Group chaired by Professor M. Jul (Denmark) to limit co-operation initially to the theme 'Thermal Treatment of Foods with priority to Preservation and Freezing'.

Also, as suggested in the Working Group's report, and in response to a recommendation of the Food Committee, it was agreed to hold a COST Seminar on 'Food Quality and Nutrition - Thermal Processing'.

The objectives of the Seminar were to familiarise research directors and those managing and co-ordinating R & D with the proposed COST projects in this area, and to define specific aspects requiring further research and development. The seminar programme, as further outlined in the Introduction, was designed to allow a consideration of the inter-related 'commodity' and 'processing' dimensions of this area.

COST Member States

Austria
Belgium
Denmark
Finland
France
Greece
Italy
Ireland
Luxembourg
Netherlands
Norway
Portugal
Spain
Sweden
Switzerland
Turkey
United Kingdom
West Germany
Yugoslavia

It is a great pleasure for me, on behalf of the Committee of Senior Officials, to thank the Food Technology Committee and its Working Group for their invaluable assistance in the preparation of the Seminar programme and to express our appreciation to the many scientists and technologists from the COST States whose shared knowledge and experience contributed to a profitable and successful occasion. We are, of course, especially grateful to the authors of the plenary papers at the Seminar and to the Chairmen, Rapporteurs and Co-ordinators of the Commodity Study Panels and of the Nutrition Coordination Group who prepared the Seminar Concluding Reports.

It is also a great pleasure to acknowledge the assistance of the General Secretariat of the Council of the European Communities and the very important contribution of the Commission. Particular mention, too, should be made of Dr. Nada Rajcan who was seconded from Yugoslavia to assist COST in this area.

Finally our special thanks are due to the Minister and to all those in Ireland who have assisted in the organisation of the Seminar and, in particular, to Dr. W.K. Downey for his major contribution and to Mr. A. Cotter who assisted in editing the Seminar Proceedings.

Dr. S. Nielsen
Chairman
COST Senior Officials

Official Seminar Opening by Mr. D. O'Malley T.D. Minister for Industry, Commerce and Energy (Ireland)

View of the auditorium

View of the auditorium

INTRODUCTION

Research and development requirements on Thermal Food Processing and its influences on Food Quality and Nutrition are outlined in the following contributions presented at the COST Industrial Food Technology Seminar held in Dublin, Ireland in November 1977.

The combined influences of the original characteristics of the major food commodities and important thermal processes on the quality and nutritive attributes of foods are examined, as well as pre- and post-manufacturing changes during storage and commercial distribution.

The seminar programme had three main components:

- Plenary session papers on specific food commodities and thermal processes
- Brief communications to Commodity Study Panels and the Nutrition Co-ordination Group
- Seminar Concluding Reports

<u>Plenary Papers</u> To set the scene for the main seminar theme on thermal processing, the initial five plenary papers emphasise the commodity aspects of thermal processing. The current focus of international R & D on the quality and nutritional parameters of <u>Meat/Poultry</u>, <u>Dairy</u>, <u>Fish</u>, <u>Cereals</u> and <u>Fruit/Vegetables</u> is reviewed with particular attention to the requirements of thermally preserved and frozen foods.

The subsequent plenary papers which constituted the major part of the seminar are devoted to the five important thermal processes of most relevance to the aforementioned five commodities namely <u>Pasteurisation-Blanching</u>, <u>Sterilisation</u>, <u>Dehydration</u>, <u>Freezing-Chilling-Thawing and Cooking</u> both domestic and institutional.

These 'state of the art' reviews consider the commodity requirements for specific thermal processes, the predisposing influences of these commodity characteristics on the quality and nutritive attributes of thermally processed foods as well as post-processing changes during storage and commercial distribution. Attention is given to those aspects of the thermal processes themselves - and to procedures for monitoring and minimising post-manufacture deterioration - which require further R & D. The two papers on each thermal process jointly consider the more important quality and nutritive aspects of the particular process in relation to the most relevant commodities.

Brief Communications - Complementary to the plenary papers, the brief communications presented by invited experts to parallel meetings of the five Commodity Study Panels and the Nutritional Co-ordination Group highlight specific quality, nutritional and safety aspects of the aforementioned commodities and/or relevant thermal processes requiring further R & D.

Seminar Concluding Reports Based on the major observations or conclusions of the Plenary Papers, Brief Communications, and the Guidelines outlined in Appendix 1, these consensus reports summarise those aspects of the five commodities and relevant thermal processing requiring further R & D. The objectives and possible benefits of the proposed research are indicated briefly as well as specific areas in which co-operative European projects may be beneficial. Additional recommendations contained in the seminar reports on the individual commodities, the overall requirements for nutritional R & D both on the food commodities themselves and the influences of thermal processing on their nutritive properties and safety are further elaborated in the Nutritional Co-ordination Group's report.

Further to this World Nutritional Research Priorities based on the recent US report entitled 'World Food and Nutrition Study - The Potential Contribution of Research' - (National Academy of Sciences, Washington DC 1977) are outlined in the address presented at the Seminar Dinner.

The contents and accuracy of the seminar contributions compiled in this publication are the responsibilities of the individual author(s). Mainly because of linguistic interpretation editorial changes were necessary in some papers. In a compilation such as this where rapid publication is an important consideration minor errors and inconsistencies are however inevitable.

Dr. W.K. Downey,* D.Sc, Ph.D, National Science Council, Dublin

*Present address: National Board for Science & Technology,
Shelbourne House, Shelbourne Road, Dublin, Ireland.

APPENDIX 1

GUIDELINES FOR THE PREPARATION OF FINAL REPORTS

Suggested working definition of Food Quality and Nutrition:

The sum of the chemical, biological and physical attributes which determine the safety, sensory appeal and nutritive functions of foods.

General checklist for content.

* Characterisation of commodities from the quality, nutritional and safety viewpoints including their biological micro-structure and influence of these characteristics on the suitability of the commodities for thermal processing.

* Influences of agricultural and aquacultural practices such as animal nutrition/soil fertility, animal/plant genetics and husbandry (intensive production systems, antibiotics, pesticides, detergents polyphosphates) and methods of harvesting/milking/pre-slaughter-holding on the initial quality and nutritive attributes of the raw-materals.

* Influences of post-harvesting/milking/slaughter practices and changes on the suitability of the commodities for thermal processing.

* Process improvement and development including plant modification, process automation, energy conservation and pollution abatement.

* Changes in the quality and nutritional attributes of foods during thermal processing including the influences of various processing parameters such as type of plant, temperatures/pressures/time, stabilising additives, inert gases, packaging etc.

* Evaluation of alternative forms of thermal processes in respect to nutritive loss, conservation of organoleptic attributes, product shelf-life, energy conservation and pollution abatement.

* Post-manufacture product deterioration during storage, transport and commercial distribution including comparison of the relative stability of products manufactured by alternative thermal processes.

* Development of methodology for monitoring changes in the quality and nutritive attributes of thermally processed foods with particular attention to the correlation between subjective visual or organoleptic assessments of quality (taste panels etc) and more objective chemical or physical methods.

* Development of rapid methods including automated instrumentation for the determination of keeping quality and/or shelf life of thermally processed foods leading to the elaboration of commercial quality control procedures and quality standards.

* Product modification and development, including packaging, new or novel products including the technological modification of the original commodities to meet normal or special dietary requirements and/or to enhance the quality and nutritional attributes of the products.

* Waste treatment disposal and utilisation including the up-grading and utilisation of 'waste' biological materials.

* Development of more effective mechanisms for the dissemination of R & D results on food quality, nutrition and safety with the objective of increasing their ability by providing policy makers, legislators, planners and industrial managers with authoritive scientific consensus on specific topics including the scientific data underlying existing and proposed food legislation.

PLENARY SESSION PAPERS

OPENING ADDRESS

Desmond O'Malley TD
Minister for Industry, Commerce and Energy, Ireland

It gives me great pleasure to extend a cordial welcome, on behalf of the Government, to the participants in this Seminar, in particular the distinguished Chairmen, to the officials of COST and the Commission of the European Communities.

We are both pleased and honoured that Dublin was chosen as the venue for this important Seminar on a subject in which we, as a food producing and exporting country, have a particular interest and already some competence which we hope to develop further. Traditionally, agriculture has been the cornerstone of the Irish economy.

With the growth of industrialisation leading to increased demand for processed foods, the food processing industry has developed from a relatively slow beginning through a period of rapid growth from the mid sixties onwards. In the period from 1960 to 1974 volume production in fact doubled.

There are, however, a number of critical problems confronting the Irish food processing industry and these must be tackled if it is to meet future challenges. These problems, many if not all of them common to our partners in COST, call for an increased technological input to the industry which in turn requires greater commitment in terms of money and manpower to research and development in food.

As Minister for Industry I am particularly interested in matters affecting the food processing industry.

I am concerned with raising the technological level of Irish industry and the food sector is one which is already heavily dependent on technology and becoming more so. The scope for increasing the technological input to the food processing industry in Ireland is considerable and there is evidence to suggest that the benefits to be derived from such an input could be very substantial.

There is the burning question of increasing employment opportunities for our school-leavers, graduates and technicians.

I believe that, given our solid agricultural base, there is considerable potential for an enhanced contribution in this regard from the Irish food processing industry.

As Minister concerned with consumer affairs I have yet a further interest insofar as the technological developments in the industry have been paralleled by inreasing awareness and concern on the part of consumers regarding the quality, nutritive values and safety of the foods they buy.

Up to recently industrial food research and development have been concerned mainly with problems of quality, taste, keeping quality etc. Apart, however, from ensuring the safety of foods by minimising contamination there has been little outside incentive for the food industry to give particular attention to the nutritional aspects of food processing or to invest in new developments aimed at optimising the nutritional values of process foods. Accordingly, it would appear that a comprehensive evaluation of the present food supply system would constitute a fruitful field for a co-operative European research programme. The initiative taken by Professor von Sydow of Sweden to have industrial food technology questions considered by COST was, therefore, timely and welcome.

Ideally a study of the total food area is warranted. Clearly however, such an undertaking would be enormous requiring a budget, scientific manpower, experience and facilities outside the scope of a single country. Such an undertaking is, however, within the compass of Europe's food research capabilities. Co-ordination of these capabilities would not only enhance the possibility of providing answers and improvements to the European food supply system but also of contributing more effectively to alleviating the worldwide problems of malnutrition. The underlying objective of the proposed COST Food Technology Programme is, of course, to provide the Scientific Community with the opportunity of undertaking such a co-operative European research programme.

Clearly many factors must be considered in guiding research into productive paths and in matching the improved options that the research makes possible within the context of overall economic development.

In particular research is needed to improve the utilisation of results that are attainable with limited resources. The identification of priorities for research is obviously a prerequisite to improved utilisation of results.

All of these desirable and indeed necessary objectives will not be achieved without greater expenditure of time, effort and funds.

Relative to other European countries this country's expenditure on food research and development is not yet as high as we would wish whether expressed as a percentage of Gross Domestic Product for agriculture plus food industries, or per capita, but it is increasing.

About half of the total is spent in the business enterprise sector, and a quarter in the Government sector including the State research institutes.

The Government's overall contribution however is greater than this because it also supports food R and D in the other sectors through grants administered by the Industrial Development Authority, the National Science Council and the grants to the Higher Education sector.

Scarcity of resources requires that whatever funds are available be spent in the most efficient and effective way possible. Better co-ordination of the national effort will be an essential national prerequisite and the Government has already initiated action in this regard through the recent enactment of legislation to establish a National Board for Science and Technology which will have such co-ordination as one of its major functions.

Collaboration in international research projects is another obvious way of reducing the cost of, and achieving more comprehensive returns from, research and development in this sector and the need for such collaboration gives special point and purpose to the work being undertaken here under the aegis of COST.

I would like to emphasise again how pleased we are to host this important Seminar and to say that I look forward with considerable interest to the outcome of your deliberations.

KEYNOTE ADDRESS

Erik von Sydow
SIK - The Swedish Food Institute
S-400 23 Göteborg, Sweden.

The average consumer spends the largest slice of her budget on food and related products. The food industry is, therefore, one of the largest industries in the western world. Raw materials, such as agricultural products and fish, are, however, not necessarily available where the consumers are. They must be transported and to allow transportation they must very often be processed in industrial operations. This leads to trade, national or international.

International trade with foods basically is of two kinds: firstly, trade with raw materials, such as agricultural products and fish, and secondly, trade with industrially processed food products, such as cheese, deboned meat and frozen peas. The latter sector has increased through the years and different systems of food distribution, legislation, quality evaluation and consumer appeal have been confronted with each other as a result of this increased trade. Although many of the problems arising are of an economical and political nature some important problem areas are to be found within food science and technology. Better understanding of such problems and solutions to the problems will promote trade of industrially processed food products and thus be beneficial to the food industry and allied industries and to the consumers.

COST stands for Committee on Scientific and Technical research. Members of COST are the Community countries and 10 other countries in Europe, altogether 19.

COST is guided by a Committee of Senior Officials and the activities are quite wide spread within the area of technical and otherwise applied research.

Sweden is not a member of the EEC but it is a member of the European Free Trade Association (EFTA) and COST. In July 1974 the Swedish Government presented the idea of having COST projects in the area of industrial food technology. The idea was well received at a Senior Officials meeting and a formal proposal was submitted by Sweden in a letter to the Secretary General of the Council of the European Communities dated December 4, 1974, which is very close to 3 years ago. Three projects were presented:

    A. Physical properties of food.

    B. Quality properties of food.

    C. Nutritional properties of food.

Concerning 'Physical properties of food' it was said in the official letter that 'knowledge of physical properties is needed for the process development and control. The change of rheological properties (ie consistency, texture etc) with temperature and flow velocity, and the change of thermal properties when mixing ingredients are important examples'.

The following activities were suggested for this sub-project:

    a. Co-ordination and structuring national efforts to collect data on physical properties of foodstuffs. A European centre should be appointed to collect, classify and make readily available data.

    b. Development and standardisation of methods for measuring physical properties of foodstuff. Testing of methods in practical work.

    c. Experimental studies with the aim to develop simple methods for measuring physical properties which correlate to quality properties.

The second subproject, 'Quality properties of food', was presented in the following way:

'One of the most important objectives of food research
is to improve knowledge about the effects of processing
on the quality of food products. However, the use of
modern large scale technology in food production raises
problems. Above all it becomes important to control
factors which are relevant to the consumer's choice of
food'.

The following activities were suggested for this subproject:

a. International evaluation of the importance of various
quality properties for different types of food. Determination of the importance of quality properties for the
consumers food choice in different regions in Europe.

b. Development and testing of simple but efficient
methods for evaluation of quality properties of food for
the benefit of the food industry and the food consumer.

The third subproject was concerned with 'Nutritional
properties of food'. This subproject was introduced in the
following way:

'The nutritional value of food is nowadays coming more
and more into focus for consumers both in developed and
developing countries. The consumers request content
declarations on food offered by retailers. Increased
knowledge about characterisation and measuring of nutritional properties of food is therefore of considerable
industrial interest.

The increasing trade with raw materials and food products,
on the one hand, and the increasing degree of processing
and refining in the food industry, on the other hand, call
for harmonised, appropriate and simple measuring methods
and for more reliable data concerning the nutritional value
of food.'

The suggested activity within this subproject was:

'Development of methods for measuring nutritional properties of different food products in a complete meal. Testing of methods in practical work.'

The Committee of Senior Officials on Scientific and Technical Research (COST) decided in April 1975 to set up an ad hoc Working Party to study, examine in detail and finalise a proposal from the Swedish delegation for setting up a COST Working Party on Food Technology with the aim of encouraging research in the food industrial sector.

This ad hoc Working Party met on 10 July, 24 October and 16 December 1975 and was attended by 14 delegations. Four study groups were set up to help in the preparation of the final report and each of these groups presented a preparatory report on one of the proposed projects.

In such a vast area as food technology with appreciable research activities going on in the 19 member countries it has been important to define certain criteria which would be the basis for project selection and project priorities. The Working Party agreed on such criteria in the course of the preparation of the report. These criteria were:

1. The project shall be an important contribution to research and development for the food industry and related industries.

2. The project shall draw specific advantage of the fact that several nations are involved.

3. The project shall stimulate industrial development and trade in Europe.

4. The project shall lead to improvements for the consumers.

5. The project shall give rapid and practical results for the industry.

In spite of considerable efforts on all sides, the size and complexity of the task and the desire of the rapporteurs to contact their colleagues in the field meant that the deadlines went beyond those originally planned by the COST Committee.

The secretariat, under the able leadership of Mr. Rateau, prepared a 'Report of the ad hoc Working Party of Experts on Food Technology' dated February 9, 1976. Although this report is an important document and an important step in the development of a COST activity it was found that the scope of the various projects was still too wide and that more specific and concrete project descriptions should be worked out. This has been carried out in two rather independent activities, one centred around physical properties and the other on a combination of quality and nutritional properties of food, the latter activity leading up to the present seminar.

First a few words about the other subproject. During the first half of 1976 an inquiry was sent out to 361 research groups among the COST countries to find out present activities in the area of physical properties of food and to find out the interest for a joint project. The result of the survey was discussed in a small Working Group and suggestions for the content of a COST project were made. The secretariat and the Working Group have worked out a final report on theme A, 'Physical properties of foodstuffs', dated March 17, 1977. The aim of the research programme is to make available to the food industry and allied industries, eg manufacturers of equipment, data on physical properties of foodstuffs such as finished food products or foods in the course of industrial processing. Naturally, research centres and departments will also have access to the results. The work is suggested to be organised by a co-ordinating committee, three subcommittees and the research laboratories themselves with the assistance of the Commission of the European Communities. The results will be stored in a data bank. Co-operation and integration with existing European and world-wide activities in this area is suggested. The project will deal with rheological properties, water activity and thermal properties and will, at least for the time being, be limited to certain food commodities (Table 1).

TABLE 1
PHYSICAL PROPERTIES OF FOODSTUFFS

```
Rheology of milk products
        of sweetened and cereal products
        of fruit- and vegetable-based products

Water activity in
        sweetened and cereal products
        fruit- and vegetable-based products
        meat products

Thermal properties
        of fruit- and vegetable-based products
        of meat products
        of fish-based products
```

As far as I understand, the preparations for this project are proceeding in a positive way. The Commission of the European Communities has made a proposal on August 11, 1977, to the Council of European Communities, which in its turn has asked for the opinion of the European Parliament, stating that the programme should preferably enter into operation in 1978. This time goal should be possible to keep.

Now back to quality and nutritional properties. The report of February 9, 1976, was discussed at the COST Senior Officials Committee meeting on March 15, 1976. At that meeting the Commission was asked to arrange meetings for the interested COST countries to enable the topics on quality and nutritional properties to be examined in parallel with the topic on physical properties. A meeting was held in July 1976 to review progress made in these areas and if possible to determine the research topics to be tackled. Due to the complexity of the area a group of five participants was asked to prepare more specific proposals. It was also decided to carry out a survey among national institutes concerning present activities and interest in co-operation. Replies were received from 12 countries. The type and amount of information varied a great deal from country to country due to different interpretation and understanding. In the 12 countries 90 research centres reacted on the inquiry. 50 of these are prepared to co-operate in a suitable COST project. More than 500 persons in these research centres are at present engaged in research work on quality and/or nutritive value of foods. A few examples of areas of common interest are shown in Table 2.

TABLE 2

QUALITY AND NUTRITIVE VALUE

Project	Countries
Heat treatment and protein quality	Finland, Norway, Italy, Spain, Belgium, Switzerland
Nutritive value of convenience foods	UK, Sweden, Germany
Quality of frozen meat products	UK, Spain

The group of five met several times under the able leadership of Mr. Mogens Jul, Denmark, with the purpose to prepare a report which after amendment by the ad hoc Working Party on Food Technology has appeared as a draft final report, 'Effects of processing and distribution on quality and nutritive value of food', dated in March 1977.

It is claimed in the report that there is a lack of research data which assembled would give a complete overview regarding quality and nutrition in relation to the whole food supply system. The reason for this, it is claimed, is that the food supply system is enormous in size and covers a multitude of food materials and processing, handling and distribution methods. It should be possible to achieve a complete overview if several nations co-operate in an international project which will be of a size too large for any single nation to carry out. This is one of the justifications for a COST collaborative study. Another justification is found as a result of the large international trade in that new processing and distribution methods are generally rapidly adopted in other countries than where they were developed. The subject has become acute because consumers are increasingly concerned with regard to the quality properties, the safety and the nutritive value of industrially processed foods and of foods received through the modern food distribution system. Such foods are erroneously believed to be of inferior quality than the untreated products.

In the draft report it is proposed that the effect on quality and nutritive value of the various stages of the food supply system be studied and compared to that of alternative systems. The results will be appropriately summarised and communicated to the industry as well as to the public, to legislators etc. The purpose is not only to make an inventory of the present situation. More important the study must include investigations into improvements of food processing and distribution methods in order to maximise quality and nutritive value.

The project will be of a 3-dimensional nature, as there are three systems of parameters:
- Processing and distribution
- Type of food, and
- Quality and nutritive factors.

The working method suggested includes a co-ordinating committee, responsible for guiding and managing the project as a whole, a co-ordinating secretary, ad hoc working groups for subareas and information exchange including status reports and project seminars, the first one starting today.

It is easily realised that the project will be more flexible if the numbers of processing parameters or commodities are not restricted. However, it has been felt that some restrictions should be applied. The risk of chaos is otherwise too great and the co-ordination procedures too difficult to carry out. Therefore, priority has been given to 'thermal treatment with priority given to freezing and preservation by the application of heat'.

The next step in the development of a COST project on food quality and nutrition is the present seminar. The object of the seminar is to define specific needs of research and development within the frame given. The design of the seminar is both interesting and ambitious and it should allow for an adequate identification of subprojects worthwhile to carry out. The actual selection can then be carried out at the ad hoc Working Party meeting Friday afternoon. The criteria given earlier for subproject selection should be kept in mind when drawing up the programme.

The meetings during this week should allow the Secretariat in Brussels, now under the able leadership of Dr. Rajcan, to prepare a final report with suggestions to the Committee of Senior Officials of COST. With a lot of work and a bit of luck it should be possible to start this important project during 1978.

SESSION I

CURRENT FOCUS OF R & D ON COMMODITY AREAS

Chairman: R. McCarrick

QUALITY AND NUTRITIVE ASPECTS OF MEAT PROCESSING AND DISTRIBUTION

Mogens Jul
Danish Meat Products Laboratory,
Copenhagen, Denmark.

For meat, like most other foods, there is a general tendency to assume that the result of industrial processing, long term preservation, and time consuming distribution of meat and meat products, will lead to a deterioration in both quality and nutritive value compared to when the products are consumed more or less direct after slaughter at the place of production, or at the most, only subjected to those preservation methods which have traditionally been used.

It is of great importance for the consumers as well as for the controlling authorities that the food industry and the food distribution system should have access to detailed data regarding these matters. Where no deterioration in quality or nutritive value takes place, it is important that this fact should be known so as to dispel suspicion with regard to the modern food distribution system. Conversely, where deterioration is found, it is important to determine what is the cause thereof and to investigate whether the matter can be rectified.

In considering the latter one must assume that undesirable changes cannot always be completely prevented. In those cases one needs to give careful consideration to the question of whether processing, handling and distribution methods which result in undesirable changes can be eliminated. The cost of this will often be that a certain meat product cannot be made available at all times for all consumers in all geographic areas. Thus, one will have to weigh the advantages of mass processing, mass distribution, against the possible undesirable effects on quality or nutritive value.

QUALITY

A review of the influence of various processing, handling and distribution methods on meat quality would almost be a complete review of all investigations in meat technology. Obviously, most investigations relating to say a new method of preservation, etc. have generally been accompanied by tests to determine the influence of the method on the various quality factors. In general, only such methods have been introduced where the influence on quality was considered minor, but in any case, the effect on quality was established.

Suffice it here, therefore, to review a few examples of the general effect of variations in processing and preservation on quality.

SLAUGHTERING TECHNIQUES

In general it is probably assumed that the process of slaughtering an animal in a modern meat plant is so similar to what takes place when an animal is killed and slaughtered on a farm that little suspicion exists for a negative influence of the slaughtering process on meat quality. As will be seen, this assumption is not always entirely justified.

Chicken

In Denmark, as in several other countries, complaints were often heard that frozen chickens were not as tender as chicken obtained directly on the farm in the fresh state. The Danish Meat Products Laboratory carried out a survey and determined that frozen chicken as obtained in the stores had varying degrees of tenderness, but not infrequently some were found which were quite tough. A similar phenomenon was not found where chickens were obtained directly on the farm. This discrepancy was ascribed by some people to the freezing process, by others the assumption was that free range chickens were more tender than those from modern intensive chicken raising units. This latter theory, which later proved to be wrong, failed to take into consideration that in practically no cases were the chickens obtained directly from farms, the result of free range poultry management. Generally they will result from the same type of intensive broiler raising as those purchased frozen in the stores.

Inspired by investigations in other countries, Simonsen (1967) carried out an investigation into the effect which the choice of scalding, temperature and time might have on the tenderness of broilers. A scalding temperature of 59°C was used and the results shown in Table 1 were obtained.

TABLE 1

SCALDING TIME AND TEXTURE OF BROILERS

Scalding time seconds	Texture	
	Breast meat	Thigh meat
43	7.0	7.2
87	5.4	6.6
180	3.1	5.2

Tenderness was measured by a taste panel, scoring 10 for very tender, and 1 for very tough

A subsequent survey of the practices in various Danish poultry slaughterhouses clearly demonstrated that toughness was always associated with scalding temperatures which were too high or scalding times which were too long. Therefore a regulation was passed specifying that broilers for freezing could not be scalded at temperatures above 60°C and not for longer than 180 seconds. After this, complaints about toughness in broilers disappeared.

One might assume that other factors in the slaughter process might have some undesirable influence on quality in comparison with the home slaughtered product. Therefore Tove Skårup (1977) carried out an experiment which led to the result shown in Table 2.

TABLE 2

FACTORY SLAUGHTERED AND HOME SLAUGHTERED CHICKEN

Group	Overall impression	Taste		Tenderness		Juiciness	
		breast	thigh	breast	thigh	breast	thigh
1	0.47	0.57	0.38	0.83	1.21	-0.57	0.59
2	0.65	0.61	0.54	0.96	1.44	-1.02	0.78

Group 1 = Slaughterhouse processing
Group 2 = Home slaughtering

The scores were given by a taste panel using a score from -5 for very bad, to +5 for very good. As will be seen from this Table there is no evidence that correctly slaughtered chickens from an industrial plant are any different in taste or texture from chickens slaughtered directly on the farm without the use of complicated processing machinery.

CHILLING

It has often been assumed that the introduction of modern chilling methods, ie, very rapid chilling in a very cold air stream, would have a negative effect on quality when compared to the old practice of letting the carcases chill by free moving air on hanging floors.

Pork

In the early 1950s a general switch over took place in Denmark from chilling hog carcases on hanging floors to quick chilling; gradually chilling in rapid moving air was developed, until today most chilling processes start with an air temperature of $-20^{\circ}C$ and an air velocity of at least 2m/second. At that time most Danish pork was processed into Wiltshire bacon for export to the UK. Many people expressed fears that modern quick chilled pork could not produce bacon of the same quality as that to which the British had become accustomed. First fears were expressed that quick chilling might result in a lower keeping quality. It was soon determined, however, that the surface bacterial count from hog carcases chilled on hanging floors was in the order of magnitude of 10^7/cm, while the surface count from similar carcases which were quick chilled were $10^4 - 10^5$/cm. Similarly, detailed tests carried out in the UK established that there was no discernible influence on the keeping quality there.

In the same way, of course, many people expressed fears that the taste of bacon from such quickly chilled carcases could not be equal to that of traditional bacon. Many samples were manufactured under careful control and evaluated and tasted in the UK. In no case were we able to establish any difference in any of the organoleptical characteristics of the two types of products.

However, the taste of bacon might not necessarily be the best criterion for comparing the two types of chilling since Wiltshire bacon is produced by curing the carcases at least six days and subsequently shipping them to Great Britain at temperatures of about 4°C. It was unreasonable to assume that the quickly chilled pork might be less tender than a slowly chilled product if consumed immediately after chilling. Zeuthen (1971) determined that immediately after chilling, quickly chilled pork was slightly less tender and less tasty than a slowly chilled product. However, if the products were aged three days no organoleptic difference was discernible and thus it was established that quick chilling may have a negative effect on quality but only when sales take place very quickly after slaughter.

Veal

With the development of modern carcase chillers, quick chilling of calf carcases was soon introduced in Denmark. This resulted in repeated complaints regarding toughness and tests at the Danish Meat Research Institute indicated that quick chilling of calf carcases could indeed produce very tough meat as indicated in Tables 4 and 5. For this reason it was recommended that calf carcases be chilled in air temperatures lower than 8°C.

TABLE 3

SHEAR FORCE VALUES (in kg) AND RESULTS OF AN ORGANOLEPTIC EVALUATION OF GRILLED PORK CHOPS AT DIFFERENT pH_2, AGED 1, 3 AND 6 DAYS

pH_2-cured grouping	Shear force days after slaughter			Organoleptic evaluation days after slaughter		
	1	3	6	1	3	6
≈ 5.5	17.2	15.1	13.4	-1.15	-1.46	-1.11
≈ 6.3	17.7	13.4	10.1	-0.58	+0.22	+1.14
≈ 6.5	21.8	14.7	10.6	-1.35	-0.15	1.37

Framed results show that pH_2 groups are significantly different. Arrows indicate a significant development during ageing. (Buchter and Zeuthen, 1971)

TABLE 4

TENDERNESS OF VEAL (SHEAR FORCE VALUE) AFTER CHILLING AT DIFFERENT TEMPERATURES FOR CALF CARCASSES

Temperature °C	Shear force value
10	8.6
6	10.5
4	13.7
Tunnel, then 1°C	24.5

(Lis Buchter, 1974)

TABLE 5

TENDERNESS OF VEAL AFTER QUICK AND SLOW CHILLING OF CALF CARCASSES

	Tenderness (average)
Slow chill 10°C for 24 h, 1°C for 24 h	2.50
Rapid chill -10°C for 4 h, 1°C for 24 h	0.43

Score: -5 = very tough; +5 = very tender

Lamb

New Zealand workers (Locker et al., 1972) determined that quick chilling of lamb may indeed produce very tough mutton. The toughness can be somewhat improved by ageing the product, but it is by no means possible to eliminate the effect of cold chilling by slow ageing. Figure 1 illustrates this. This discovery led to a practice in New Zealand whereby lamb has to be conditioned, ie, held at elevated temperatures for up to several days prior to chilling or freezing.

It was because of the difficulties in maintaining such large stocks of slowly chilled carcases etc. that the modern practice of electrical stimulation, with its complete elimination of cold shortening, was introduced.

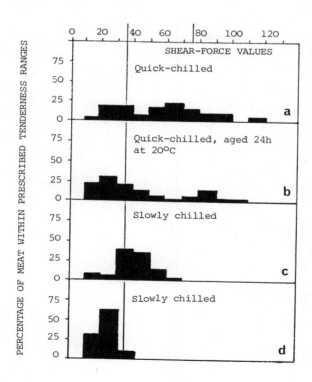

Fig. 1 Quick chilling of beef

Beef

Comparatively few defects have been determined as a consequence of too quick chilling of beef carcases. The main reason for this is that the carcases are so large that even a quick chilling heat penetration to the interior parts of the musculature is so slow that only the surface layers can be classified as quickly chilled. Also the advantages of quick chilling are fewer because the reduction in total chilling time is rather insignificant. Nevertheless, it is generally agreed that beef should not be chilled at temperatures below $4^{\circ}C$.

Chicken

Broilers are often chilled by passage through a cold water bath prior to packaging and distribution either in the fresh or frozen state. Some countries maintain that especially for broilers to be marketed in the fresh state, an air chilling is preferable. A survey carried out by the EEC Commission (1977) indicated the following bacterial counts on chickens which had been air chilled as opposed to water chilled in various commercial installations.

TABLE 6

SURFACE BACTERIAL COUNTS ON WATER CHILLED AND AIR CHILLED BIRDS
(Log total count before and after chilling of chickens)

	Immersion chilling			Air chilling	
Slaughter-house No.	Before	After	Slaughter-house No.	Before	After
1	5.15	4.51	1	4.64	4.54
2	5.16	5.06	2	4.66	4.73
3	5.00	5.14	3	5.84	5.65
4	5.25	5.05	4	6.39	6.12
5	5.25	4.99	5	5.26	5.15
			6	5.25	5.37
Average	5.16	4.95		5.34	5.26

As will be seen from the above, water chilling, because of its washing effect, results in a lower surface count than air chilling.

However, it is normally known that air chilling results in a slight water loss while water chilling results in a water uptake. It could be feared that the higher water content of the water chilled bird could result in a lower keeping quality. Nevertheless it should be remembered that the water activity is determined by the ratio between dissolved molecules and the total amount of water present. Since the amount of dissolved matter is very small compared to the amount of water present, a few per cent more or less water has no determinable effect on the water activity. Thus, as could be expected, no difference has been demonstrated between the keeping quality of water chilled and air chilled birds in a survey carried out by Thomson et al. (1975).

One might also have assumed that the water uptake - and possibly the leading effect of water chilling - might result in an inferior, possibly bland, taste. Figure 2 shows this not to be the case as long as efficient chillers with lower water uptake are concerned.

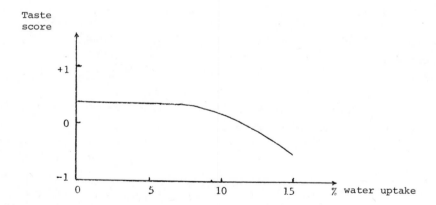

Fig. 2 Taste - Danish taste panel tests on broilers chilled in air and in water, with various percentages of water uptake, gave the above results (0 = normal; -1 = barely noticeable less desirable quality).

Conclusion

Consumers have no taste advantage using air chilled birds instead of water chilled. For the prepared product on the plate they pay about 10 - 12% more for the air chilled product.

Prices

It is often assumed that water chilling is popular in the industry because of the yield increase due to water uptake. Conversely, consumers object to it because of the fear of 'buying water for the price of chicken'. However, both are correct only if poultry prices per kg were independent of chilling method. Danish workers have collected representative prices in some countries and calculated, on the basis of known water uptakes and cooking losses, the price a consumer will have to pay for:

1. 1000 g of eviscerated chicken, ie, before any shrinkage, water uptake, etc. (probably the most logical comparison)

2. 1000 g of chicken cooked and served on a platter (probably the best psychological comparison)

The price calculations below show clearly that water chilling is economically advantageous to the consumer. Only at water uptakes of about 15% would the break even point be reached. The reason is the much higher cost of air chilling.

In the USA both fresh and frozen birds are water chilled. Here, frozen birds are more expensive than fresh birds; this suggests that with air chilling frozen birds would become quite expensive.

Danish experiments suggest the following yield of 1000 g chicken purchased:

Weight	Air chilled 1 - 2% water loss	Water chilled 5 - 7% water gain
Eviscerated weight	1020 g	950 g
At purchase	1000 g	1000 g
At plate	800 g	750 g

Observed prices are as follows (Feb 77)	Fresh birds (air chilled)	Frozen birds (water chilled)
Denmark, D.kr	15.80	13.50
Fed. Rep. of Germany, DM	6.50	5.00
UK, p	108	86
	(water chilled)	(water chilled)
USA (wholesale) $	0.97	1.08

Based on these data the following prices are calculated per kg

	Eviscerated (before chilling)	Cooked (on plate)
Denmark, D.kr		
fresh, air chilled	15.49	19.75
frozen, water chilled	14.21	18.00
Fed. Rep of Germany, DM		
fresh, air chilled	6.17	7.88
frozen, water chilled	4.95	6.21
United Kingdom, p		
fresh, air chilled	106	135
frozen, water chilled	91	115
USA (wholesale) $		
fresh, water chilled	1.02	1.14
frozen, water chilled	1.14	1.44

TABLE 7

RESULTS OF THE EVALUATION OF THE FROZEN SAMPLES FROM THE SIX MODEL EXPERIMENTS

Freezing unwrapped pork sides

Model exp. No.	Product initial temp. °C	Air temp. °C Initial	Air temp. °C Final	Air velocity m/sec Initial	Air velocity m/sec After the surface temp. has reached -6°C	Freezing time (h)	Freezing loss (%)	Visual judgements of samples
1	7	-25	-31	2.7	unchanged	3.33	0.8 - 1.3	5 - 10% of exposed meat surface discoloured
2	11	-25	-31	5.7	unchanged	2.60	0.75 - 1.2	50% of exposed meat surface discoloured
3	10	-25	-30	1.0	unchanged	4.50	0.9 - 1.3	5% of exposed meat surface discoloured
4	12	-25	-30	1.0	2.7	4.00	0.9 - 1.35	5 - 10% of exposed meat surface discoloured
5	8	-30	-37	1.0	2.7	3.70	0.6 - 1.3	5% of exposed meat surface discoloured
6	7	-20	-27	1.0	2.7	5.00	0.8 - 1.6	5 - 10% of exposed meat surface discoloured

FREEZING

Popularly it is assumed that freezing, frozen storage and consequent thawing will result in relatively minor changes in product quality, especially if freezing is rapid and storage takes place at low temperatures. A closer study will reveal that many more factors need to be taken into consideration.

Effect of freezing on organoleptic qualities

In some cases no difference can be found in taste between a frozen and thawed product in comparison with an unfrozen product.

In other cases freezing may have a deleterious effect, eg, on appearance. Thus Clemmensen and Zeuthen (1974) found, as indicated in Table 7 that hog carcasses show considerable quality deterioration if the freezing has taken place at very low temperatures and very high air speeds, while moderate temperatures and speeds at the beginning of the freezing process resulted in a fully acceptable product.

Effect of freezing on microbiology

It is generally known that freezing and thawing a product will have a deleterious effect on the keeping quality of the defrosted product. However, as Figure 3 shows, Simonsen (1960) was unable to determine any difference in microbiological keeping quality of chopped meat which was stored at $20^{\circ}C$ either in the unfrozen or defrosted state.

Freezing and refreezing

It is often assumed that freezing a product which has already been frozen will result in a quality deterioration. However Table 8 demonstrates that broilers can be frozen, defrosted and refrozen several times without any discernible effect on cooking yield, taste or tenderness.

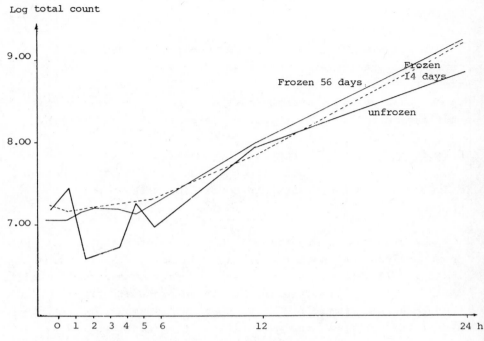

Fig. 3 Effect of freezing on the quality of chopped meat

TABLE 8

EFFECT OF REFREEZING BROILERS

Frozen, defrosted	Cooking loss	Taste	Tenderness
Once	27.1%	+0.3	-0.4
Twice	25.2%	+0.1	-0.1
Three times	23.1%	+0.3	-0.3
Four times	24.2%	+0.5	+1.2

Score: -5 = very poor; +5 = very good

Freezer storage temperature

It is often assumed that when a frozen product is kept at a very low freezer storage temperature, quality will be maintained. However, Roy Nilsson (1973) demonstrated the reverse may be true. In the case of bacon he determined that the product kept better at -10 than at -20°C as indicated in Figure 4.

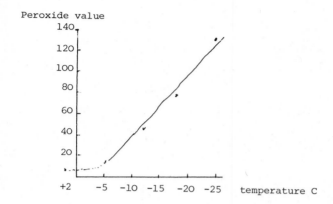

Fig. 4 Freezer storage of bacon

In general, however, it is true that lowering the temperature will increase the keeping time for a frozen product. This has led to so-called TTT expression, ie, the keepability of a frozen product is assumed to be characterised by its time temperature tolerance. However, many other factors are involved, principally those referred to as PPP, ie, product, process and packaging.

Effect of product quality of frozen products

Figure 5, quoted by Andersen et al. (1966) indicates that frozen pork liver in a good microbiological state will keep much longer during freezer storage than a questionable product.

Package

Similarly, Andersen et al. quote one example, illustrated in Figure 6, where a frozen broiler kept much better in a package of good quality than in a package of poor quality. This difference is remarkable because both types of packages were such which were at that time found in commercial use.

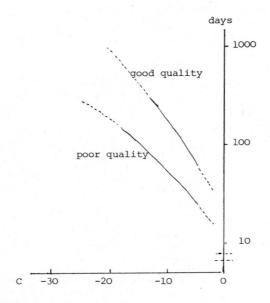

Fig. 5 Frozen pork liver

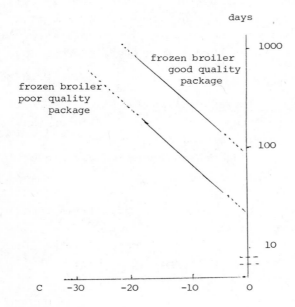

Fig. 6.

Effect of process on frozen products

Andersen et al. (1966) also quote an example where a cooked pork product kept much longer than the corresponding uncooked product. It is assumed that cooking has destroyed some enzymes which otherwise would be active, even during the low freezer storage temperatures.

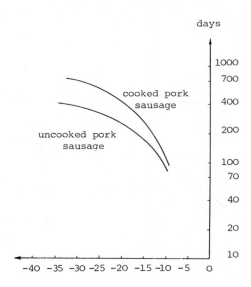

Fig. 7

Distribution of frozen foods

It is often assumed that frozen products keep reasonably well in normal distribution systems, since low temperatures are maintained throughout transport and store display. Bøgh-Sørensen (1977) has demonstrated that this assumption is often not justified. In particular, the storage temperatures in the retail display cases varied widely - a maximum of $-4.5^{\circ}C$ to a minimum of $-35^{\circ}C$. Testing the chickens which were offered for sale in normal retail stores, it was found that only 49% of these could be termed as having a normal quality. Thirty-three per cent were found to be of acceptable quality, while 17% were found non-acceptable by trained personnel. This survey, and many similar surveys carried out in other countries, clearly demonstrate that most of the effort which goes into the improvement of frozen foods needs to take the difficulties of maintaining desired temperatures in retail cabinets into consideration. Such efforts must be directed, through the better use of packaging materials, the improved formulation of the product, etc. towards the improvement of the ability of the product to withstand the unfavourable conditions during store display.

PASTEURISATION

Most heat preserved meat products are preserved by pasteurisation. This is the case for such products as hams, luncheon meat, roll products, etc. These products are normally cured with salt and nitrite and the cooking is not much more than that which is carried out when the products are prepared for direct use. Therefore the heat preservation process has little, if any, effect on quality. Nevertheless it is well-known that such products as canned cooked hams, often have to undergo a slightly more thorough cooking process than that required for regular cooked ham. The result may be a higher cook-out of jelly and a correspondingly drier ham. This phenomenon can often be offset by the addition of polyphosphates in the curing process. Many people assume that the addition of such water binding substances may result in a decreased quality. However, as Table 9 shows, a taste panel preferred cooked ham which had a water uptake during the curing process of about 16 - 20%. Since these products may be sold at considerably lower prices than those which have been prepared without any increasing curing yield, it seems that the combined addition of polyphosphate and use of heat pasteurisation may actually result in a cheaper and more palatable product.

TABLE 9

SENSORY EVALUATION OF HAM WITH DIFFERENT LEVELS OF ADDED WATER

Added water (%)	Average general impression
3	8.3
15	8.1
17	8.1
19	6.7

Score: 10 = excellent; 1 = unacceptable

CANNING

Cooking meats to complete shelf stability often results in a considerably more thorough cooking process than would be used when the meat is prepared in the home. Consequently the meat will often loosen and change somewhat in taste. Thus, in general the quality of meat and shelf stable canned meats is lower than it would be in the corresponding product prepared at home. This quality loss is unavoidable when a shelf stable product is desired.

Today shelf stability is often not required by the consumer who may often prefer a meat dish which is prepared by a pasteurisation process, eg, hot filled in a casing and maintained at refrigeration temperatures until consumed. Such products will retain quality much better and have a sufficiently long keeping time for the normal distribution system.

The Danish Meat Products Laboratory recently carried out a quality test of pasteurised, refrigerated meat dishes and frozen dishes, where the refrigerated dishes were given the highest score for taste. From a bacteriological point of view the frozen meals showed a higher total count than the refrigerated ones.

TABLE 10

NUTRITIVE VALUE OF MEAT PRODUCTS

	Raw ham	Pasteurised ham	Sterilised ham	Cooked pork sausage	Cooked salami
% Protein	17	20.2	18.3	18.1	17.5
% Fat	23	1.8	12.3	44.2	25.6
% Carbohydrate	-	0.1	0.9	-	1.4
% Water	59.2	73.6	65	34.8	51.0
% Ash	0.8	4.3	3.5	2.9	4.5
Calories/100 g	270	102	193	476	311
Vitamin A (IE/100 g)	-	<10	-	-	-
Vitamin C (mg/100 g)	-	13	-	-	-
Thiamin (mg/100 g)	0.54	0.4	0.53	0.79	0.25
Riboflavin (mg/100 g)	0.25	0.2	0.19	0.34	0.24
Niacin (mg/100 g)	4.8	3.9	3.8	3.7	4.1
Calcium (mg/100 g)	10	5	11	7	10
Iron (mg/100 g)	2.5	0.8	2.7	2.4	2.6

TABLE 11

VITAMINS OF THE B-COMPLEX IN RAW, FROZEN AND COOKED BEEF STEAKS (µg/g on fat-free dry basis)

Sample	Thiamin		Riboflavin		Nicotinic acid	
	Steer A	Steer B	Steer A	Steer B	Steer A	Steer B
Unfrozen, raw	2.8	3.8	5.7	7.5	303	277
" cooked	1.8	2.3	5.5	6.8	265	265
Frozen to -18°C	2.4	3.7	5.8	7.0	284	244
" immediately cooked	1.5	2.0	4.8	6.7	254	242

TABLE 12

PERCENTAGE RETENTION AFTER COOLER STORAGE AT 4°C

	Thiamin	Riboflavin	Niacin
Minced pork loin			
after 2 weeks	93	104	97
after 8 weeks (putrefied)	89	141	105
Cured pork loin			
after 2 weeks	95	105	90
after 8 weeks	91	120	93

NUTRITIVE VALUE

It is characteristic that any changes in technological procedures have generally been preceded by an investigation of the effect which the change might have on food quality. In contrast, few experiments have ever been carried out prior to any technological change to establish whether the change would have any effect on nutritive value. The reason probably is that consumers are generally assumed to attach a great deal of importance to the organoleptic quality of the product while they have little knowledge of its nutritive value and display little active interest in this regard. In spite of this, much data is available regarding the nutritive value of meat products as related to the various technological treatments (Tables 10, 11 and 12). However, these have been carried out mainly in an effort to monitor the content of the various nutrients in the daily diet and the data has had little effect on decisions with regard to choice of technologies.

Lee et al. (1950) give values for some of the vitamins of the B-complex in beef steaks from two beasts, before and immediately after freezing, in the raw and cooked states. Although there appear to be slight losses due to freezing, these are small compared with the variability between animals and the effect of cooking.

Changes in vitamins are small. High bacterial counts on meat may raise the vitamin content due to bacterial production of B-vitamins (Rice et al., 1946, 1949).

Except for riboflavin, freezer storage does not seem to have any significant effect on vitamins.

The data for riboflavin suggests that freezer storage should not be extended beyond 6 months. Other investigators found more favourable results, eg, Harris et al. (Table 13).

TABLE 13

RETENTION OF VITAMINS IN PIG MEAT AFTER FREEZER STORAGE

Storage time	Storage temperature	Percentage of initial content after slaughter		
		Thiamin	Riboflavin	Niacin
24 hours	$-18°C$	79	117	78
48 hours	$-18°C$	84	94	78
3 months	$-18°C$	81	62	85
6 months	$-18°C$	68	66	90
24 hours	$-26°C$	97	115	87
48 hours	$-26°C$	97	109	82
3 months	$-26°C$	78	75	93
6 months	$-26°C$	83	79	98

COOKING

All the above data indicate the retention of vitamins in the stored product. This may not be completely relevant because the pig meat is practically always eaten cooked. Thus it may be more appropriate to look at the relative retention in the cooked pig meat after various treatments. Lehrer et al. quote the following percentages of retention after various treatments of pork loins.

TABLE 14

PERCENTAGE RETENTION OF VITAMINS IN PORK LOINS

Treatment	Retention of initial content (%)		
	Thiamin	Riboflavin	Niacin
Frozen and thawed	85	109	81
Frozen, stored 3 or 6 months	78	71	92
Freshly cooked	44	66	83
Frozen, stored, cooked after thawing	36	52	59
Frozen, stored, cooked without thawing	51	63	52

REFERENCES

Andersen, Jul and Riemann, 1965. Industriel Levnedsmiddelkonservering, Teknisk Forlag, Copenhagen.

Buchter, L. 1974. Proceedings of the 21st Easter School in Agricultural Science, University of Nottingham.

Buchter, L., and Zeuthen, P. 1971. Proceedings of the 2nd International Symposium, Condition Meat Quality Pigs, Zeist, Pudoc, Wageningen, 247.

Bøgh-Sørensen, L. 1977. The quality of frozen chicken in Denmark, Report No. FJ-64, Danish Meat Products Laboratory, Copenhagen. (mimeographed).

Clemmensen, J. and Zeuthen, P. 1974. Meat Freezing - Why and How? Meat Research Institute Symposium, Langford.

EEC-Commission 1977. 'Study for the evaluation of the hygienic problems related to the chilling of poultry carcasses', Commission of the European Communities, Information on Agriculture No. 27.

Harris and Karmas, 1975. Nutritional Evaluation of Food Processing, Second Edition, Westport, Connecticut, The AVI Publishing Company, Inc.

Lee, F.A. et al. 1950. Effect of freezing rate on meat, Food Research, $\underline{15}$.

Lehrer, W.P. et al. 1951. Effect of frozen storage and subsequent cooking on the vitamin content of pork, Food Research, $\underline{16}$.

Locker et al. 1972. A new concept of processing beef and lamb, Mirinz, 257.

Nilsson, R. 1973. Livsmedelsteknik, $\underline{1}$.

Simonsen, B. 1960, 1967. Unpublished experiments from the Danish Meat Products Laboratory, Copenhagen.

Skårup, T. 1977. Factory slaughtered versus home slaughtered chickens, Thesis, Department of Food preservation, The Royal Veterinary and Agricultural Univ. Copenhagen, Denmark.

Thomson et al. 1975. Bacterial counts and weight changes of broiler carcasses chilled commercially by water immersion and air-blast, Poultry Science, Vol. 54, No. 5.

MILK PROCESSING - R & D FOCUS AND NEEDS

W.K. Downey* and P.F. Fox**

*National Science Council
**University College, Cork, Ireland.

ABSTRACT

Based on a limited survey of publications in dairy processing the current focus of international R & D is indicated. Quality and also some nutritional aspects of dairying requiring further R & D are discussed with particular attention to milk as a raw material for processing.

FOCUS OF DAIRY PROCESSING R & D

While it is hardly possible to indicate briefly the current focus of international R & D in milk and dairy products the following semi-quantitative patterns may be discerned by scanning Dairy Science Abstracts (Dairy Science Abstracts).

Between January 1975 and June 1977 over 10 000 scientific papers, reviews, articles and other technical publications on milks and dairy processing/products were abstracted. These constituted about 60% of total publications referred to in Dairy Science Abstracts. The others not included were abstracts of articles on husbandry and milk production; physiology and biochemistry (mammary gland and lactation); mastitis and other diseases; economics, (legislation and standards); dairy research and education and annual reports.

As expected, the bulk (94 - 97%) of the annual abstracts on milks and dairy processing/products are concerned with bovine milk, its processing and products. Though constituting the next most important category, abstracts of publications on human milk are appreciably less numerous representing only 2% of the total for the three year period or about 50 - 100 publications per annum. Publications on milks from other species including buffalo, sheep and goat are each little more than 0.4% of the total.

Differentiation of the 10 000 publications on milks and dairy processing/products according to discipline/specialisation (Figure 1) reveals the highest concentration in dairy chemistry which alone accounts for one third of the total publications. Next are dairy microbiology and technology, each accounting for more or less one fifth of the abstracts and then dairy engineering with about 15% of the total. It is interesting that the relative distribution of abstracts between these four disciplines/ specialisations is rather constant over the three year period. Even more important, if not disappointing, is the observation that little more than one tenth of the 10 000 abstracts are

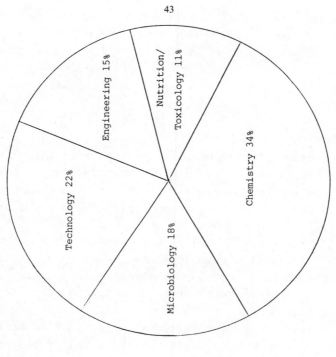

Dairy Science Abstracts
Distribution According to Discipline/Specialisation
1975-1977: Total 10,000

Fig. 1.

concerned with nutrition/toxicology. Indeed, as explained later the actual number of publications specifically concerned with the nutritional attributes of milk and dairy products may be little more than half of those classified under the heading nutrition/toxicology.

Dairy Chemistry

As mentioned above publications on various facets of dairy chemistry constitute the largest segment accounting for one third of the annual abstracts or more than 3 500 over three years. Of these, publications on analytical methods are consistently the most prevalent (Figure 2).

The analytical methodology used in dairy chemistry with some exceptions could be appreciably improved. Thus, the observed increase in the percentage of publications devoted to chemical analysis from 25% in 1975 to 29% in 1976 and up to 32% in 1977 (7 months) is most welcome. As further elaborated subsequently it presumably reflects the widely felt need for improved methodology not only in chemistry but in all areas of dairy science. Abstracts on the related topic of the composition of milk species and dairy products including studies on trace elements and other minor constituents represent 12 - 15% of the annual total. Together with the previously mentioned methodology studies they account for up to 40% and more of the total chemistry publications.

This is roughly of comparable magnitude to the percentage of abstracts devoted to the four major dairy constituents (proteins, lipids, lactose and salts) which jointly account for 36 - 38% of the annual chemistry publications. Since these include most of the fundamental dairy chemistry research it is welcome to note that the proteins receive the most attention (20% of abstracts) especially the casein followed by lipids (average 13%) with lactose and milk salts accounting for less than 3 and 2% respectively. Enzymology the other important area of concentration of basic research accounts for an average 14% of the annual abstracts. Studies on the native

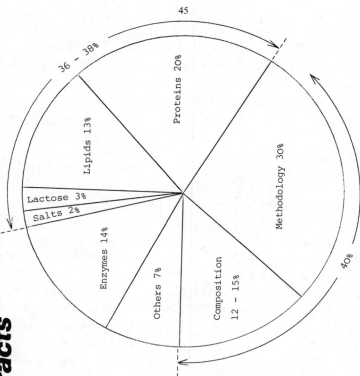

Fig. 2.

milk enzymes (4.7%) closely followed by rennins (4.3%) are the major interest together with lactase (1.3%). Of the twenty enzyme activities reported in bovine milk most attention has been focused on the lipolytic enzymes resulting from their involvement in promoting hydrolytic rancidity or lipolysis. (discussed later).

Dairy Microbiology

Close on 2 000 microbiological publications were abstracted between January 1975 and June 1977. Studies on the culturing of bacteria and other micro-organisms were the main theme of the publications, accounting for almost 40% of the annual total (Figure 3). As with chemistry, it is noteworthy that publications on methodology were also prominent representing over 20% of the total. These two categories jointly account for roughly 60% of microbiological publications abstracted between 1975 and 1977. Other significant categories include publications on microbiological spoilage (12%) and food-borne diseases (8%) which together constitute some 20% of the total. Although difficult to quantify publications on the more fundamental aspects of microbiology (8%) including characterisation studies appear to be appreciably less popular than in dairy chemistry.

Dairy Technology

Of the 2 300 dairy technology publications on milk and dairy products abstracted between 1975 and 1977 it had not been expected that those on cheese would outweigh the abstracts jointly devoted to the other conventional products.

As illustrated in Figure 4 abstracts on cheese technology represent 25 - 30% of the total. On the other hand, butter, milk powders, ice cream and milk (pasteurised and UHT) each constitute only about 4 - 6% of the total dairy technology publications. Taken together with cream (2%) these five products which include such dominant and large volume items as butter, milk powder and liquid milk account for little more than 20% of the total and appreciably less than cheese alone.

DairyMicrobiology Abstracts
Distribution According to Topic
1975-1977: Total 2,000

Fig. 3.

Dairy Technology Abstracts
Distribution According to Product
1975-1977: Total 2,300

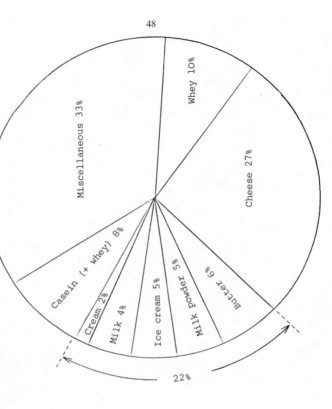

Fig. 4.

Moreover, as a percentage of the annual total, technology publications on butter, milk powder and cream have each declined a little in recent years. The relatively high proportion of abstracts devoted to cheese reflects the large number of cheese varieties and the relatively high technological requirements of cheese manufacture as well as the desirability to broaden the range of dairy products. This is further evidenced by the increase in publications on whey technology and other dairy products from 7 and 27% respectively in 1976 to 12 and 33% in 1977. Nonetheless the question arises of whether such a polarisation of technological input in favour of cheese and away from other conventional large volume products is desirable. As discussed subsequently there are well recognised technological problems confronting the other conventional products where a satisfactory solution is urgently required to arrest either declining consumption or rapidly accumulating stocks.

Dairy Engineering

Publications on dairy engineering exceeded 1 500 over the period under consideration. As expected they are predominantly concerned with plant design, improvement and development which account for up to two thirds of the annual total (Figure 5). Publications on farm buildings as a percentage of the total, seem to have increased from 9% in 1975 to 17% in 1977 (average 12%) and there has been a corresponding reduction in abstracts on cleaning and sterilisation from 8 to 3% (average 7%) while those devoted to transport have remained relatively constant (average 5%). The traditional food processing industries including dairying are frequently implicated in water pollution, and the industry is incurring rapidly increased cost in all European countries in controlling pollution. Since these costs may be most easily reduced through new and improved technology it may be noted that consistently less than 10% of dairy engineering abstracts are devoted to effluent treatment. The aforementioned increased emphasis on technology aimed at increased whey utilisation has together with developments designed to minimise product loss (especially in drying) the added advantage

Dairy Engineering Abstracts
Distribution According to Specialisation
1975-1977 Total 1,500

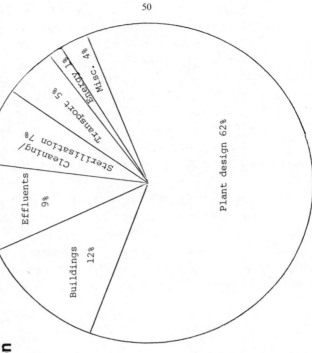

Fig. 5.

of reducing pollution due to dairy processing. Perhaps the most noteworthy indeed disappointing feature of dairy engineering publications is the relatively low number directly concerned with energy.

Out of a total of more than 1500 dairy engineering abstracts over the period only 20 or a little more than 1% seem to be specifically concerned with energy considerations. Bearing in mind escalating energy costs and the predominance of sequential heating and cooling operations in almost every facet of milk handling and processing (from milking and farm bulk collection through pasteurisation, factory holding and product manufacture to commercial and retail cold storage and distribution) dairy engineering innovations aimed at energy conservation including waste heat utilisation would appear to be a fruitful field for R & D.

Nutrition/Toxicology

Turning next in this survey of dairy science and technology to abstracts on nutrition/toxicology, these as already indicated (Figure 1) constitute the smallest specialisation amounting to about 1 200 in total or slightly more than 10% of the total publications over the period. The percentage of the abstracts devoted to nutrition/toxicology has increased from less than 10% in 1975 to near 12% in 1976 and over 13% in 1977. However, over half the publications classified under nutrition/toxicology (Figure 6) are concerned with the presence in milk and dairy products of pesticides (21%), heavy metals (11%) and other contaminants (18%) and are accordingly more concerned with quality control, inspection and trade certification requirements. Thus the percentage of milk and dairy publications directly concerned with their nutritional attributes may not be appreciably more than 5% of the total and over half of these are specifically concerned with human milk and infant feed formulations. In overall, therefore, only about 2% of all publications abstracted between January 1975 and June, 1977 or little more than 100 publications per annum jointly cover such important aspects of milk and dairy products as basic

Nutrition/Toxicology Abstracts

Distribution According to Specialisation

1975–1977: Total 1,200

Fig. 6.

nutritional research, composition, fortification and nutritional stability in decreasing order of occurrence. Thus, despite its obvious importance, publications on the nutritional stability of milk and dairy products, which is of particular relevance to the seminar, are rather low in number amounting to about 10 or less per annum.

Dairy Processing Publications and R & D Expenditure/Return in COST and Other States

Finally, it may be interesting to consider briefly the distribution of dairy processing publications between COST States. In particular it may be useful to explore possible relationships between expenditure on food science and technology and publication output which may not only provide an index of relative investment in different sectors (as above) but also a partial measure of return for expenditure.

Of the 4 500 processing publications from 33 major dairying countries cited in the 1976 Dairy Science Abstracts, COST States account for about 40% (Figure 7). This is roughly of comparable magnitude to the combined output of the USA (21%) and USSR (16%) which are the dominant contributors. COST States together with the US and USSR jointly account for over three quarters of the 1976 abstracts on dairy processing.

Within COST States both the UK and Germany are the dominant sources of dairy processing publications and together with France account for well over half the COST total for 1976 (Figure 7). The number of publications from other COST States ranges from 20 - 140 per country and may be compared to such major dairying countries as Canada, New Zealand, Australia and India whose publication output range from 80 - 140 per country (Figure 7).

Because of the dramatic differences between countries in the gross domestic product (including that of the agriculture and food industries) and of course population, it would be more realistic to compare numbers of publications per million

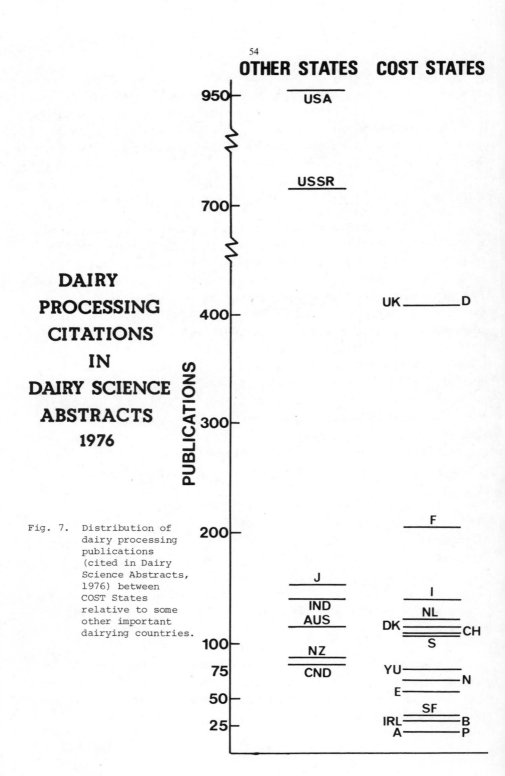

Fig. 7. Distribution of dairy processing publications (cited in Dairy Science Abstracts, 1976) between COST States relative to some other important dairying countries.

population or ideally numbers of publications as a function of relative expenditure on dairy science and technology.

On comparing COST States in terms of dairy processing publications per million of population, Denmark topped the scale in 1976 with a high output of 23 publications per million (Figure 8) followed by Switzerland and Norway and then Sweden, Ireland and the Netherlands with close on 10 publications per million population. Other COST States range from about 2 to 10 publications per million population. This may be compared to India, Japan, USSR, Canada, USA and Australia but not New Zealand, which even for a major dairying country had the exceptionally high figure of 30 dairy processing publications per million population in 1976.

The rather wide spread between countries in the number of publications per million population presumably reflects the relative expenditure in dairy science and technology. Unfortunately the appropriate R & D statistics are only available for EEC countries and even they are rather incomplete. Thus, in considering the analysis depicted in Figure 9, it should be borne in mind that the number of dairy publications (1976) per million population (1974) is expressed as a function of total food R & D expenditure (1973/74). Notwithstanding these discrepancies, an interrelationship is discernible which concurs to some degree with the relative publication output in the different countries and may account for the typically high output in Denmark (Figure 8). Such an inter-relationship, if confirmed, would not only demonstrate the critical dependence of scientific output on investment but also would validify the use of publications as an indicator of relative investment in different sectors of dairy science and technology.

While probably involving excessive extrapolation the inter-relationship may also be interpreted as suggesting that the average cost of R & D leading to a typical dairy science publication is of the order of 0.125 m $ within EEC States. Alternatively an expenditure of 1 million $ in dairy science

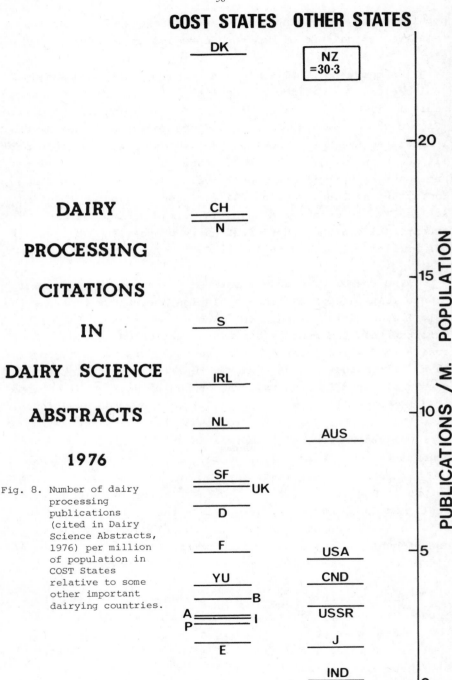

Fig. 8. Number of dairy processing publications (cited in Dairy Science Abstracts, 1976) per million of population in COST States relative to some other important dairying countries.

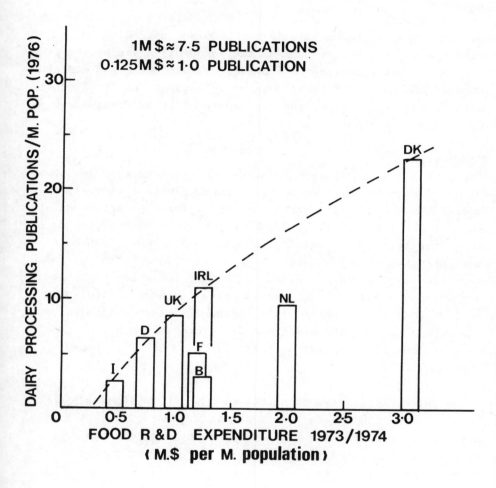

Fig. 9. Number of dairy processing publications (cited in Dairy Science Abstracts, 1976) per million of population in EEC countries relative to total food R & D expenditure (1973/74) in each country.

and technology might be expected to generate roughly 7.5
publications. Clearly these estimates are at best first
approximations which, in addition to the aforementioned
discrepancies in the available data, do not take many important
parameters into account such as the large differences in costs
between member countries. Nevertheless, with existing budgetary
constraints it has become important to estimate the total costs
of research publications and its presentation is designed to
prompt consideration of such estimates.

Conclusions

There is little need to stress that the cursory analysis
of dairy science and technology presented in this paper is at
best rudimentary. Definitions of the different disciplines/
specialisations are rather arbitrary and the classification of
abstracts into the selected categories has in some instances
been subjective. Nonetheless it does provide a useful semi-
quantitative 'bird's eye view' of dairy science and technology
and while few statisticians would agree, as a crude indicator
of gaps and/or imbalances, it is perhaps no less informative
than the more conventional procedure of measuring R & D
performances in different sectors simply in terms of money
expended.

While recognising that extreme caution is necessary in
interpreting the limited and semi-quantitative survey of the
literature summarised in Figures 1 - 9, nonetheless the one
overriding conclusion emanating from the study of particular
relevance to the seminar is:

> the relatively low international priority afforded to
> R & D on the nutritional functions of milks and dairy
> products including such important parameters as
> nutritional stability and general health considerations.

Other deductions which may be made from the limited
survey are more tenuous and they may be better presented in the
form of questions:

Is the concentration on dairy chemistry excessive relative to other disciplines/specialities or should it be directed more to methodology?

Should more emphasis be given to basic microbiology?

Is the concentration of dairy technology on cheese excessive in view of the major problems confronting other conventional large volume products?

Is there a need for further R & D in energy conservation and technological innovations aimed at controlling effluent pollution?

Is it reasonable to estimate the R & D costs of a dairy processing publication as $0.125 m and to suggest that an investment of $1m might support R & D leading to 7.5 publications?

QUALITY R & D REQUIREMENTS

Having attempted to indicate semi-quantitatively the current focus of international dairying R & D attention will now be directed to R & D requirements, mainly in milk itself, but also in dairy processing. Some areas which seem to be receiving insufficient support have already been alluded to, notably nutrition. Consideration, however, will initially be given to R & D problems relating to product quality where the authors can draw more readily from experience.

Physico - Chemical Properties of Milk

Fat Globule Membrane Elucidation of the structure and nutritional functions of milk fat globules, (especially their membranes) and casein micelles are the two major aspects of milk requiring further research. These are the dominant constituents of milk. They are largely responsible for the colloidal nature of milk, largely determine its processability and are in turn altered to a greater or lesser extent by thermal processing. Both components are the subject of extensive investigation and are perhaps the major focus of fundamental dairy research. There are, however, many important facets of each requiring answers.

For example, mention may be made of the paramount importance of the fat globule membrane in protecting milk fat triglycerides from hydrolysis by the lipolytic enzymes (Downey and Cogan, 1975). Also, as discussed subsequently, it is uncertain whether or not the propensity of milks to develop lipolysis reflects the integrity of the fat globule membrane. Further to this questions arise, as further elaborated later, concerning the proposed involvement of what appear to be blood lipoproteins in spontaneous lipolysis (Driessen and Stradhenders, 1974). Assuming that these detergent-like substances do actually potentiate spontaneous lipolysis, do they penetrate the fat globule membrane and if so how do they link the casein micelle bound lipolytic enzymes to their triglyceride substrates within the fat globule? These examples illustrate the important role of the fat globule membrane in determining the ultimate quality of dairy products.

Casein Micelles

From the processing and nutritional viewpoints casein micelles are perhaps the most important ingredients of milk. The precise nutritional function of casein micelles is uncertain although they are generally presumed to facilitate ingestion of calcium and phosphorus. During the last decade considerable progress has been made towards elucidating the structure of casein micelles and a number of structural models have been postulated (Rose, 1969). While these differ significantly in many important features, casein micelles may be envisaged as more or less spherical colloidal particles (diameter 300 - 3000$\overset{o}{A}$) in which the caseins are arranged in an open network linked by the milk salts and resulting in a porous sponge-like structure with an overall negative charge.

In some of the models β-casein is considered to play an important structural role and it has been depicted as constituting the matrix or frame-work which directly or indirectly integrates the other casein components into intact micelles. Recent results however are contrary to this hypothesis (Downey and Murphy, 1970). Rather the β-casein appears to be mainly micellar packing material which is associated with the αs and κ-casein framework through hydrophobic bonding so that the bulk of the β-casein can be removed without micellar disruption.

Further to this a differential packaging of the β-casein within the intact micelles may be postulated (Downey et al., 1973). Preliminary results suggest that about 10% of the total β-casein is loosely associated with the micellar surface and this is released simply by cooling milk. The bulk (50 - 70%) of the β-casein seems to be more firmly bound and is perhaps located within the micellar pores where it is entrained by the colloidal calcium phosphate. The remainder of the β-casein seems to be more strongly bound to αs and κ-caseins and is only released when the soluble milk salts are removed. Additional research might be undertaken to test this hypothesis (Downey, 1978).

Another aspect of special interest is the role of colloidal calcium phosphate in maintaining the structural cohesion of the casein micelles. In one of the previously mentioned micellar models, casein micelles are depicted as collodial calcium phosphate linked submicelles, each consisting of an inner core of αs + β-casein surrounded by an outer layer of κ-casein (Rose, 1969). According to this model removal of colloidal calcium phosphate should yield submicelles with an αs/β-casein ratio approaching that of the original micelles. Contrary, however, to what had been expected on removal of colloidal calcium phosphate the casein micelles are disputed to yield apparently discrete κ, αs + β-casein rich particles in order of decreasing molecular size (Downey et al., 1973). Based on these observations it would appear that colloidal calcium phosphate not only supplements the micellar integration of β-casein as mentioned above, but even more importantly it appears to be critical in linking αs and κ-caseins into intact micelles (Downey, 1978).

Accordingly, the possibility might be examined of whether casein micelles would be better depicted as agglomerates of submicelles composed mainly of αs-caseins which are integrated into intact micelles by colloidal calcium phosphate mediated κ-casein linkages with the β-casein differentially packaged within this micellar framework where it is mainly held by hydrophobic bonding supplemented by colloidal calcium phosphate.

Structural studies concerning the juxta-location of the milk lipolytic enzymes within the casein micelles might also be considered (Downey and Murphy, 1975).

Finally, particular attention should be given to establishing the effects of thermal processing on the structure of the micelles. To date, such studies have been mainly focused on studying the heat induced complexing of κ-casein with β-lactoglobulin.

Organoleptic/Sensory Attributes and Stability of Milk and Dairy Products

The European food supply system is increasingly dependent on technology. Dairy production is heavily dependent on intensified production systems involving genetic selection, extensive use of fertilisers, pesticides, antibiotics, detergents and many other technological innovations. Moreover, the raw milk is stored for longer periods and then subjected to processing sometimes involving advanced technological processes such as UHT which has become increasingly popular in many European countries. The final products are then stored and transported for extended periods under more diverse conditions than hitherto, both at the wholesale and retail levels and often also in the homes where refrigerators and freezers have become common household appliances.

Paralleling these technological developments consumers in many European countries are becoming more concerned with regard to the quality and nutritive attributes and indeed the safety of foods produced by intensive farming systems and then subjected to increased industrial processing. Such foods are often held to be less attractive in respect of their sensory properties and of reduced nutritional quality and to contain harmful residues. <u>It should be stressed that such views are seldom sustained by scientific evidence and are with some possible exceptions unjustified.</u> Nonetheless greater consumer awareness has increased the demand for foods of consistently high quality and with the growing complexity of the food supply system it is essential to ensure that the initial quality of foods is preserved for the longer periods and the more diverse conditions frequently encountered in international trade.

Accordingly, there appears to be a need for further research aimed at identifying and more importantly quantifying the relative importance of the multitude of factors which collectively determine the initial quality and influence the storage stability and ultimately the shelf-life of both conventional and modified/new dairy products. The proposed studies should give particular attention to lipid oxidation (oxidative rancidity) and lipid hydrolysis (lipolysis) which in common with other fat containing foods are two of the paramount causes of off-flavour development in milk and dairy products especially high fat products like butter. (Downey and Cogan, 1975; Wilkinson, 1964; Foley, 1978; Downey, 1975 and Cogan, 1976). The other critical parameter of special relevance to thermally processed dairy products such as evaporated and dried milks and also UHT products especially cream, is the protein content and heat stability of the milk as well as the balance between soluble and colloidal salts. These compositional factors profoundly determine the suitability of milk for thermal processing and critically influence the quality and stability of the final products. Though not immediately relevant to the seminar, the quality of cheese and other fermented products is similarly controlled by the bacteriological and compositional quality of the raw material and of course the presence of antibiotic and other residues (discussed later).

Milk is more variable in composition than the raw materials used in most other sectors of the food industry (McGann and O'Connell, 1971 and McGann, 1975). Apart from its nutritional implications this poses major problems for dairy processing. Compositional variations in milk are major impediments to thermal heat processing. The heat stability of milk (Fox and Morrissey, 1977) is notoriously unpredictable and depends on a host of compositional factors.

Until recently numerous studies failed to correlate heat stability with any of a wide range of compositional factors. It now appears, however, that at least in the case of bulk factory milk most of the variation (85%) may be accounted for by variations in urea concentration which seems to be related to nutrition of the animal (Muir and Sweetsur, 1976 and Muir and Sweetsur, 1977).

Although perhaps less critical, variation in composition and physico-chemical properties of raw milk also creates problems for other sectors of the dairy industry. For example variations in rennet coagulation time, rennet curd strength and syneresis characteristics pose problems in cheese manufacture (Ernstrom and Wong, 1974). Changes in fatty acid profile lead to variations in the plasticity and spreadability of butter (Foley, 1978). Moreover, inconsistencies in the keeping quality of butter and other products reflect compositional changes in the milk as subsequently discussed (Downey, 1975 and Cogan, 1976).

In aiming therefore to identify and quantify the relative importance of the multitude of factors which collectively control the quality and stability of thermally processed dairy products the proposed study should give special emphasis to the initial quality of the raw milk which for all practical purposes predetermines the quality of the final products. The ultimate objective might be to develop improved criteria for selecting milk for processing by indicating the suitability or otherwise of different milk supplies for particular processes. In other words, improved compatibility between raw material quality and particular processes leading to better quality control, improved product quality and stability. As an example of what might be envisaged for thermally processed dairy products, the conclusions of a recently completed comprehensive study (Downey, 1975 and Cogan, 1976) of the controlling parameters in butter quality are schematically summarised in Figure 10 (Downey, 1978). This shows that salt content and pH, the essential compositional differences between the four butters examined, are the two critical factors which distinguish between the susceptibility of salted and unsalted sweet and ripened cream butters to post-manufacture lipid auto-oxidation, lipolysis and associated microbial growth. However, both factors counteract each others' potentiating influences on these quality defects when they are either simultaneously increased or decreased and the combination of the two factors, viz. low pH (4.75) and high salt content (2%), most conducive to lipid auto-oxidation is inhibitory to lipolysis and associated microbial growth, conversely pH6.6 and no added salt are favourable to lipolysis and microbial growth but not lipid auto-oxidation.

Over and above these intrinsic compositional parameters are the pre-disposing influences of both packaging and seasonality which critically determine the propensity of each of the types of butters and indeed other dairy products to develop lipid auto-oxidation, pre- and/or post-manufacture lipolysis and microbial growth (Downey and Cogan, 1975; Downey, 1975 and Cogan, 1976).

Unless butter is packaged immediately following manufacture (ex-churn), rather than following cold storage in bulk (ex cold-store), the risk of post-manufacture lipolysis and associated microbial growth are both greatly enhanced. Moreover, there is a ubiquitous risk of photo-oxidation unless light impervious packaging materials are used.

These somewhat generalised conclusions may in fact be even further simplified

Copper induced lipid auto-oxidation is the primary cause of flavour deterioration in <u>salted ripened cream butter</u>

Lipolytic defects are the primary cause of flavour quality impairment and deterioration in <u>unsalted sweet cream butter</u>, <u>unsalted ripened cream butter</u> and <u>salted sweet cream butter</u> in order of decreasing risk.

<u>Lipid Auto-Oxidation</u>

In addition to metal induced lipid auto-oxidation and resultant flavour impairment, bovine milk may develop <u>spontaneous oxidised flavour deterioration</u>. This phenomenon has been most extensively studied in the US and various reports are present in the literature indicating that while metal (copper) - induced oxidised flavour in milk is a chemical oxidative process, spontaneous oxidised flavour development (without added copper) is an enzymic reaction involving xanthine oxidase (Aurand et al., 1967). While the proposition has not however received general acceptance it would seem to merit further investigation.

Lipid oxidation, as previously mentioned, seriously limits the storage potential of most fat containing foods including dairy products (Wilkinson, 1964). It has been a major stumbling block in the manufacture of whole milk powder (Pyle, 1975) and it is also a critical factor in limiting the shelf life of UHT cream (Downey, 1968). Flavour defects due to lipid auto-oxidation also arise in butter (Figure 10) but they are likely to be less common than lipolysis in that they are confined mainly to one type of butter <u>viz</u>. salted ripened cream butter and even then only in butter manufactured from winter milk, because of its higher copper levels (Downey, 1975 and Cogan, 1976). Indeed from a quality control viewpoint, it might be said that the copper content of salted ripened cream butter is not entirely disadvantageous. In that it is the critical determinant of auto-oxidation, the level present can provide a useful, <u>albeit</u> approximate, predictive index of keeping quality (Figures 11a, 11b) (Downey, 1975).

BUTTER QUALITY – CONTROLLING PARAMETERS

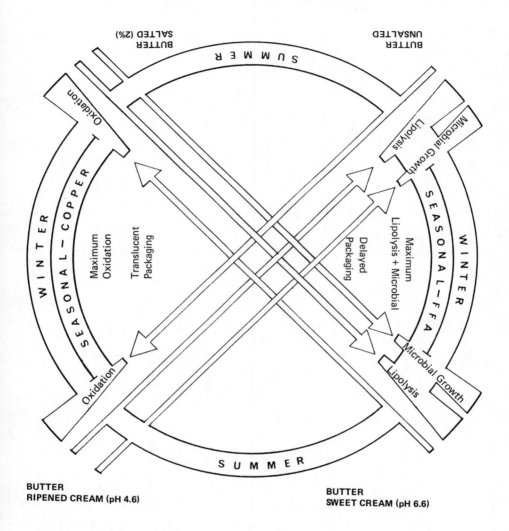

Fig. 10. Schematic representation of the interactive influences of the intrinsic pH value and salt content of the butter and the predisposing influences of both packing and seasonality of milk composition on the propensity of sweet cream butter (salted and unsalted) and ripened cream butters (salted and unsalted) to develop flavour impairment due to either auto-oxidation, pre-manufacture lipolysis or post-manufacture lipolysis and microbial growth (Downey, 1978; Downey, 1975; Cogan, 1976).

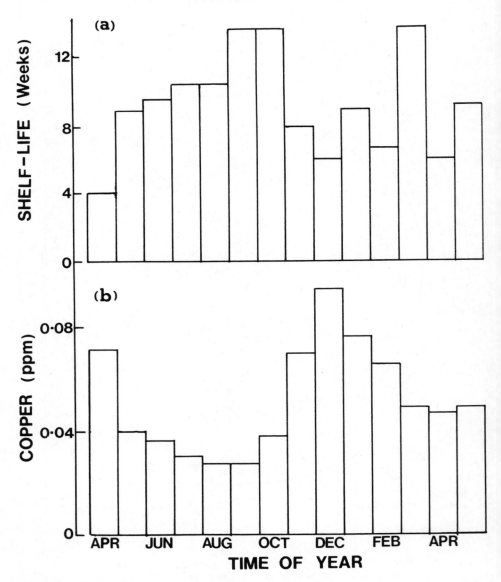

Fig. 11a. The inverse relationship between the seasonal trends in (a) the shelf-life at 15°C and (b) the intrinsic copper content of ripened cream butter (slightly salted: packed ex-churn) (Downey, 1975).

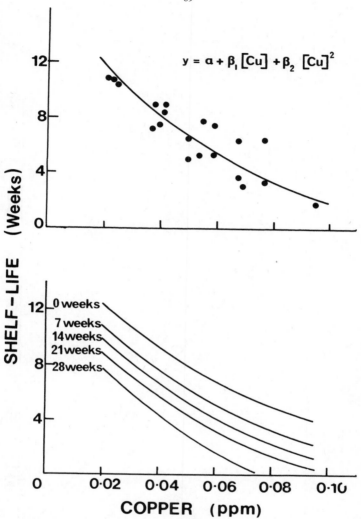

Fig. 11b. Polynomial regression equation depicting the inverse relationship between the shelf-life at 15°C and the intrinsic copper content of ripened cream butter (slightly salted: packed ex-churn). (Downey, 1975).

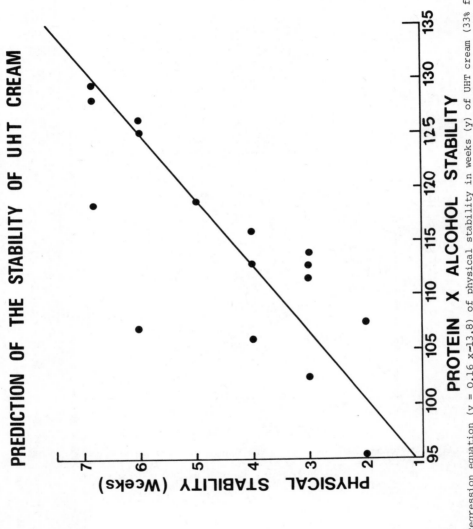

Fig. 12. Regression equation ($y = 0.16\ x-13.8$) of physical stability in weeks (y) of UHT cream (33% fat) held at 18°C relative to the product (x) of protein concentration x alcohol stability (Downey et al., 1970).

It is hardly necessary to emphasise the usefulness of keeping quality prediction tests, even if only guidelines, to the dairy processing industry, a further example of which is given in Figure 12 for predicting the physical stability of UHT cream (Downey et al., 1970). Unfortunately, there are too few such tests and instead, the industry is largely dependent on testing of control samples. Thus, it is recommended that the development of keeping quality/stability prediction tests, including more meaningful accelerated deterioration tests, should constitute a high priority for future dairy research. In this regard the need for a rapid method for monitoring the moisture distribution in butter - which largely determines its microbiological keeping quality - comes immediately to mind.

The development of an in-process control method (perhaps based on infra-red scanning) to replace the current post-hoc subjective procedure, but which is nonetheless a critical butter grading test, would make a major contribution to stock control of butter.

Lipid Hydrolysis

Flavour impairment and deterioration in butter and other high fat dairy products, due to lipolysis, is more complex than lipid auto-oxidation, in that there are multiple causative factors. Lipolytic enzymes from at least three origins may contribute to an elevated free fatty acid level in dairy products and there are a number of unresolved questions on each requiring both fundamental and applied research. The relative importance of the different causes is a continuous source of debate and from an organoleptic viewpoint it may be important to distinguish between the causes, as explained subsequently (Downey and Cogan, 1975; Downey, 1978; Downey, 1975; Cogan, 1976 and Downey, 1974).

Lipolysis in Milk

Bovine milk contains a significant level of native lipolytic enzymes (Downey, 1974). Yet normal milk does not develop lipolysis. The fat globule membrane protects the

triglycerides from attack by the lipolytic enzymes whose
activity is further impaired by their association and/or
occlusion within the casein micelles, the large casein particles
responsible for the colloidal nature of milk. Thus, the native
milk lipolytic enzymes are well partitioned from their triglyceride
substrates (Downey and Murphy 1975). Hence, little or no
lipolysis occurs in normal milk since for all practical
purposes the lipolytic enzymes cannot make contact with the
triglyceride substrate.

However, if the raw milk or cream is subjected to excess
agitation or turbulence the fat globule membrane may be
disrupted resulting in the enzymes having access to the triglycerides which are accordingly hydrolysed. Accumulation of the
reaction products, especially the free fatty acids (FFA) as
well as the mono and diglycerides, confers a rancid or
lipolysed flavour on the milk which may in turn be transferred
to products manufactured from the milk. This is known as
<u>induced lipolysis.</u>

It is generally accepted that excessive agitation of
raw milk and/or cream is the major cause of induced lipolysis
(Downey, 1974). Farm handling of milk, especially milking, is
considered to be mainly responsible for the excess agitation
(Downey and Cogan 1975). It may also arise subsequent to the
farm handling and some investigators hold the view that much
of the agitation occurs during the creamery handling of milk.
The relative significance of both potential causes may vary
with different circumstances and these would need to be
investigated.

Agitation on the farm may be minimised by ensuring that
milking machines are well serviced, maintained and correctly
operated so as to eliminate excess air intake, air leakages and
other causes of turbulence. Further to this, low-line milking
installations which may be more expensive but less prone to
agitation are preferred in some European countries. The
advantages of low-line milking installation are not however

universally accepted. To overcome the problems of induced lipolysis the International Dairy Federation's Expert Group on Lipolysis recommended in 1975 (Downey and Cogan, 1975) that special attention should be given to developing milking machines and installations which cause minimum agitation of milk. Accordingly, this may be recommended as a fruitful area for further R & D in dairy engineering.

Even in the absence of agitation, milks from individual cows sometimes develop excessive free fatty acids. This is termed <u>spontaneous lipolysis</u>. Among the physiological factors considered to pre-dispose milk to this quality defect which is extremely difficult to control, stage of lactation, the nutrition and individuality of the cow, manner of feeding, and milking technique are most frequently cited (Downey and Cogan, 1975). Spontaneous lipolysis is perhaps the least well understood facet of lipolysis and it is thus recommended for further research (Downey, 1974). There are few aspects of spontaneous lipolysis about which an unequivocal statement may be made, apart from the fact that it occurs. Even then the incidence of the phenomenon is uncertain and difficult to establish since individual cows maintained under apparently the same conditions vary markedly from day-to-day in the susceptibility of their milks to spontaneous lipolysis (Downey and Cogan, 1975). In considering remedial action to arrest lipolysis, it is obviously important to know whether the problem is due to induced lipolysis which may be alleviated by attention to mechanical factors or whether it reflects some physiological circumstance such as an inadequate energy content in the diet.

To distinguish between these alternatives, further fundamental research should be undertaken with the objective of defining the intrinsic difference(s) between milks which exhibit spontaneous lipolysis and those exhibiting only induced lipolysis. Differences in the complement of lipolytic enzymes present in the two types of milk is unlikely and attempts to examine by electromicroscopy whether the fat globule membrane in spontaneous lipolysed milk may be different (less intact?) from normal milk have not been successful (Downey and Cogan, 1975).

Recently it has been postulated that the propensity of individual milks to develop spontaneous lipolysis is determined by the level of phospholipid containing substances present in the milk (Driessen and Stradhenders, 1974 and Downey, 1974). It is suggested that in certain physiological conditions these substances infiltrate from the blood into milk. To confirm the hypothesis it would be necessary to establish a correlation between the incidence of spontaneous lipolysis and the level of phospholipid containing substances, possibly lipoproteins, present in individual milks. With existing methodology this may not be easy. The difficulty may be further compounded by the aforementioned day-to-day seemingly erratic fluctuations in the propensity of milk from an individual cow to develop spontaneous lipolysis. Even if the hypothesis is accepted it is difficult to envisage the precise mechanism by which the detergent-like substances actually potentiate lipolysis. Do they become attached and penetrate the fat globule membrane, and if so, how do they link the substrate to the lipolytic enzymes which are bound to the particulate casein micelles? (Downey and Murphy, 1975). Alternatively, they may become attached to any free fat present in the milk and thereby enhance the activity of the milk lipolytic enzymes whose activity against long chain triglyceride substrates is known to be enhanced by lipoproteins. Thus, the propensity of milk to develop spontaneous lipolysis would be related to the quantity of free fat present which among other things may be influenced by agitation of the milk. In these circumstances the distinction between spontaneous and induced lipolysis would be mainly one of degree (Downey, 1974).

Lipolysis in Dairy Products

As already indicated enzymes from at least three sources may contribute to elevated free fatty acid levels in dairy products (Downey, 1978). In addition to the lipolysis of milk

(Downey, 1974) and uptake of the free fatty acids by the products (termed post-manufacture lipolysis), lipolysis may also develop in the dairy products themselves during storage (Downey, 1975 and Cogan, 1976). <u>Post-manufacture lipolysis</u> may arise either due to microbial growth in the actual products or the activity of heat-resistant (survive HTST pasteurisation and also UHT sterilisation) microbial enzymes produced in the bulk cooled milk/cream by psychrotropic bacteria (Cogan, 1977).

The ability of phsychrotropic bacteria, especially the *Pseudomonas* genus, to produce in bulk cooled milk heat-resistant enzymes which may concentrate in the cream and remain active in products, notably butter and cheese, but also in various thermally processed products such as condensed and dried milks and UHT products, is obviously of considerable commercial significance, not just because of the risk of lipolytic defects in the products but perhaps even more so proteolysis leading to product instability. There are, however, many uncertainties regarding heat-resistant psychrotropic enzymes requiring further research and indeed the significance of the potential risk during commercial storage of dairy products is the subject of continuing debate (Cogan, 1977).

<u>Phychrotropic bacterial enzymes and product quality</u>. Compounding the adverse effects of mechanical agitation and turbulence in milking machines (already discussed) the attendant widespread use of farm bulk refrigeration tanks where raw milk may be stored for 2/3 days at $2 - 4^{\circ}C$ coupled with prolonged holding of the milk at the factory prior to pasteurisation (based on Cogan, 1977) have further enhanced the risk of quality impairment of dairy products. During the 2 - 3 days or longer which may elapse between milking on the farm and processing in the factory there is ample opportunity for psychrotropic bacteria to grow in the cooled milk and produce enzymes, including lipase and proteinases. Though the bacteria themselves are inactivated, their enzymes are reputed to be remarkably heat stable and capable of surviving normal pasteurisation

temperatures (72°C - 15/20 sec.). For example, the data summarised in Table 1 (taken from review submitted to the IDF Lipolysis Group by Dr. T.M. Cogan) indicates that the time required for 90% inactivation of these microbial enzymes especially the proteinases is considerably longer than that employed in normal HTST pasteurisation. As a further instance of the remarkable heat resistance of some microbial enzymes it may be noted that a proteinase from Pseudomonas MC 60 is 400 times more heat-resistant at 149°C than spores of PA 3679, one of the common organisms used to evaluate time/temperature treatments in the canning industry.

TABLE 1

TIMES REQUIRED FOR 90% INACTIVATION OF BACTERIAL ENZYMES (COGAN, 1977)

Organism	Enzyme	Time (min)
Ps. fluorescens 22F	Lipase	4.80
Ps. fluorescens 31H	Lipase	1.67
Alcaligenes spp 23a 2	Lipase	1.60
Ps. fragi 14-2	Lipase	0.27
S. marcesens D2	Lipase	0.38
Achromobacter spp 23 O	Lipase	33.0
Ps. putrefaciens R48	Lipase	0.74
Al. viscolactis 23 al	Lipase	0.58
Ps. fragi 3	Lipase	33.0^b
Pseudomonas 21B	Lipase	170^a
Pseudomonas 21B	Proteinase	250^a
Pseudomonas MC 60	Proteinase	420^a
Ps. fluorescens P26	Proteinase	350^a

[a] calculated and/or extrapolated

[b] estimated since the inactivation curve is slightly curvilinear.

Reports are contained in the literature of quality deterioration in various dairy products attributed to heat-resistant bacterial enzymes (Cogan, 1977). Flavour impairment of cheese and gelation of UHT products are two important quality defects often associated with such causes. The data is rather limited (Table 1) but microbial proteinases seem to be more heat-resistant than the lipases and hence a greater risk to product quality.

There is, however, insufficient information on the propensity of the dominant psychrotropic bacteria present in milk microflora, such as the gram negative bacteria of the genera *Pseudomonas*, *Achromobacter* etc., to produce heat-resistant lipolytic and proteolytic enzymes and the heat-resistance of the enzymes themselves is poorly quantified. State-of-the-art reviews currently being prepared from the International Dairy Federation Lipolysis Group recommend that further research should be undertaken on heat resistant microbial enzymes, their contribution to the quality deterioration of dairy products during storage and the quantity of the enzyme which must be present to induce defects. The proposed research should establish the heat resistance of psychrotropic dairy bacteria and their enzymes under conditions comparable to those prevailing in milk and at temperatures employed in HTST pasteurisation and UHT sterilisation. Also lower temperatures should be investigated since there is a suggestion that some bacterial enzymes may be more labile at $50 - 60^\circ C$. If confirmed it would be advantageous in that the enzymes could be inactivated at temperatures which would have little effect on product quality. Particular attention should be given to establishing whether the enzymes are produced during the exponential growth phase, which would be most adverse from the viewpoint of product quality and perhaps also the raw milk (Cogan, 1977). If, however, the enzymes are produced only during the stationary phase of growth, there would be less risk to the quality of products since then the milk would have deteriorated beyond processability. Also the stability of the heat resistant enzymes during storage of dairy products should be considered.

especially products such as butter which are stored under freezing conditions for long periods. Further to this, the possibility of enzymic deterioration in frozen dairy products, concerning which there are some suggestions in the literature, might merit consideration.

Such studies on the more fundamental aspects of microbiology which, as mentioned earlier, may not be receiving sufficient attention (Figure 3) would help to establish whether, as frequently postulated, heat-resistant bacterial enzymes are in practice a serious risk to dairy product quality. Clearly, it would be most disquieting if the increasing use of refrigerated farm bulk tanks to improve the microbiological quality of the milk itself is inadvertently resulting in dairy produce with reduced keeping quality.

Microbial Growth and Lipolytic Deterioration With milk of good (5×10^3 <counts /ml) or reasonable ($<10^5$ counts/ml) quality, lipolysis due specifically to microbial growth is unlikely to constitute an important cause of flavour impairment of milk itself during storage at $6°C$ or less for up to 3 days, even with temporary increases in temperature when fresh milk is introduced at successive milkings (Downey, 1974). The intrinsic milk lipolytic enzymes are the primary cause of milk lipolysis as already explained. In dairy products however microbial growth is frequently associated with post-manufacture lipolysis and quality deterioration as illustrated in Figure 10. As previously mentioned microbial growth in butter is largely governed by the size of the water droplets and when these are reduced below a critical limit growth is considerably retarded. Butter which has been packaged from cold storage tends to have larger moisture droplets than butter packaged immediately following manufacture and is accordingly more susceptible to microbial growth and quality impairment (Downey, 1975 and Cogan, 1976). The marked differences between microbial growth in butter packaged ex-churn and ex-cold store reaffirms the aforementioned importance of developing a rapid method for monitoring the moisture distribution of butter. Further to this the development of improved analytic methods is a general requirement of almost all facets of dairy science and technology.

Development of Analytical Methodology

Inadequate analytical methodology is arguably the major impediment in almost all facets of dairy science including the assessment of the quality and nutritive attributes of milk and dairy products. With some notable exceptions the existing methodology is not adequate for the increasingly complex product and in-process control which is a prerequisite of modern food technology. Discrepancies between analyses from different control laboratories are common even with the rudimentary compositional tests. Conclusions as to the current quality and nutritional functions of the products which can be drawn from the analyses are inevitably the subject of debate and even more importantly methods for predicting keeping quality and nutritive stability are often unreliable and too few in number as previously stated. The apparent increase in the annual percentage of dairy chemistry abstracts devoted to analytical methodology (Figure 2) which also constitutes a significant proportion of dairy microbiology abstracts (Figure 3) presumably reflects the widely felt need for improved methodology.

Methodology for Monitoring Quality Determination

Further research is, therefore, recommended aimed at developing more meaningful methodology for monitoring changes in the quality and nutritive attributes of milk and dairy products during thermal processing, storage and distribution. Such methodology is obviously a pre-requisite to undertaking the previously proposed comprehensive research programme aimed at identifying and quantifying the relative importance of the multitude of factors which collectively control the quality, nutritive attributes and stability of thermally processed dairy products. Further to this the development of automated instrumentation for monitoring changes in the quality and nutritive attributes of dairy products would greatly facilitate in-process control. The success already enjoyed by the automated methods for establishing milk composition, counting bacteria in milk and related procedures based on measuring bacterial metabolites, attestify to the requirements for automatic analytic control in the dairy industry.

The development of improved analytic methodology is certainly an area where European cooperation could be beneficial as evidenced by the invaluable work of the International Dairy Federation (IDF) in the formulation of standard methods for <u>quality</u> assessment and which are extensively used by the dairy industry. To complement this work by alleviating the long delays in elaborating standard methods which sometimes arise due to incomplete data and inadequate experience on the relative merits of alternative and/or new procedures, the proposed European cooperative research should concentrate on the more fundamental facets of methodology including modification and evaluation (initially for research) of techniques used in other fields notably biochemistry.

Recognising the widely differing sensory appreciation of food quality not only between countries but even among individuals from one family it is not envisaged that the proposed research should aim to establish universally accepted chemical/physical indices of good food quality. This is hardly possible. Extensive research has already been undertaken aimed at identifying the specific compounds and mixtures responsible for the unique flavour of milk, dairy products and various other foods. The practical use of such information in defining the intrinsic quality of a particular food is uncertain. Instead the objective of the proposed research should be to develop analytical methods for <u>monitoring changes in the quality and nutritive attributes of milk and dairy products</u> resulting from newly adapted agricultural production practices (some examples of which have already been mentioned in regard to lipolysis), thermal processing, storage and distribution with the longer term aim of improved production and process control and ultimately predicting the keeping quality and nutritive stability of milk and dairy products.

In addition to the quality prediction tests already mentioned (including the moisture distribution in butter) particular attention should be given to establishing the statistical correlations between subjective visual and organoleptic assessment of quality (taste panel etc) and more objective chemical and/or physical analytical methods. A number of relatively simple well known tests are extensively used in research and quality control to monitor, for example, lipid autoxidation and hydrolysis the two previously discussed paramount causes of flavour impairment. It is however surprising to note that the statistical basis for grading products on the basis of these tests is at best tenuous and often non-existent.

Correlation between Organoleptic and Chemical Methods

Flavour impairment in dairy products due to lipid autooxidation is generally measured by one of many modifications of the traditional Thiobarbituric acid (TBA) or Peroxide (PV) tests. As indicated in Table 2 there is a good statistical correlation between these chemical indices of lipid autooxidation in UHT cream and taste panel assessment (Downey, 1968). The problem is that corresponding correlations for many other dairy products are not well documented in the literature. For example, like UHT cream, butter with a peroxide value greater than 2.0 is generally taken to have developed excessive levels of lipid auto-oxidation. This concurs with the author's experience with salted sweet cream butter (Downey, 1975) though the statistical basis for the correlation is not well documented in the literature. It is hardly satisfactory however to have the same tolerance figure for ripened cream butter. Its characteristic '<u>lactic flavour</u>' masks the onset of oxidised flavour (and also lipolytic flavour), so that peroxide levels of about 2.0 are usually associated more with a bland flavour rather than the typical oxidised flavour which is only manifest at higher peroxide levels (Downey, 1978). In overall, therefore, the peroxide and thiobarbituric tolerances corresponding to unacceptable levels of lipid oxidation in different types of butter and

TABLE 2

FREQUENCY DISTRIBUTION OF UHT CREAMS ACCORDING TO FLAVOUR SCORE AND CHEMICAL ASSESSMENT (Downey, 1968).

(a) Thiobarbituric acid (TBA) ranges ($P < 0.01$: $x^2 = 400$)

Flavour scores	$\leqslant 0.08$	0.08-0.16	$\geqslant 0.16$	No. of samples
Acceptable	181 (90%)	49 (44.5%)	18 (6.8%)	248
Doubtful	13 (6.5%)	45 (40.9%)	48 (18.2%)	106
Unacceptable	7 (3.5%)	16 (14.6%)	198 (75%)	221
Total	201 (100%)	110 (100%)	264 (100%)	575

(b) Peroxide ranges ($P < 0.01$: $x^2 = 71$)

	< 2.0	> 2.0	No. of samples
Acceptable	118 (43.5%)	3 (4.9%)	121
Doubtful	66 (24.4%)	2 (3.3%)	68
Unacceptable	87 (32.1%)	56 (91.8%)	143
Total	271 (100%)	61 (100%)	332

butter oil (where the peroxide tolerances are frequently questioned) and other dairy products ought to be statistically reassessed.

Similarly, with lipolysis the statistical basis for grading products according to FFA levels needs to be re-examined and again having the same tolerances for different types of butter is not satisfactory. Indeed, recent observation seriously questions the validity of the FFA methods currently used to measure lipolytic flavour defects in butter and presumably other dairy products (Downey, 1978).

Free fatty acid levels in butter are generally estimated by titrating butter oil prepared from the butter. Based on these procedures good quality butter has an FFA level less than 20 units (1 unit = 1 mg NaOH/100 g butter oil) while values above 40 units are indicative of excessive lipolysis. Recently, however, it has been observed that salted sweet cream butter which develops high post-manufacture FFA levels (>40 units) tends to exhibit a more pronounced lipolytic flavour than freshly manufactured butter with similar FFA levels, due to pre-manufacture lipolysis (Downey, 1975 and Cogan, 1976). It is tempting to suggest that the organoleptic differences between the butters despite their comparable FFA readings may be due to preferential loss (because of greater water solubility) of the more strongly flavoured shorter chain fatty acids during separation of the lipolysed milk, churning of the cream and butter washing. Thus, the ratio of short to long chain fatty acids in the butter with elevated FFA levels due to pre-manufacture lipolysis may be sufficiently lower to confer on it less of a lipolytic flavour than a butter which develops post-manufacture lipolysis and contains the full complement of liberated short and long chain fatty acids (Downey, 1978). While the organoleptic discrepancy between butters with elevated FFA levels due to either pre- or post-manufacture lipolysis has not been unequivocally confirmed, the observation is clearly of considerable commercial significance especially if it also applies to other dairy products. Its mention underlines the necessity of establishing more precise statistical correlations between organoleptic assessment and standardised methods of FFA determination, not only in different varieties of butter and other dairy products, but also a different tolerance might be reasonable for drinking and manufacturing milk.

NUTRITIONAL R & D REQUIREMENTS

Turning finally to the nutritional attributes of milk and dairy products consideration will be briefly given to the possible safety aspects or potential health hazards arising from contamination of milks with environmental contaminants.

Persistent Insecticides and Other Chlorinated Residues

Bovine Milk and Dairy Products. Organochlorines are the principal insecticidal contaminants of food - between 70 and 85% of the dietary intake of pesticides in the US is due to chlorinate residues of which DDT is the most prevalent, constituting over 75% of the daily intake. Extensive surveys have been undertaken to establish the levels of these residues in milk and dairy products from major dairying countries. Compilation of the results reveals that the average level of DDT and other organochlorine residues reported in milk and dairy products from major dairying countries are generally less than the Codex Tolerance Limits (Downey, 1972). While the maxima levels detected in products from a number of countries do exceed the tolerance limits, samples containing such high levels are infrequently encountered. Excluding these atypically high samples it is pleasing to record that the organochlorine residues in bovine milk and dairy products manufactured in Europe are generally less than the Codex Tolerance Limits and accordingly pose no obvious hazard to man. Further to this and of particular relevance to the seminar, it should be added that both domestic and commercial thermal processing of foods reduce the organochlorine level. Thus thermal processing which is often criticised for impairing the flavour and also the nutritive value of food is an added safeguard in this regard. The adventitious removal of organochlorine residues from milk by thermal processing is schematically summarised in Figure 13.

Apart, therefore, from accidental contamination, it may be concluded that there is little immediate risk to human health from organochlorine contaminants in bovine milk and more especially thermally processed products. There is, however, a related aspect requiring further research.

Animal feed from pesticide treated crops is the principal origin of bovine milk organochlorine residues. It should be noted firstly that milk is the primary excretory by which lactating animals eliminate ingested organochlorines and secondly that during transfer from feed to milk low dietary organochlorine levels may be concentrated as much as 20 - 30 fold. For example feed levels of DDT comparable to those specified in EEC regulations have been shown by some investigators to yield milk fat residues of comparable magnitude to the Codex Tolerance Limits. In view, therefore, of the many uncertainties involved in predicting milk residue levels from those present in animal feed, further research might be prudent in order to establish with certainty the total concentration of each organochlorine which can be tolerated in animal feed without impairing the current purity status of bovine milk (Downey et al., 1975).

Human Milk

In contrast to bovine milk, the levels of organochlorines and other chlorinated residues in human milk may be a matter for concern. Compilation of the published literature reveals that the average levels of DDT and other chlorinated residues in human milk are about 5 - 10 times higher than the corresponding bovine milk, (reflecting the terminal position of man in the food chain) (Downey et al., 1974). Moreover, within the last decade other chlorinated residues, notably α-BHC, β-BHC and HCB have been detected in human milks from many countries. Most concern has, however, been aroused by the

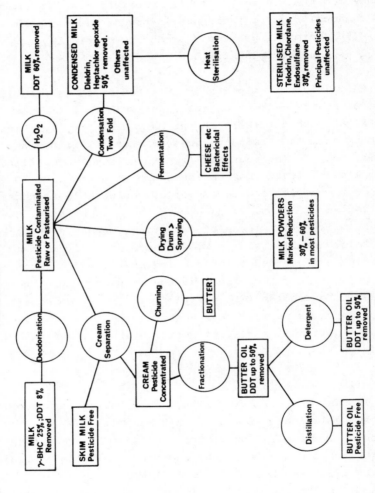

13. Schematic representation of the step-wise removal of organochlorine pesticides from milk with increasing thermal processing (Downey, 1972).

detection in human milk from some countries - of the highly toxic and persistent polychlorinated biphenyls (PCB's) which appear to be rapidly overtaking DDT as the most abundant of the chlorinated pollutants in the global ecosystem.

In the absence of tolerance limits for human milks, the extent of the risk to which infants breast-fed milks containing high levels of chlorinated residues are exposed is open to conjecture. It is, however, generally accepted that infants whose enzyme systems are still in the development stage are more susceptible than adults to certain chemicals. Accordingly, it may be noted that the average DDT and aldrin/dieldrin content of human milk from many European countries exceeds the Codex Tolerance Limits for bovine milk and the WHO/FAO acceptable dairy intake by factors ranging from 2 - 5 generally, but by as much as 10-fold in some countries (Downey et al., 1974). By definition of course this does not imply that breast-fed infants are in imminent danger. Rather it highlights the requirement for further research aimed at establishing internationally accepted tolerance limits for pesticides and other residues in human milk.

PCB Contamination of Food

PCB's are universal, and together with DDT residues constitute the most abundant chlorinated hydrocarbon pollutants in the global ecosystem.

The detection in human milk of PCB's which are chiefly used as dielectric and industrial fluids in transformers, capacitors, hydraulic systems etc. but also as plasticisers in adhesives, textiles and packaging material, does raise the question of food contamination by these highly toxic residues. Although the likelihood of cow's milk becoming contaminated is high, only one instance seems to be reported in the literature and this arose from silo-paint. Perhaps the most imminent danger as far as food is concerned, including dairy products, arises from the leaching out of PCB monomer and other chemicals into high fat products (butter, margarine, lard etc).

in which the chemicals are highly soluble.

Recognising the hazards of PCB pollution, member countries of OECD decided by common accord to control the production and sale of chemical products known as Biphenyls or Diphenyl Polychloro (PCB) except in closed systems such as condensers and transformers. The agreement recommends that the decision be applied first to the elimination of the use of PCB's from such things as coolants in the food industry.

Biodegradable Insecticides

One of the major recommendations of various government reports from the UK., US and other countries concerning pesticides is the necessity for further research on the mechanism of toxicity, metabolism and more especially the detoxication systems in mammals and other species (Downey, 1972). Thus, while the current shift to biodegradable insecticides is welcome, the more widespread use of these highly toxic chemicals in agriculture and public health raises a number of questions to which clear-cut answers are not available (Downey, 1978). Regarding detoxication systems the possibility might be investigated of whether in conjunction with phosphatases and the pseudo-enzymic activity of serum albumins, the A-type carboxylic ester hydrolases constitute a universally distributed 'hydrolase detoxication system' analogous to the microsomal mixed function oxidases which catalyse the activation and detoxication of insecticides by a wide variety of reactions including deamination, N-dealkylation, desulphuration, oxidation, hydroxylation, epoxidation, etc. A-type esterases are ubiquitous in nature, their natural substrates are not known but they can split organophosphates, carbamates and other extraneous esters.

Another aspect requiring further attention is the possible effects of organophosphates and carbamates on non-target enzymes. These insecticides are relatively specific enzyme inhibitors. They exert their insecticidal functions by inhibiting cholinesterases, of which acetylcholinesterase is the most critical being located in the nervous system where it hydrolyses acetylcholine a few milli-seconds after it is released and has executed its function in transmitting nerve pulses. However, they are not absolutely specific, organophosphates and carbamates also inhibit other enzymes of the carboxylic ester hydrolase family which include lipolytic enzymes important for example in fat digestion and also in milk fat synthesis.

In recommending, therefore, the more widespread use of organophosphates and carbamates in dairying it should be remembered that the key lipolytic enzymes involved in mammary gland milk fat synthesis are totally inhibited by relatively low concentrations (10^{-8}-$10^{-5}\underline{M}$) equivalent to a few ppm of these insecticides. While the concentration of organophosphates and carbamates in bovine tissues following treatment of animals with these compounds is unlikely to be sufficient to inhibit the lipolytic enzymes involved in milk fat synthesis nevertheless the possible long-term effects arising from continued application of the insecticides should not be disregarded. Also the possibility exists of developing laboratory screening methods for grading individual organophosphates and carbamates according to their relative biodegradability by examining their rates of hydrolysis by A-type carboxylic ester hydrolases.

In overall, therefore, it may be concluded that organochlorine insecticides used in agriculture and their residual levels in bovine milk and dairy products pose no obvious hazards to adult man. The elevated levels of chlorinated residues in human milk is, however, disquieting. Moreover, fear of persistence per se must not however induce premature widespread use of alternative insecticides of relatively higher toxicity and concerning which there are many unanswered questions.

Antibiotic Residues

The widespread use of penicillin and other antibiotics in bovine therapy and disease control coupled with failure by some farmers to withhold milk from treated cows for the specified period are the major causes of antibiotic residues in milk and dairy products (Bulletin de l'Office International des Epizooties, 1971). Because of possible health implications together with the inhibitory effect of even trace residue levels on the fermentation processes involved in the manufacture of cheese, yoghurt etc, bovine milk is routinely screened for antibiotics and other residues. The general consensus seems to be that while individual milks may contain antibiotic residues when these are diluted in bulk milk there appears to be no major risk to man.

Dissatisfaction is however sometimes expressed with the present screening methods which are rather tedious and may not be sufficiently sensitive especially in regard to some of the more recently developed antibiotics. Accordingly, further research may be required to develop rapid automated antibiotic screening procedures which would not only be helpful to the industry but would also safeguard the consumer.

Heavy Metals

During the last decade the environmental consequences of non-organic micro-pollutants have received increasing attention. In 1974 the EEC compiled an inventory of the available data concerning heavy metal pollutants in the environment (Non-organic Micro-pollutants of the environment, 1976). Of interest to the seminar are the levels (mg/kg w.w) of arsenic (0.001 - 0.1) cadmium (0.01 - 0.1, up to 1.0), copper (0.01 - 1,0, up to 10.0), iron (1.0 - 10.0), lead (0.1 up to 1.0) and mercury (0.001 - 0.01, up to 0.1) detected in foods. The levels of cadmium, lead and mercury in some samples of drinking water exceeded the WHO (1972) international standards. Because of the lack of standards for general foods, it is difficult to comment on the possible health risk associated with the levels detected in foods. However, the EEC report points out that the mercury levels in some food, notably certain fish, could lead to a dietary intake in excess of the provisional tolerable dietary intake recommended by FAO/WHO.

Heavy metal contamination of milk and dairy products has been extensively studied in most European countries. As with pesticides, compilation of the existing data (Heavy Metals and Other Elements in Milk and Milk Products, 1977) already in progress by the Internation Dairy Federation is a prerequisite to assessing the extent of the contamination and the potential health risks. Further to this there is a critical need for improved methodology for the detection of heavy metals in milk and dairy products. Also attention should be directed towards establishing further tolerance criteria for heavy metals in milk and dairy products analogous to the Codex Tolerance Limits for organochlorine residues. Some progress has indeed already been made in regard to lead levels in canned infant food formulations.

In 1971/72 it was observed that both bovine and human milks are roughly 25 and 50% respectively lower in lead than many canned infant food formulations. Accordingly, a diet composed of such canned products would deliver much larger dosages of lead to an infant than either cows' or human milk. In view of the serious cumulative consequences the FDA established a standard of 0.25 ppm as the maximum for lead in evaporated milk and infant formulation. In the meanwhile there has been a conspicuous reduction in the US in the lead levels in infant feed formulations attributed to improvements in canning technology.

NUTRITIONAL R & D PRIORITIES

Finally, to consider briefly nutritional R & D priorities; it will be recalled that the overriding conclusion from the survey of dairy publications is the relatively low international priority afforded to R & D on the nutritional functions of milk and dairy products especially on such important parameters as nutritional stability where the number of publications per annum is less than 10.

The recent report by the US National Research Council entitled '<u>World Food and Nutrition Study</u>' (World Food and

Nutrition Study, 1977) examines in detail priorities for nutritional research. Some European countries have also published policy statements on nutritional research. While under-nutrition may be relatively rare in Europe today, there is increasing evidence that suboptimal nutrition may be of great importance overall in relation to health and it is clear that too little is known of what constitutes optimal. In considering priorities for nutritional research in dairy products it may be useful to consider briefly the recently published UK priorities (Joint Statement of Policy for Biomedical Research, 1977).

UK Priorities for Nutritonal Research

1. Nutritional Surveillance and Nutrient Requirements for Health

Little is known of the intake of energy, protein, individual amino acids, minerals, trace elements and vitamins which is optimal for health or development at different ages, in different developmental stages or in recovery from disease or trauma. Hence the requirements for various nutrients recommended at present are based on scant evidence and a high priority is therefore attached to research in this area. In-depth surveys are needed of groups nutritionally at risk and the development of improved analytic methods and techniques for such surveys is one of the highest priorities for research. High priority is also attached to research on the factors that determine the selection of foods and their relative importance and on ways of modifying undesirable patterns.

2. Nutrition and Disease

Obesity is associated with diminished performance and increased morbidity and mortality from various diseases. Research on obesity should give a high priority to improved methods of measuring fatness; epidemiology of obesity especially in children; identification of the causative factors including variation in energy requirements between individuals; effects of obesity on morbidity from different conditions and on treatment of obesity by diet, drugs and psychosocial measures.

It is important to determine the role of nutritional factors in for example, cardiovascular disease, diabetes mellitus, cancer, osteoprosis and femoral neck fracture and diseases of the large bowel. The significance for health of the fibrous part of diet has not been fully established and further studies of the chemical and physiological characteristics of fibre will be necessary in order to interpret clinical results.

The relationship between ageing and biochemical efficiency should be investigated with particular attention to establishing whether different diets influence the rates of ageing and whether nutritional requirements vary as the rate of metabolic processes change with age.

3.   Food Quality and Safety
Studies are needed of the effects of processing food in various ways on its nutritive value which may also be affected by changes in agricultural practices. High priority should be given to the development of short-term tests for assessing toxic hazards.

4.   Metabolism of Nutrients and Whole Body Studies
The solution of nutritional problems relating to human health will significantly depend on enlarging the present understanding of the underlying physiological as well as biochemical and pharmacological principles. Emphasis should be placed on the need for reliable whole-body studies which should be correlated with studies of the changing requirements of the individual at different ages for energy, protein, amino acids, minerals, trace elements and vitamins.

These comments which are abbreviated from the Joint Statement of Policy for Biomedical Research - Nutrition - by the Medical Research Council and Health Departments (Joint Statement of Policy for Biomedical Research, 1977) clearly refer to general R & D requirements in nutrition. Some of the comments, notably point 3, are of particular relevance to the seminar and it was felt that their presentation might be useful in preparing the seminar reports. In regard to dairy products specifically it may be interesting to consider their significance relative to the decreasing consumption of milk and dairy products in Europe, which may be associated with possible health implications. Is there sufficient scientific evidence to advocate that a reduction in the intake of fats (both animal and vegetable) may be a practical method of controlling excessive calorific intake? On the other hand does sufficient information exist on the nutritive attributes and importance of milk and the influence of thermal processing?

Effects of Thermal Processing

From the nutritional viewpoint the proteins and minerals are the important constituents of milk together with vitamins, fat and carbohydrate. As shown in Table 3 industrial thermal processing, apart perhaps from roller drying, causes no serious loss in either the biological value of milk protein or the overall nutritive value of milk (Wanner, 1974 and Porter, 1975). While there are appreciable losses of vitamins B_{12}, C, folic acid and B_1, especially with high temperature long heating processes, in a mixed diet milk is not an important source of these vitamins. With high temperature short time processes such as UHT coupled with the use of stainless steel equipment the vitamin losses are appreciably reduced. It should, however, be mentioned that with certain thermally processed products such as carmelised roller dried milk powder the nutritive value is significantly reduced - for example up to 75% of the lysine is unavailable or destroyed. Also, some reports suggest 10% reduction in the biological value of evaporated milk due to destruction of available sulphur-containing amino acids. In general the more modern thermal processes including instantisation and UHT sterilisation cause little loss of nutritive properties except perhaps vitamin C unless excessive oxygen is eliminated.

In overall therefore the influence of thermal processing on the nutritive value of milk has been extensively studied.

TABLE 3

HEAT INDUCED NUTRIENT DESTRUCTION DURING FOOD PROCESSING (Warner, 1974; Porter, 1975).

Process	Milk (Nutr. val.)	Protein (Biol. Val)	NUTRIENT DESTRUCTION (%)					
			B_1	Biotin	B_{12}	C	Folic Acid	Lysine Avail.
Pasteurisation	Negligible	Negligible	3.20	0	10	50	–	–
Sterilisation	Small	Small	33	0	90	50	50	–
Evaporation	Small	Small	40	10	90	60	60	20
Roller drying	Small	Slight red.	15	10	30	30	30	35
Spray drying	None	None	10	10	30	20	20	4

Although perhaps an oversimplification it would appear that the nutritive losses from industrial thermal processing are not serious and may in fact be less than could be encountered in domestic cooking due to over-heating of milk and dairy products.

Recently attention is being given to functional modification of milk proteins especially casein with a view to improving their physical characteristics. Research should be undertaken on the nutritive value of these modified milk proteins and on their possible toxicity.

Aspects of Milk Requiring R & D

Regarding milk itself it might be interesting to investigate whether the presence in bovine milk of large colloidal calcium caseinate complexes or casein micelles is advantageous from the viewpoint of human nutrition. Moreover, it might be mentioned that a recent report suggests that animal proteins including skim milk, casein and lactalbumen may raise the blood cholesterol level of rabbits while the converse may be true of plant proteins such as soya protein (Anonymous, 1977). This observation, if confirmed, would substantiate the concern frequently expressed regarding the excessive consumption of animal foods in many European countries and accordingly merits further investigation especially since it is also reported that milk proteins lower cholesterol levels (Howard, 1977).

Another aspect of bovine milk which may merit further research is its allergesic effects and also that of dairy products notably cheese where protein degradation products such as histamine and tyramine have been implicated in certain reactions. Bearing in mind that up to 5% of the US population are reputed to manifest various forms of food allergies (McGovern and Zuckerman, 1956) it might be interesting to establish the incidence of milk allergies and to identify the intrinsic factor in bovine milk responsible for eczema which appears to be absent from goat milk.

Milk is an important source of vitamins A and D and a
recent Canadian report (Nutrition Canada National Survey, 1970 -
1972) indicates that children's diets may be inadequate in
vitamin D and calcium arising from which milk is now fortified
with vitamin D. Also preliminary results suggest that in
Ireland vitamin D intake by children may not be adequate and a
similar situation may exist in other European countries (Cremin,
personal communication). Although prohibited in some European
countries the possibility of fortifying milk and more especially
skim milk with vitamin D should be considered. Further to this,
present methods for estimating vitamin D are rather difficult
and this is another instance of where further research on
improved methodology would be advantageous.

Finally, mention must be made of the major nutritional
question regarding milk and dairy products namely their
purported contribution to coronary heart disease. Because of
its widespread and apparently increasing prevalence, research
on coronary heart disease has been greatly intensified in
recent years. Although extremely difficult to indicate briefly
the current scientific consensus on this emotive topic it
would appear that total calorific intake or energy balance is
a critical factor in coronary heart disease and when this is
in excess of requirements the quality and quantity of the fat
intake becomes significant (Connolly, 1977). Extensive
research is being undertaken in almost every European country
on the multitude of pre-disposing factors to coronary disease.
The question therefore arises of whether it would be beneficial
to coordinate the results of this research and whether the
COST framework is an appropriate mechanism for doing so? It
certainly would have many advantages, not only in pooling
scientific manpower and resources but it also might provide a
useful mechanism for epidemiological surveys etc.

DAIRY TECHNOLOGY

Having considered some of the major technological problems
currently affecting the quality of conventional dairy products,

mention should be made of technological innovations required to improve existing products and develop modified/new dairy products. This is obviously an immense field and is somewhat outside the scope of this paper. However, the following topics have been singled out as high R & D priorities.

Improvement of the Organoleptic Qualities of Instant Whole Milk Powder

This product is now produced commercially but its flavour is inferior to that of good-quality liquid milk (Pyle, 1975). The availability of a high quality instant powder would greatly expand the potential market for dairy products.

Continued Development of UHT-Products

These products are particularly suitable for developing countries and are enjoying considerable success in Europe presently because of their convenience (Burton, 1977 and Regez, 1977) UHT products are potentially an important means of conserving energy through more rationalised distribution systems and elimination of refrigeration at various levels of distribution.

Cheese Manufacture

Cheese is the major growth product of the dairy industry and this should be encouraged. Among the major problems confronting the cheese industry requiring research are:

> increased mechanisation leading to a reduction in manufacturing costs (Crawford, 1976) the biochemistry of cheese ripening so as to produce cheese of consistently high quality and permit developments in what might be referred to as chemical or biochemical methods of cheese manufacture through the use of acids (or acidogens) and isolated enzyme systems (Fox, 1977).

Butter Technology

Milk fat is facing increasing competition from non-dairy alternatives as evidenced by the declining butter consumption in all European countries (Figures 14a, 14b) (EEC Dairy Facts and Figures, 1976). In addition to cost and nutritional considerations poor spreadability is an important contributor to declining consumption. To alleviate this disadvantage it is recommended that research should concentrate on improving the technology of blending milk fat with polyunsaturated oils (Foley, 1978 and Zillen, 1977).

Membrane Technology

Arising partly from increased cheese production and pollution abatement, whey processing has assumed major significance. Membrane technology has made significant contributions in this regard (Evans and Glover, 1974; Cotton, 1974 and Donnelly et al., 1974). However, development of profitable outlets for the products ie whey protein concentrates and lactose needs further research.

R & D on membrane technology should be intensified with the objective of adapting the process to a greater range of products. Concentration by reverse osmosis or ultrafiltration should be investigated as alternatives to thermal evaporation in the manufacture of milk concentrates both from the point of view of energy saving but also toward increasing the compositional and functional range of such products.

Novel Dairy Products

Milk and dairy products have been used traditionally as ingredients for non-dairy foods both at domestic and industrial level. However, until recently and only now in a few specific cases, notably milk powders, dairy products were not produced to meet specific industrial requirements (Fox, 1978). This is in contrast to the endeavours of competing industries, eg fats and oils, starch and other hydrocolloids, soya protein, who attempt to produce a range of tailor made products. The paucity of such developments in the dairy industry is due in part to restrictive legislation but also to a prevalent attitude that milk and its products are sacrosanct.

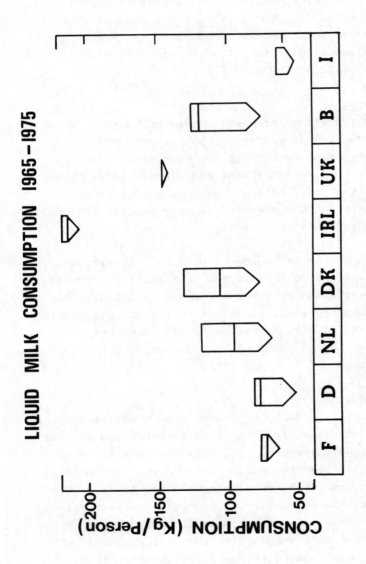

Fig. 14a. Declining consumption of milk in EEC member states (EEC Dairy Facts & Figures, 1976).

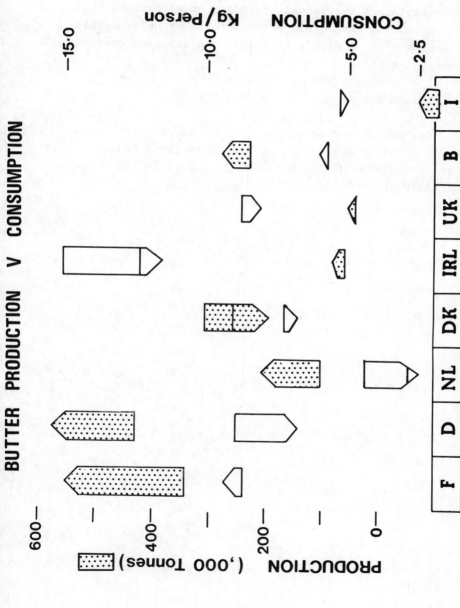

Fig. 14b. Disparity between declining butter consumption and increasing production of butter in many EEC member states (EEC Dairy Facts & Figures, 1976).

Research aimed at the development of functionally defined dairy products is recommended, with particular emphasis on milk proteins which even ten years ago were not considered as food materials. Due largely to R & D undertaken in Australia and New Zealand food grade milk proteins of well-defined properties are now widely available and enjoying considerable commercial success.

ACKNOWLEDGEMENTS

The advice and assistance of Mr. A. Cotter (National Science Council) and Dr. F. Cremin (University College, Cork) in providing some of the material contained in the paper is gratefully acknowledged.

REFERENCES

Aurand, L.W., Chu, T.M., Singleton, T.A. and Shen, R. 1967. J. Dairy Sci., 50, 465.

Anonymous, 1977. Nutrition Reviews, 35, (6), 1148.

Bulletin de l'Office International des Epizooties, 1971. 75, (9-10).

Burton, H., 1977. J. Soc. Dairy Technol., 30, 135.

Cogan, T.M., 1976. Butter Quality - Chemical and Microbiological Properties of Salted and Unsalted Lactic and Sweet Cream Butters, Dairy Research and Review Series No. 8; published by An Foras Taluntais, Dublin, (ISBN 0-905442 - 0-905442 - 03 - 2).

Cogan, T.M., 1977. Ir. J. Fd. Sci. 1, 95.

Cremin, F. Personal Communication.

Connolly, J.F. 1977. The prevention of coronary heart disease; published by An Foras Taluntais, Dublin.

Crawford, R.J. 1976. J. Soc. Dairy Technol., 29, 71.

Cotton, S.G., 1974. J. Soc. Dairy Technol., 27, 121.

Dairy Science Abstracts, 1975-1977. Commonwealth of Dairy Science and Technology Reading, England.

Downey, W.K. and Cogan, J.M. 1975. Proc. of the Lipolysis Symposium - Cork, Annual Bulletin, Doc. No. 86, International Dairy Federation, Brussels.

Driessen, F.M. and Stradhenders, J. 1974. Neth. Milk Dairy J., 28, 130.

Downey, W.K. and Murphy, R.F. 1970. J. Dairy Sci., 37, 361.

Downey, W.K., Kearney, R.D. and Aherne, S.A. 1973. Proc of the Fed. European Biochem. Soc.

Downey, W.K. 1978. D. Sc. Thesis, National Unversity Ireland.

Downey, W.K. and Murphy, F.F. 1975. Proc. of the Lipolysis Symposium - Cork, Annual Bulletin, Doc. No. 86, International Dairy Federation, Brussels, p. 19.

Downey, W.K. 1975. Butter Quality - Oxidative and hydrolytic rancidity in salted sweet cream and slightly salted ripened cream butter Dairy Research and Review Series No. 7; published by An Foras Taluntais, Dublin, (ISBN 0 - 901317 - 85 - 3).

Downey, W.K. 1968. J. Dairy Res. 35, 429.

Downey, W.K., O'Sullivan, A.C. and Keogh, M.K. 1970. In: Ultra high temperature processing of dairy products; published by the Society of Dairy Technology, London.

Downey, W.K., 1974. 19th International Dairy Congr. New Delhi. Vol, 11. p. 323.

Downey, W.K. 1972. Pesticide Residues in Milk and Dairy Products Annual Bulletin (Part 2), International Dairy Federation, Brussels.

Downey, W.K., Aherne, S.A. and Flynn Mary P., 1975. J. Dairy Res. $\underline{42}$, 21.

Downey, W.K., Flynn, Mary P., Aherne, S.A. and Kearney, R.D. 1974. 19th International Dairy Congr. New Delhi. Vol IE, p. 189.

Donnelly, J.D., O'Sullivan, A.C. and Delaney, R.A.M. 1974. J. Soc. Dairy Technol., $\underline{27}$, 128.

Ernstrom, C.A. and Wong, N.P. 1974. Milk clotting enzymes and cheese chemistry. In Fundamentals of Dairy Chemistry Ed. by B.H. Webb, A.H. Johnson and J.A. Alford. Published by Avi Publishing Company Inc. Westport, Connectitcut, USA.

EEC Dairy Facts and Figures; 1976. Published by the Economic Division, Milk Marketing Board, Surrey, England.

Evans, E.W. and Glover, F.A. 1974. J. Soc. Dairy Technol, $\underline{27}$, 111.

Fox, P.F. and Morrissey, P.A. 1977. J. Dairy Res. $\underline{44}$, 627.

Foley, J. 1978. J. Soc. Dairy Technol. $\underline{31}$, (1). 21.

Fox, P.F. 1977. DIrect Acidifaction of Dairy Products, 61st Annual Session International Dairy Federation.

Fox, P.F. 1978. The use of milk and dairy products in non-dairy foods. In: Factors affecting the yields and content of milk constituents of commercial importance. Monograph edited by J.H. Moore et al., published by International Dairy Federation, Brussels.

Heavy Metals and Other Elements in Milk and Milk Products, 1977. A. Document 37, International Dairy Federation, Brussels.

Howard, A.N. 1977. International Conference on Arteriosclerosis (abstract) Milan, 9. 34.

Joint Statement of Policy for Biomedical Research - Nutrition, 1977. by The Medical Research Council and Health Departments, UK.

McGann, T.C.A. and O'Connell, J. 1971. Ir. Agric. and Cream Rev., $\underline{24}$ (11) 5.

McGann, T.C.A. 1975. Ir. Agric. and Cream Rev. $\underline{28}$, (10), 11.

Muir, D.D. and Sweetsur, A.W.M. 1976. Dairy Research $\underline{43}$, 495.

Muir, D.D. and Sweetsur, A.W.M. 1977. J. Dairy Res. $\underline{44}$, 249.

McGovern, J.P. and Zuckerman, J.I. 1956. Nutrition in Paediatric Alergy. In: Borden's Review of Nutrition Research, $\underline{17}$, (3), 27.

Non-organic Micro-pollutants of the Environment. Report of Working Group of Experts - Rapporteur: J. Bouquiaux, EEC Report No.V-F-1966/1974e, Vol. 2, Luxembourg.

Nutrition Canada National Survey, 1970-1972. A report by Nutrition Canada to the Dept. of National Health and Welfare.

Porter, J.W.G. 1975. Milk and Dairy Foods. From the Value of Food Series ed. by P. Fisher and A.E. Bender.

Pyle, J.R. 1975. Instant milk powders and the principles of high-fat powder manufacture, In: Winter School of Dairying Ed. by B.C. Baker, L.L. Muller and A.T. Griffin, Australian Society of Dairy Technology.

Rose, D. 1969. Dairy Sci. Abstr., 31, (4), 171.

Regez, W. J. 1977. J. Soc. Dairy Technol. 30, 164.

Wilkinson, R.A., 1964. Theories of the mechanisms of oxidised flavour in dairy products div. Dairy Res. CSIRO Internal Report No. 4, Melbourne.

World Food and Nutrition Study, 1977. The potential contributions of research, The National Academy of Sciences, Washington, D.C.

Wanner, R.L. 1974. Effects of Commercial processing of milk and milk products on their nutrient content. In: Nutritional Evaluation of Food Processing, Ed. by Robert S. Harris and Harry von Loesecke; published by The Avi Publishing Company, Inc., Westport, Connecticut, USA.

Zillen, M. 1977. Proc. on Seminar on New Dairy Foods, Killarney, Doc. 16, International Dairy Federation, Brussels.

R AND D NEEDS FOR FISH AND FISH PRODUCTS

J.J. Connell
Assistant Director, Torry Research Station,
Aberdeen, Scotland.

ABSTRACT

1. *The main stimulus for further R and D on quality and nutritive value is the need to utilise optimally all old and new materials. This is due to the severe and increasing pressure on conventional aquatic resources.*

2. *Fish is particularly susceptible to physical, chemical and microbiological deterioration; thus, there are many points during harvesting, processing and distribution where quality and nutritive value can suffer.*

3. *The active international collaboration, particularly within Europe, which exists on methods for assessing quality illustrates the relative importance of this topic.*

4. *Aquatic resources vary within and between species in their intrinsic quality more than many foods and this causes difficulty in the maintenance of uniform and high quality during processing.*

5. *Main recommendations for further research are:*

– Ways of improving or manipulating the quality of the underutilised edible materials which are becoming available.

– Study of the course and causes of quality deteriorations in recovered, edible material and in new resources, and of means for controlling these deteriorations.

– Evaluate the magnitude and significance of quality and nutritive value loss during those handling, processing and distribution systems which have been less intensively studied.

– Study the incidence and control of the growth of micro-organisms, particularly pathogens, in raw material and products.

- *Formulate quality assessment and grading schemes for attributes and products not yet examined.*

- *Examine and codify the effect of intrinsic variations in quality on processing characteristics and product quality.*

INTRODUCTION

This paper examines, first, the main reasons why R and D on the quality and nutritive value of fish and fish products is needed by industry, consumers, Government and users of fish, and, secondly, the kind of R and D believed to be necessary to fulfil those needs. In most cases R and D is already being carried out in the areas identified, often in one or other of the laboratories in COST countries; international collaboration is already occurring on some projects. Although national circumstances dictate R and D programmes, which consequently differ from one country to another, nevertheless certain common problems are apparent and these will be identified under the headings which follow. These headings are placed according to my view of their descending order of importance. I have tried to avoid needs and research topics which are of interest to only one country. The preparation of this paper has been aided by recent discussions held under the auspices of the West European Fish Technologist Association (WEFTA - an annual forum for discussions between Government laboratories); the views expressed here are, however, entirely my own.

OPTIMAL UTILISATION OF AQUATIC RESOURCES

A very important requirement for R and D stems from the need to make full and efficient use of aquatic resources of protein and edible oil. Most of the world's wild fishery resources of traditionally-used species are fully or overexploited and in the face of a continued general increase in demand attention is increasingly being paid to making use of what was hitherto discarded or ignored or to upgrading low grade material (Burgess, 1977). It should be noted that fish farming can at best in the foreseeable future make only a relatively small contribution to supplies in Europe. The effort to increase utilisation has thrown up many problems most of which involve processing technology and the creation of new products, but there are in addition some important questions to be answered about quality and nutritive value.

With the decreasing supply of traditional species such as cod, haddock and herring, it has become essential in most countries to recover as much edible material from them as possible. An important development here has been the introduction of machines capable of stripping flesh in the form of mince from parts of the fish previously discarded (Conference, 1976; Technical Seminar, 1974). Unfortunately the quality of the recovered material has hitherto often been poor; colour tends to be too dark; texture is sometimes too firm or granular, sometimes too sloppy; spoilage rates tend to be high; the material may contain unacceptably high proportions of connective tissue, bone or other non-fleshy parts. If the fullest possible use of this recovered material is to be achieved it is necessary, therefore, to discover either conditions of machine operation under which products of different acceptable qualities can be obtained or the means by which quality can be improved after recovery. In pursuance of the first aim it would be useful to relate different qualities to different types of machine at different machine settings with different input materials. As an example of the second aim, the application of bleaching agents has been investigated in order to achieve a desirable whiteness in the recovered material. Although bleaching can be successful it brings with it dangers to nutritive value: sensitive amino acids in the protein and vitamins may be oxidised. In addition, toxic components may be formed. Thus, we need to know conditions under which bleaching can be achieved without rendering the recovered flesh nutritionally unsatisfactory. In a somewhat different direction, a good deal remains to be found out about the sensory and microbiological quality of these materials and how they limit storage under different conditions of chill and frozen storage. It has been observed, for example, that the texture of certain kinds of recovered flesh deteriorates very rapidly under normal commercial conditions of frozen storage. As a result, large sums of money have been lost. The reasons for this textural behaviour and solutions for its control are emerging from research but the picture is far from complete.

One advantage of recovered minced flesh is that it can be re-formed, mixed with modifiers or improvers or otherwise manipulated into a variety of guises having different appearance and texture. Systematic work is still in its infancy and a first step is to describe the qualities of the products which result. Ultimately is might be possible at will to tailor such fish flesh into several different desirable forms.

Rather similar considerations apply to species not, or little, used directly for human food. Into this category would fall capelin, sand eel, blue whiting, squid, small crustacea, argentines, horse mackerel and others discarded as by-catch. These resources are very large in quantity but tend to comprise species which are small in size, bony, high in lipid and shell content, and awkward in shape. Thus, filleting, cleaning, cutting and incorporation processes normal for traditional species are often difficult or impossible to apply. One means of overcoming these difficulties is to use a flesh-stripping machine of the kind just mentioned, usually after preliminary gutting and cleaning operations. The resulting mince or paste can suffer from the same type of quality problems and accordingly a similar but much wider area of research has been opened up, only relatively few parts of which are as yet being worked on.

There is also the question of the effects on quality of handling before processing of such under- or non-utilised species. The patterns of quality change and the limits of edibility during chilling in ice, or refrigerated sea water with or without carbon dioxide, and during freezing and frozen storage have not yet been fully delineated. This information is a basic requirement in the design of practical systems of handling and processing. For example, if a species can only be kept after capture for three days in melting ice, this places rather severe limitations on the type of boat and the location and type of processing and distribution system.

ENSURING NUTRITIVE VALUE AND SAFETY TO THE CONSUMER

1. Nutritive value

Some of the components of fish are among the most unstable of any in foods and much is known about the relative loss or destruction of nutrients during handling and processing. The effects of heating on fish and shellfish have, for instance, been reviewed recently (Aitken and Connell, 1977). The results in the main show that no serious effects on nutritive value occur during processing for human consumption. In some instances, however, there are indications of large changes as, for example, the recent finding that cooking fish in oil causes a fairly marked reduction in available lysine (Tooley and Lawrie, 1974). It would obviously be worthwhile following up this observation. Extensive oxidation of the polyunsaturated lipids characteristic of fish often occurs during heating and frozen storage, but the nutritive implications have not been fully investigated. Furthermore, there is a need to understand more about the effects of drying and smoking on protein availabilities in fish products.

Apart from the loss of a relatively small proportion of nutrients in drip or through leaching, few undesirable effects of chilling are expected. Freezing and frozen storage has been little studied from this point of view. Drip losses can be extensive and there is some evidence for loss of protein amino acids (Kolakowski et al., 1972). Dimethylamine and formaldehyde are known to accummulate during the frozen storage of some species but the nutritional consequences have not been explored. Therefore, from a number of points of view it would be useful to have more definitive information on the effects of frozen storage.

Rather severe and prolonged heating occurs during the usual preparation of animal feeds from fish (fish meal and oil) with varying degrees of damage to lysine and methionine. Meals with lower available lysine fetch lower prices. Recent research has gone a good way to revealing the nature of the underlying reactions but a full understanding is still awaited. Further work, also, is necessary in order to demonstrate how it is possible to produce meal economically of consistently high nutritional value.

Considerable interest surrounds the role of fish oils in human nutrition. As just mentioned these contain high proportions of polyunsaturated fatty acids. There is evidence that diets containing this class of acids could be beneficial in reducing the incidence of heart disease (Royal College of Physicians, 1976). Somewhat in contrast, concern has been expressed recently over the possible harmful effects of the docosenoic acids (for example cetoleic acid) in natural and processed fish oils. A similar acid, erucic acid found in rape seed oil, has been shown to cause heart lesions when fed under certain conditions (for example, Beare-Rogers, 1975). Research so far has shown that fish oils are unlikely to be a cause for concern at normal levels in the diet but further work is in progress. The International Association of Fish Meal Manufacturers (IAFMM) have been active in fostering collaboration on this subject and in my view the involvement of COST would not be warranted.

2. Safety and Hygiene

An important consideration in relation to fish quality is contamination with pollutants. This is a topic which, however, conceivably lies outside the terms of this Seminar. It will not be discussed further beyond pointing out that a good deal of research has been afforded it and much will continue. In addition, some aquatic organisms occasionally used inadvertently for food are intrinsically toxic, but this topic is of little general concern to COST countries.

The incidence to damage of health caused by micro-organisms in fish and fishery products is relatively low. Nevertheless, two pathogens present in fish have been the focus of recent research in COST countries. The first is *Vibrio parahaemolyticus* which causes a mild form of food poisoning. Until recently it was thought not to be present in temperate sea water but field surveys have now established that is is fairly widespread round the southerly reaches of the UK coastline. The second organism is *Clostridium botulinum* which is the causative agent of the rare but often fatal disease of botulism. Outbreaks of botulism have been associated in recent years with a number of different imperfectly heat processed and stored fish products. Research is currently being carried out in at least five laboratories in COST countries on measures to evaluate and control the risk. Some international exchange has occurred in this research area and it is likely that progress would be facilitated by further collaboration.

The major agents of fish spoilage are, of course, microorganisms and this fact justifies further research directed towards understanding their activities, incidence and control. Pasteurisation combined with chilling of shellfish such as crab and shrimp is a useful method of processing but it has been attended by a number of unexpected failures. Research is needed, therefore, to define the exact parameters of time, temperature and packaging which will ensure consistent freedom from spoilage. During normal handling, chilling and further processing, the microbial population changes in character and numbers. Information on these changes is essential in order to set the limits to microbial incidence which can be achieved in good manufacturing practice. It is from such information, also, that microbiological standards are formulated. All major methods of handling, processing and distribution have not been investigated and so research is still needed to fill the gaps.

From time to time concern is expressed about the likelihood of toxic components developing during the heat processing of fish. Oxidation of lipids have already been referred to. Further examples are nitrosamines from the relatively high concentrations of amines present in many fish - it will be recalled that nitrosamines were first discovered as carcinogens in a fish product (fish meal) - and polynuclear-aromatic hydrocarbons deposited during the smoking process. These concerns have generated a large body of research and it is possible that the further work which will undoubtedly continue would benefit by closer international collaboration.

METHODS FOR EVALUATING QUALITY AND NUTRITIVE VALUE

The need for scientific methods of evaluating quality is as self-evident for fish as it is for any food. Indeed, in view of the much greater number of species and public concern about the freshness of fish, it could be argued that the need is greater for this commodity. Much research effort has been and is being spent on methods for fish and fishery products which can be grouped as (1) sensory, (2) non-sensory related to sensory attributes or suitability for processing, (3) microbiological, (4) nutritional.

(1) Several schemes for assessing quality sensorially have been devised for different purposes and more are under development. Recently there have been some moves towards getting agreement on schemes which can be used internationally. The stimulus here has been the need to harmonise grading standards. Thus, the EEC have an agreed scheme for the quality grading of wet fish at first sale. Also, a successful collaboration has taken place under the aegis of WEFTA which resulted in a scheme for the sensory assessment of whole frozen fish. There is a need for the extension of this collaboration to cover all industrially important products but especially filleted fish. Such collaboration requires not only discussion but mutual laboratory investigations.

(2) A small number of non-sensory tests usable both in research and in industrial quality control or surveillance are available but there is a need for more. The goals of satisfactory, causally-related non-sensory methods for assessing aroma, flavour and texture in spoiling and deteriorating products have been elusive and are likely to remain so. For this reason research towards these goals might well benefit from a concerted collaborative effort. Other pressing needs are for unambiguous methods for assessing (a) the fish flesh, bone and parasite

contents of products, (b) the proportions of glaze and of minced flesh in products containing mixture of mince and fillets, (c) the identity of species in products containing mixtures of species, (d) the degree of oxidation of fish lipids.

(3) More rapid, cheaper and replicative methods for microbiological enumeration suitable for fish and fishery products are, as for any commodity, still required. Collaborative research in this area is, however, being satisfactorily organised by the International Commission on Microbiological Specifications and a COST initiative would not be warranted.

(4) The need for better nutritive methods specific to fish is confined to fish meal but again there is already a mechanism for international collaboration in the IAFMM.

THE INFLUENCE OF INTRINSIC QUALITY ON PROCESSING

Uniquely among major commodities, most fish is still hunted and the hunter has, to a large extent, to accept the nature of what he catches. This means that the intrinsic quality of fish raw material in many instances varies more uncontrollably than for other foods. The processor then has the added task of turning a very variable raw material into as uniform an end product as possible. This task may involve the selection of suitably uniform batches of raw material from a variable mass and the continuous modification of processes. It is made easier by the possession of as much systematic information as possible on these variations and their effect on processing characteristics and storage behaviour.

The types of intrinsic quality which are subject to variation are: size; shape; colour; water, protein and lipid content; ultimate pH; degree of softness and of 'belly-burst'; stage of gonad development; incidence of pathogens. As an extreme example, lipid content can vary within one species from 1 to 25%, with marked effects on processing and storage capability. Thus, the quality characteristics of a canned product made from batches whose lipid contents are this disparate will be quite different; indeed, it is well known in some cases that it is impossible to manufacture a satisfactory canned product from raw material whose lipid content exceeds a certain value. Similarly, the frozen storage life of fish with high lipid or low pH is known to be less than that with low and high values, respectively.

A comprehensive description for the industrially important species of intrinsic variations and their consequences for handling and processing is not yet available. Because the results would be of widespread benefit work towards such a description would seem to be a suitable subject for international collaboration.

GENERAL QUALITY CHANGES DURING PROCESSING AND DISTRIBUTION

Fish is one of the most perishable foods. It must inevitably feature prominently in any general survey of quality change during processing and distribution which is designed to identify where excessive quality loss occurs and to try to reduce it.

Several surveys of this kind on fish and other foods have been reported for different sectors of the processing and distribution system. In some cases severe loss of quality has been shown. As these systems change with the introduction of new technological developments it is inevitable that new surveys will be mounted. For example, little up to date definitive information is publicly available on the eating quality and nutritive value of frozen fishery products sampled in retail shops, home freezers or after cooking. Since research of this kind tends to be time consuming and costly there would be obvious value in sharing the effort on a collaborative basis. The same holds for systematic laboratory studies designed to discover the effects of time and temperature on quality.

REFERENCES

Aitken, A. and Connell, J.J. 1977. 'Fish' in 'Chemical and Physical Effects of Heating on Foodstuffs' (Priestley, R., ed.), Applied Science Publishers Ltd., London - to be published.

Beare-Rogers, J.L. 1975. in 'Modification of Lipid Metabolism' (Perkins, E.G. and Witting, L.A. eds), pp 43-50, Academic Press, London and New York.

Burgess, G.H.O. 1977. Proc. Nutr. Soc. 36, in press.

Conference on the Production and Utilisation of Mechanically Recovered Flesh (1976), Ministry of Agriculture, Fisheries and Food, Aberdeen, Scotland.

Kolakowski, E., Fik, M. and Karminska, S. 1972. Bull. Intern. Inst. Refrig. Annex 2, pp 65-72.

Royal College of Physicians 1976. Prevention of Coronary Heart Disease. J. Roy. Coll. Phys., 10, April.

Technical Seminar on Mechanical Recovery and Utilisation of Fish Flesh. 1974. National Marine Fisheries Service, Department of Commerce, Boston, USA.

Tooley, P.J. and Lawrie, R.A. 1974. J. Fd. Technol. 9, 247-253.

RESEARCH AND DEVELOPMENT REQUIREMENTS ON SOME ASPECTS OF THE QUALITY AND NUTRITIVE VALUE OF THERMALLY PROCESSED CEREALS

Christiane Mercier and J. Delort-Laval

Institut National de la Recherche Agronomique
Centre de Nantes
44072 - Nantes, France

ABSTRACT

Variation in raw material quality and development of new industrial processes, consideration of hygienic and nutritive value of cereal products, lead to the proposal of the following topics for R and D in Western Europe:

- *Genetic research to breed varieties of wheat better adapted to industrial requirements which involves development of adequate European tests for the machinability of the dough and for the breadmaking.*

- *Evaluation of the technological aptitude of cereals taking into account all the characteristics of the raw material (genetic origin, growing and harvesting conditions, storage) and their properties for a definite process (bread, biscuits, pastries, etc).*

- *Technical aspects of new treatments of non baking cereals and their effects on the organoleptic and nutritive properties of the end-products.*

- *Contamination of cereal raw materials and end-products by mycotoxins during storage. Behaviour of additives (antimicrobial, texturising agents) during technological processing.*

- *Development of methodology to evaluate sensory properties and nutritive value of cereal products containing unusual components such as bran, protein, modified starch.*

Other topics are submitted for consideration to the following panels:

- <u>*Chilling and freezing*</u> *: chilling during rest after dough fermentation in French breadmaking.*

- *Cooking* : *evaluation of cooking quality of pasta with relation to raw material and industrial treatment.*

- *Dehydration* : *effect of storage and drying treatment of cereals on the technological properties of cereals for starch extraction and breadmaking.*

About 70 million tons of cereals are collected each year within the 9 countries belonging to the EEC, including soft and durum wheat, barley, maize, rye, oats and rice. (Table 1, Cordonnier, 1977).

Forty-two million tons are used by the food industry; in this amount are included milling by-products, exports of flour and cereals used for alcoholic beverages, only briefly discussed further in this report (Table 2).

The distribution among the different countries shows that Italy and the United Kingdom process the maximum of cereals followed by Germany and France.

Human consumption of the different cereals in the 9 countries of the European Community, expressed in kg/capita/year, is presented in Table 3.

TABLE 3

HUMAN CONSUMPTION OF CEREALS IN EEC IN Kg/Capita/Year

Cereals	Countries								
	EEC-9	Denmark	Germany	Netherlands	Ireland	UK	France	UEBL	Italy
Total	115	87	88	85	123	115	96	100	180
Soft wheat	86	54	58	74	113	93	85	93	115
Rye	5	18	17	5	-	-	-	-	-
Durum wheat	16	-	5	-	-	1	10	4	57
Barley	0.5	-	-	-	-	-	-	-	1
Oats	1	10	2	-	3	3	-	-	-
Maize	7	5	7	3	4	17	1	2	6
% Baking cereals	77	83	85	93	92	81	88	93	64

It can be seen that the baking cereals (soft wheat and rye) are the major cereals used in the human diet in all 9 countries, representing an average percentage of 77%. Durum wheat follows. Barley, oats and maize, whose use is not based on their baking

TABLE 1

USE OF CEREALS BY FEED AND FOOD INDUSTRIES WITHIN EEC PER YEAR (Mean of 2 years, 1972-1974, expressed in 1 000 tons)*

Cereals / Countries	Soft wheat (1)	Durum wheat (3)	Barley (2)	Maize (3)	Others (3)	Total	% of every country
Germany (D)	4 900	330	3 350	2 750	1 700	13 030	19.2
France (F)	6 900	540	1 900	3 400	250	12 990	19.1
Italy (I)	7 000	3 200	700	3 650	100	14 650	21.6
Netherlands (NL)	1 550	-	550	2 550	450	5 100	7.5
Belgium - Luxemburg (UEBL)	1 250	30	1 000	1 200	1 000	4 480	6.6
United Kingdom (UK)	6 750	100	4 300	2 800	450	14 400	21.2
Ireland (IRL)	450	-	650	150	50	1 300	1.9
Denmark (DK)	350	-	1 050	250	250	1 900	2.8
USE EEC-9	29 150	4 200	13 500	16 750	4 250	67 850	100.0
COLLECT. EEC-9	30 530	2 700	18 770	11 000	3 740	66 740	

(1) Incuding exportation (2) Including malt for exportation (3) Home consumption

* Eurostat 1975 - 1976

TABLE 2

USE OF CEREALS BY FOOD INDUSTRIES WITHIN EEC PER YEAR (Mean of 2 years, 1972-1974, expressed in 1 000 tons)*

Cereals Countries	Soft wheat (1)	Durum wheat (3)	Barley (2)	Maize (3)	Others (3)	Total
Germany (D)	4 038	325	2 334	830	1 223	8 750
France (F)	5 569	521	876	678	39	7 683
Italy (I)	6 780	3 192	265	883	13	11 133
Netherlands (NL)	1 135	3	222	329	75	1 764
Belgium – Luxemburg (UEBL)	1 154	35	735	528	10	2 462
United Kingdom (K)	5 080	109	1 973	1 725	175	9 062
Ireland (IRL)	333	1	181	29	21	565
Denmark (DK)	268	–	190	18	130	606
EEC-9	24 357	4 186	6 776	5 020	1 686	42 025

(1) Including export flour   (2) Including export malt   (3) Home consumption

* Eurostat 1975 – 1976

properties, correspond to a small percentage of the cereals in the human diet. When expressing the national consumption relative to the total EEC use of cereals for food, Italy is the country which consumes the maximum of soft (29%) and durum (77%) wheat, whereas Germany is the main user of rye (83%). With respect to the other cereals, the United Kingdom is the main consumer of maize (52%) and oats (45%) (Table 4).

TABLE 4

THE PERCENTAGE OF CEREALS USED BY THE MAIN CONSUMER COUNTRIES WITHIN EEC

Cereals	Country	%
Soft wheat	Italy	29
Rye	Germany	83
Durum wheat	Italy	77
Oats	United Kingdom	45
Maize	United Kingdom	52

When considering cereals as a commodity for human consumption, two main points have to be taken into account:

1) The variability of the composition is much higher than in the large cereal-growing areas such as the US, Canada, Australia and the USSR: however, it must be observed that the French cereal is different since France is the main exporter of cereals within the EEC.

2) As human food, cereals are not usually eaten in the unprocessed state. They have to be treated to improve the digestibility of their starch and protein and therefore their nutritive value.

Until recently, cereals have been processed mainly through the following conventional technologies (Mercier, 1977).

The major amount of <u>soft wheat</u> is used within the EEC for bread production. The consumption of biscuits, 'biscottes' and various pastries is increasing and small quantities are transformed into malt, gluten, starch and alcohol.

Rye is also used in breadmaking, especially in Germany and in France for a typical pastry called 'pain d'épice'.

Durum wheat is mainly transformed into semolina and used in the preparation of pasta and noodles.

The most important use for barley for human consumption in Europe is in the form of malt used in the brewery industry. In Great Britain and Ireland, malt is also used in the preparation of whisky.

A large proportion of maize is converted into semolina in the EEC countries, except in Ireland and Luxemburg. This semolina is then used either as such (polenta in Italy), or in the brewery as gritz, used eg for the preparation of whisky in Ireland and in cornflakes for breakfast cereals. The other technology, the importance of which is increasing, is the transformation of extracted maize starch into different products used in food, such as pregelatinised starch, modified starch, glucose syrup, dextrose, isoglucose.

During the past few years, the use of cereals for food has not increased, in fact, the consumption of flour and pasta has decreased within the European Community. This phenomenon can be attributed partly to the stagnant population within the EEC and partly to a great social evolution in which the feeding habits of the population have been changing. On the one hand, the consumer who has little time to spend on food preparation in our present society, is asking for fast, ready-made and long-storage foods such as frozen products. More diversification in the appearance, presentation, flavour and colour of those products seems also to be fashionable. On the other hand, the same consumer, the potential buyer on the market, has a great concern for the organoleptic properties and the safety and nutritive value of the industrial processed foods. Such foods are often believed to be of inferior quality compared with the 'old' conventional products and they are considered to be less attractive with respect to their sensory properties and of reduced nutritional or hygienic quality.

All these demands lead either to the modification of existing technology and/or to the development of new technology for cereals.

It is therefore important to acquire a sufficient knowledge of the compostion of cereals and of the effects of technological treatment in order:

- to achieve a better control of existing technologies with regard to the sensory and nutritive value quality of the end product;
- to develop new processes to satisfy the consumer's requirements stated above by a diversification of the end products.

NUTRITIVE VALUE OF RAW MATERIAL

Nutritive value of cereals as raw materials is dependent on their total composition, which varies, to some extent, with genetic origin, soil and climatic conditions, farming techniques, harvesting and storage conditions.

Cereals represent an important energy source in the human diet due to their high starch content (60 - 90%). The proteins (8 - 15%) are essentially recognised for their functional properties in technology and constitute only a limited part of the dietary protein, except perhaps in Italy where cereal consumption is high and animal protein intake lower than in the other EEC countries.

Lipids, vitamins and minerals are in such small quantities that they cannot be considered important enough to be involved in the quality and the nutritive value of cereals taking into account the high degree of diversification in the human diet in Europe. However, we cannot overlook the importance of maize germ oil. Besides the quantitative aspect, it is important to note that the consumption of cereals includes a fibre fraction which has aroused considerable interest in recent months.

The presence of fibre in a diet, as a ballast can change the digestive pattern by its own properties; consequently, not only does it reduce the energy content of the diet, but it also plays a role in the nutritive value of food bringing a certain 'comfort' through the digestion which is more needed by people in sedentary occupations.

The role recently attributed to wheat bran, as the best ballast improving the digestive transit, is accepted by most of the gastroenterologists. It seems well recognised that the hemicelluloses and cellulose of wheat bran possess the capacity to bind the bile acids, possibly by hydrophobic linkages (Burkitt and Trowell, 1975) whilst on the other hand rye bran, with a higher content of cellulose and hemicellulose, does not appear to be as efficient. It is therefore urgent that a study of the mechanism and action of bran in the digestive system through the biochemical structure of cellulose and hemicellulose, should be undertaken.

So, as a traditional food of general acceptability in the EEC population, cereal products (and especially bread) could be a privileged component of qualitative enrichment of the human diet. The development of <u>bread enriched with bran</u> or obtained from a higher extracted flour containing more bran, is certainly a new and very useful area, since all the European countries tend to consume too much refined carbohydrate, leading to changes in transit pattern and digestion. On the other hand, although cereals are not recognised as a significant protein source, <u>protein enrichment</u> of baked cereal products should not be overlooked: addition of purified milk protein, vegetable protein extracts and/or cultivation of micro-organisms on cereal substrates, might allow the development of new foods. Some of them could be of immediate interest in under-nourished populations and, in the long term, might involve a larger proportion of the consumers, even in Europe, where there is no evident substitute up to the present, to animal protein in the human diet.

The conditions of <u>storage</u> of wheats (moisture content and temperature) are very important for the quality of flour, through the <u>growth of micro-organisms</u>, the increase of the enzymatic action and the alteration of the gluten quality. These factors are all involved in the quality and the nutritive value. The development of mycotoxins especially resistant to thermal processes, leads to a new field of research in relation to the safety of the final product.

In this respect, de-hydration, essential to prevent the deterioration of cereals grown and harvested under bad weather conditions, deserves some attention. The use of solar energy has to be taken into consideration as well as the recent flash-heat treatment which reacts on the micro-organisms. The effects of dehydration on the technological properties (baking ability, extractability of main constituents) and the <u>nutritive value of the cereal based foods</u> are submitted for consideration to the panel on 'dehydration'.

2. TECHNOLOGICAL VALUE OF BAKING CEREALS

Since baking cereals, soft wheat and rye, represent the basis of the diet within the EEC, breadmaking is certainly the process which needs the most research and development (Buré, 1975)

It should be noted that the type of bread in Europe differs largely from country to country. The raw material as well as the processing are both involved in the quality and the nutritive value of bread.

2.1. <u>Influence of genetic origin</u>

The different breadmaking processes and their industrialisation imply a greater homogeneity in the varieties of soft wheat There is also a need to breed varieties of cereals better adapted to industrial requirements. In France, there is a great need to breed varieties of 'hard' wheat which could avoid the necessity of importing American hard wheats. The recent development of high yield varieties, such as Clement and Maris Huntsman, has shown that it is difficult to improve both quantity and quality. Research therefore needs to be done on the development of good baking quality wheat. It is somewhat unfortunate that the objectives of the wheat breeder to achieve the highest yield of the grain, of the miller to maximise his flour production and of the baker to obtain the greatest production of bread, are often mutually conflicting when the quality of the loaf is to be maintained. In the context of the European Community, these problems necessitate a more positive collaboration, research and understanding, between the various interests involved (Greenwood, 1975).

Owing to the differing breadmaking techniques in Europe, the only test at present proposed concerns the machinability of the the dough. Without any complete baking test, which will be the only valuable one to detect all the characteristics of baking wheat, important economic problems could occur rapidly in the European cereal trade. The baking tests should serve as the European reference method from which other quick tests would thus be adapted according to the national breadmaking conditions.

2.2. Industrialisation of breadmaking processes

The rapid evolution of industrial processes in breadmaking, connected with a reduction in cost and labour, leads to significant changes in the processing conditions (Doublier, 1977).

Different factors, applied to raw materials, to the breadmaking itself and to the storage of bread, permit the process to be industrialised: the protein baking quality of a soft wheat can be improved by using a mixture of various baking flours, some enriched protein flour obtained by new techniques such as air-classification, and/or by adding the gluten fraction extracted from wheat; during breadmaking, the addition of agents capable of modifying the functionality of proteins, permits the use of different grades of baking wheats. However, those additives are dependent on the legislation applicable in the countries concerned, for example, even though some additives are commonly used in some countries (English breadmaking), they are prohibited in the French process.

The improvement of the final stability of bread, which is partially possible by using anti-staling agents such as monoglycerides, modified starch, etc, and the preservation of the hygienic quality of bread during long storage, are certainly important and new steps to be studied in commercial breadmaking.

At the present time, millers and bakers choose their commercial wheats with a test based only on the identification of varieties of wheat or mixture of wheats; this test does not allow for the differentiation of technological capacities for definite applications.

There is, in this field, a need to develop some diversified technological tests which would take into account all the characteristics of the raw material (genetic origin, growing and harvesting conditions, storage and its application for a given purpose (bread, biscuits, pastries, etc)).

The industrialisation of breadmaking may lead to a different product in terms of appearance, organoleptic and hygienic properties: this is the case for the so-called 'intensified' processed French-bread. Although the loaf is similar to the conventional bread, its volume is doubled and the crumb is different in respect of texture, colour and flavour. The staling is even faster; the crust is more important and the Maillard products formed are the flavour precursors but the effect of those products on the nutritive value of the bread is not yet known. The use of microwave techniques in the UK prior to baking in order to reduce the α-amylase activity, seems to allow the baker to handle various wheats for breadmaking.

The modifications observed in the appearance and organoleptic properties of the processed product, are worth considering in the choice of a technological treatment, since they greatly involve the consumer acceptability of the new product. Bound to this acceptability, the safety properties must be taken into account. The behaviour of contaminants, such as preservative agents of wheat and bread and improver additives during the industrial process, has to be studied in terms of their eventual degradation into different metabolites. This hygienic capacity deserves a special consideration in order to maintain 'l'image de marque' or the high reputation of bread.

2.3. Pastification

The industrial production of pasta requires a regular semolina from durum wheat in order to obtain a consistent final pasta, (Feillet, 1977). Here again, as with soft wheat, there is a need for genetic improvement of varieties and for the development of technological tests showing their potential aptitude to pastification.

The main quality factors of noodles are their colour and behaviour during cooking. The enzymatic mechanism involved in the colour formation and stability is known but research is still needed in order to study the influence of protein and starch properties on the preparation of the dough and on the cooking quality of pasta.

3. TECHNOLOGICAL VALUE OF NON-BAKING CEREALS

Although maize and rice are a basic food in large areas of the world, their consumption in Western Europe is somewhat limited and plays only a small part in the human diet. Various technological treatments (wet and dry milling, heat, extraction) are applied in order to obtain a large range of diversified products. Heat-treated cereals are intended for direct consumption; processed extraction products are often mixed with other food materials in order to give end-products with improved organoleptic and functional properties.

3.1. New thermal processes

Diversification of cereal products implies, for the food industry, the creation of new processes. In order to be competitive with the usual technology, these processes have to be continuous, fast and of low cost, including, of course, reduction of energy consumption. Innovation should be directed not only to controlling a given process, but also in finding products in form, texture and presentation to appeal to the customers' requirements. With a commodity such as a cereal, heat treatments tend to improve the digestibility of starch and the mechanical effect added in flaking, popping, extruding, expanding, results in fast and continuous processes where the raw material is not first constituted into a dough.

Flaking is the customary treatment for breakfast foods, corn, oats and rice flakes.

Extrusion cooking is a recent food technology where considerable efforts have been made with the object of developing new products such as snacks, baby foods, breakfast cereals and modified starches (de la Gueriviere, 1976).

The end-products have an expanded texture mainly due to starch gelatinisation. A great variety of them have already been obtained by incorporating different flavours or coatings (lipids, cheese, chocolate) and by changing empirically the physical parameters of extrusion (temperature, time, mechanical treatment). But there is little real knowledge about the mechanisms involved in the process.

Research and development are especially required on the following topics:

- measurement of the real values of the physical parameters inside the different types of extruders;

- evaluation of the raw material;

- study of the texture of end-products and of the factors which determine it;

- evaluation of the nutritive value, especially when this technology is used to prepare baby foods and dietetic products, particularly, almost nothing is done about the effect of the processes on proteins (intrinsic and added).

3.2. Processed extraction products

The products prepared from extracted maize starch are increasingly used in food technology. Pre-gelatinised starch, modified starches, products of enzymatic and chemical hydrolysis, such as dextrins, glucose syrup, dextrose, isomerised glucose, are commercially available and utilised in a wide range of foods, on account of their functional properties: gelification, viscosity, stability at various temperatures. The food industry provides these new products and public research institutes are usually not involved in their production. However, those institutes will have to take part in the evaluation of the nutritive and hygienic properties of these food components. Their digestion and absorption may be modified by the treatment and they may act upon the transit of the whole diet. Since growing attention is given to the digestion of the carbohydrate fraction, an effort could be made to increase our knowledge of the digestion behaviour of modified starch and its derivatives.

CONCLUSION

The need to industrialise the cereal processing for human food implies some adjustment of the raw material quality (variety, growing, harvesting and storage conditions) to the continuous parameters of the industrial treatments.

In order to stimulate the cereal consumption needed for the equilibrium of the diet, it is necessary to maintain the high quality of cereal products in conventional uses. The only way to reach this goal is the critical evaluation of the processes.

This evaluation assumes the existence of a common agreement between the different countries on the development of the methods to estimate the technological aptitudes of baking cereals.

The new technologies, developed empirically must be studied on basic research level. It would be interesting to study the physical association and the chemical interaction between carbohydrates, lipids and proteins. Special attention has to be given to investigation on the sensory estimation, the nutritive value and the safety of processed foods in order to meet the consumer's requirements.

Everything cited above leads to the development of estimation procedures for the new products resulting from physico-chemical, microbial or enzymatic reaction applicable to cereal products. It implies also the control of contamination which can occur during the production and storage of cereals and during treatment. This particular problem must be studied in all the raw materials involved in human food.

The possibility of qualitative enrichment of cereal products by addition of cell-wall constituents (bran), protein or essential amino-acids, must be considered as long-term R and D topics.

Finally, a great deal more work needs to be done to study the physico-chemical structure of starch and cereal protein in order to understand their behaviour during thermal processes.

REFERENCES

Buré, J. 1975   Cereal Foods Worlds, 20, 39-45
Burkitt, D.P. and Trowell, H.C. 1975   Refined Carbohydrate Foods and Disease, p. 356, Academic Press, London and New York.
Cordonnier, J.M. 1977   Perspectives Agricoles, 5, 64-72.
Doublier, J.L. 1977   Blé et Pain, monograph APRIA
Eurostat 1975-1976   Cereal production and use in the EEC countries.
Feillet, P. 1977   Ann. Nut. Diététique, in press.
Greenwood, C.T. 1975   Cereal Food Worlds, 20, 23-28
de la Guerivière, J.F. 1976   Ind. Alim. Agric., 5, 587-595
Mercier, C. 1977   Science et Avenir, 18, 26-32

NUTRITIONAL AND QUALITY ATTRIBUTES INVOLVED IN THERMAL PROCESSING OF FRUIT AND VEGETABLES

C. Cantarelli
Istituto di Tecnologie Alimentari, University of Milan,
20133 Milan, Italy

ABSTRACT

The available information regarding the composition of fruits and vegetables is considered in order to give a general evaluation of the topics of research actually developed in the field of thermic preservation and related processes. Both domestic and institutional food processing have to be considered.

With few exceptions, the most important characteristics of fruit and vegetables are colour and structure because of the hedonistic orientation of the demand for these products. We must also consider that fruits and vegetables are necessary in the diet because of their contribution in vitamins, minerals and fibre, and that severe losses of these nutrients can occur during processing.

As far as thermal treatments are concerned, the important topics of interest seem to be the following:

The maintenance of the attributes of colour and structure

The evolution of available and non-available glucides

The reactions of proteins with sugars and with phenolics

The thermal resistance of anti-nutritional factors (proteins, phenolics, glucosides, alkaloids) taxonomically present in many species.

The loss of minerals and organic acids important for their role in the saline equilibrium

The conditions leading to a loss of vitamins

From a general point of view an improvement of cytochemical research on the phenomena accompanying cell disruption or disorganisation and on the means for their control, is important. A primary goal must be an increase of bioavailablity of some nutrients, particularly the minerals, including trace elements.

1. THE ROLE OF FRUITS AND VEGETABLES IN NUTRITION

The psychological gratification derived from the organoleptic characteristics of fruits and vegetables may be the main reason for their consumption by humans. Nutrients such as glucides, minerals and vitamins do not play a major role except in the case of oleaginous fruits and pulses, rich in proteins and lipids. In industrialised countries the energy and protein contribution of vegetables represent only about 10% of the total, while their lipids account for only 1%. On the other hand, vegetables supply the total need of vitamin C, more than 50% of vitamin A, and 20% of iron, magnesium, calcium and potassium (Aykroya and Doughty, 1964; White and Selvey, 1976).

The psychological appeal of fruits and vegetables due to their intrinsic characteristics (wholesomeness, sweetness, fragrance), as well as to their image of prestige, orients their consumption to quality rather than quantity. (Claudian and Serville, 1975).

Other important aspects are the physiological ones without a nutritional contribution, represented by the positive effects of the high content of fibre and inorganic cations. On the other hand, a great number of anti-nutritional compounds is widely distributed in vegetable species, such as specific and non-specific enzymatic inhibitors, alkaloids and other toxicants (lathyrogenes, glycoalkaloids, favism factors), chelating substances (oxalates, phytates), glucides (flatulence factors, low digestible starch), bitter and astringent components.

From a nutritional standpoint, a very relevant aspect is the very high water content of fruits and vegetables which causes the ingestion of considerable quantities of them; consequently nutrients such as proteins, even diluted, can, in some cases, represent a far from negligible source of nourishment.

A list of different vegetable foods by Rizek et al. (1975) based on a global evaluation of the nutritional value, is presented in Table 1.

2. UTILISATION OF FRUIT AND VEGETABLES

With regard to the consumption of fruits and vegetables, two aspects are relevant:

a) Their traditional consumption as fresh and raw products

b) The remarkable increase of their consumption due to the above mentioned psychological reasons.

On this subject it is interesting to compare the data of recent European and USA surveys. In Italy and in France fruit is eaten at almost every meal and in Europe the consumption of vegetables is linked with that of meat. In the USA from 1925 to 1971 the consumption of vegetables rose only 7% with a 57% reduction in the consumption of fresh products and a 350% increase of processed products, which results in a decreased consumption of nutrients (Rizek et al. 1975).

Generally the total consumption of fruits and vegetables is broadening in Europe and in affluent countries with a wide increase of market and farming areas.

The trend is towards an increasing production of processed vegetables, therefore genetic and technological research is being developed to improve quality and reliability of preserved vegetables. For these reasons it is interesting to examine any alteration of natural characteristics brought about by thermic treatments with special reference to the nutritional aspects.

3. THERMIC TREATMENTS

We can summarise the treatment of vegetables as preparatory and preserving operations.

Preparatory operations:

a) Treatment of the product as it is

b) Morphological modifications (or liquid separation)
   - peeling, grinding, slicing
   - comminution
   - squeezing and expression

c) Chemical modifications
   - blanching
   - chemical peeling
   - fermentation (pickling, alcoholic)
   - electrolyte extraction

Thermic treatments in fruits and vegetables processing can be listed as follows:

a) Heating
   - home cooking
   - blanching
   - canning
   - storage
   - hot drying
   - specific treatments (roasting, conching)

b) Cooling
   - preservation of raw products in controlled atmosphere and temperature
   - refrigeration
   - deep-freezing
   - freeze-drying

These treatments involve changes in structure, colour and flavour with a loss of nutrients and a formation of artefacts. We must point out that modifications of organoleptic characteristics due to industrial treatments are particularly important for vegetables if compared with other foodstuffs.

A completely different processing of fruits and vegetables is the extraction of proteins from oleaginous seeds, pulses and leaves, involving special thermal treatments.

4. STRUCTURE MODIFICATIONS DUE TO THERMIC TREATMENTS

Since morpho-structural characters are very important for the acceptance of this kind of food, many researches have been devoted to the study of the thermic conditions causing modification of these sensitive products (Weier and Stocking, 1952).

Histological, chemical and physical data describe various phenomena which occurred following thermic treatments as due to:

a) The anatomy of the vegetables

b) The rheological behaviour of the product

In thermic treatments the preservation of the structure is substantially dependent on the intercellular constituents of the median lamella (pectins, hemicellulose), and, for the pulpy fruits and tubers, on the type of starch.

There is a relationship between the intercellular adhesion measured by mechanical parameters, and the amount of polyuronides, the degree of pectin methoxylation, the amount of hemicellulose and amylose, and the divalent cations (Linehan and Hughes, 1969).

Heating acts on the polyuronides of the lamella causing both physico-chemical modifications (the so-called 'melting') and enzymatic ones (by activation of depolymerising enzymes).

The hemicellulose of cell walls is modified in the same way. Heating effects and technological treatments are important from the nutritional standpoint because they modify:

a) The size of cellulosic fibrils and the state of peptic polymers forming the food ballast

b) The level of pectin methoxylation, with methanol release (Van Buren et al., 1960; Doesburg, 1966)

c) The final concentration of calcium

Another aspect to take into consideration is the callose formation due to fruit bruising and heating (Currier, 1957; Dekazos and Worley, 1967). This glucan plays an interesting structural role, being stable to heat; its metabolic fate has not yet been taken into account with regard to animal nutrition.

Comminuted nectars and juices present interesting rheological characteristics depending on the size of dispersed particles and on proteic and pectic colloids. Heating treatments modify the rheological behaviour of the product and affect its palatability. The product can be treated in two different ways: cleared or thickened by the addition of extraneous clouding agents; neither treatment has yet been nutritionally investigated.

During freezing we can observe a cellular disorganisation with mixing of the cytoplasm with the vacuolar content and damage of the intercellular adhesion, this latter producing a true structural collapse (Monzini et al., 1975; Ulrich, 1977).

This effect increases the one caused by heating in the blanching. On the other hand, freezing without blanching compromises the quality of the product due to latent enzymatic activity.

A great number of studies have been devoted to establishing the conditions which minimise the side effects of freezing of fruits and vegetables, including a genetical approach to this problem. The consequences of denaturation by freezing will be discussed further.

The crioconcentration of the juices is a special method of freezing, affecting the rheological behaviour. The characteristics of viscosity and 'pulpiness' of concentrated, cut-back and diluted juices, which result from different thermal treatments are interesting both from the technological and the nutritional points of view.

5. DISCOLOURATION

Colour is an essential characteristic for the acceptance of vegetables which raises complex problems of stability. Discolouration is perhaps the phenomenon receiving the most active experimental interest.

The pigments involved are chlorophylls, carotenoids and polyphenols; their alteration follows different kinetics influenced in various ways by temperature, oxygen level, pH and other cellular constituents.

The degradation of chlorophyll to pheophytin is a heat induced metamorphosis of the chloroplasts to a smectic mesophase which occludes the chlorophyll.

Haisman and Clarke (1975) demonstrated that reaction velocity depends on the hydrogenium ions transfer through the interface. To avoid discolouration a slight alkalinisation is helpful; surface-active agents are also effective because they lower the potential on the lipid-water interface, reducing hydrogenium ions transfer. Thermic treatments are carried out on this basis to avoid discolouration. The alkalinisation effects and the type of treatment have, nevertheless, to be considered from a nutritional and toxicological point of view. Modification of carotenoid pigments due to thermic treatment has been widely observed during processing and storage.

Peculiarities of different products influence the stability of the pigments, in particular β-carotene stability. To a certain extent this is due to liposolubility of these pigments and to their location in tissues and organs. For instance, different losses have been observed, under the same operating conditions, for beans, peas, maize, carrots and spinach (Lee, 1958).

The loss of carotenoids in freeze-dried products is related to the A_w level, as Labuza (1972) demonstrated for lipid oxidation. Colour and vitamin A loss is an interesting critical point of the new preservation processes based on intermediate moisture levels. Regarding polyphenols, in particular anthocyanins, we can say that their location in vegetables causes problems of colour stability and protein availability. The loss of anthocyanins is due to hydrolysis reactions (the anthocyanidins are very unstable), oxidations and complex formation. Data by Dalal and Salunkhe (1964) related to canned fruit preservation are reviewed. Since anthocyanins have positive pharmacological effects, their stabilisation can have a nutritional interest (Gabor, 1975).

Other discolouration phenomena concern polyphenols, other than anthocyanins and flavons, such as cathechins and polymeric flavanes and phenolic acids (Mathew and Parpia, 1971).

Browning, due to phenolase activity and auto-catalytic oxidation of these polyphenols, is a crucial problem in the processing of fruits and vegetables such as canning and the production of juice and fermented beverages.

Genetic research in order to obtain products low in phenolics and phenolases, treatments to inactivate enzymes or to subtract phenolics and complexing metals, has been developed to avoid this discolouration phenomenon. Therefore some of these processes present toxicological problems (such as the use of SO_2, acidulants, chelating and adsorbing compounds). More generally a crucial problem in nutritional research is presented by polyphenols-proteins reactions.

6. FLAVOUR CHANGES

The sapid and volatile substances as a whole form another essential organoleptic characteristic of vegetables, in particular of fruits; differences between their natural flavour and the flavour of processed products cover a wide area of analytic research. Some aspects could have a nutritional implication, one of them is the release of carbonyl compounds, due to cell disruption, reacting with amino acids and phenolics (Marchesini, 1973).

7. NUTRIENTS MODIFICATIONS BY THERMAL TREATMENTS

The nutritional characteristics of vegetables are, as mentioned above:
a) Simple and complex available glucides, present in high concentration in any species
b) Lipids, present in slight concentration except in oleaginous fruits and seeds
c) Proteins in small concentration (1 - 2%) except in pulses
d) Minerals, of which vegetables are an irreplaceable source, particularly calcium but also magnesium, potassium and iron
e) Vitamins, particularly ascorbic acid, β-carotene, thiamin and other B group factors

Besides these elements which have a direct nutritional value, vegetables contain, according to their taxonomical characteristics, different organic acids and polyphenols of physiological interest. Of the above mentioned anti-nutritional factors, some are thermolabile while others are modified during preservation treatments (Jaffe, 1973).

8. LOSSES DURING COOKING

Vegetables show weight variations according to cooking techniques, from blanching to home cooking, and according to different species (Péquignot et al. 1955). These variations range from losses of 30 - 40% in home cooked vegetables; to losses of 50% in frying, and to a slight weight increase for beans, Brussels sprouts and boiled potatoes. Losses of nutrients which are soluble in cooking water have been evaluated in detail. These factors are essentially minerals, simple glucides, amino acids and aliphatic acids (Cain, 1975).

Other losses of solubles take place during the storage of canned products, depending on temperature; glucides, pigments and polyphenols are particularly affected.

9. GLUCIDES ALTERATIONS

Thermic treatment modifies the reactivity of various glucides. The disruption of simple glucides and reduction of oligosaccharides to the chain of Maillard reaction due to heating have been amply demonstrated even at subcellular level. Refrigeration does not cause alteration of these glucidic fractions. As far as quality is concerned, the progressive shift from canning to freezing is due to the differences in the modification of glucidic constituents.

Complex unavailable glucides, particularly pectins and some oligosaccharides, also undergo degradation during heating and preservation. Dalal and Salunkhe (1964) estimated, in canned cherries, a 90% loss of pectins in 16 weeks at 100 - 120°F. At the same time they observed an increase in simple or complex reducing glucides, in acids and in hydroxymethylfurfurol, leading to a sharp alteration of flavour (burt flavour). In different processing conditions these autocatalytic alterations are associated with severe enzymatic modifications.

Finally, the glucides responsible for flatulence, which are present in legumes, must be considered. Many attempts have been made to eliminate them, both technological (germination, fermentation, ultrafiltration, alkaline extraction) and genetic. Thermic treatment of vegetables does not seem to modify these glucides.

10. LIPIDS AND RELATED COMPOUNDS

Lipids, besides being scarcely represented (except for the oleaginous species) do not have a relevant significance because vegetable processing does not usually reach auto-oxidation and cyclisation temperatures. The only exceptions are frying, in which the cooking fats are prevalently modified, and the roasting of coffee, cocoa beans and nuts. Therefore roasted products must be protected by antioxidants and inert gases. A systematic survey on deterioration of these products which are widely used in confectionery should be nutritionally valuable. Fat rich fruits such as olives and avocado show a relative stability in their lipid content (Cantarelli, 1969); this is caused by low saturation of fatty acids, richness in natural phenolic antioxidants, and absence of lipoxydase in the mesocarp.

Lipoxygenase is a critical enzymatic activity of many seeds which can be avoided by blanching. This enzyme is still active at a very low A_w because of capillary water (Brockmann and Acker, 1977; Klop, 1974). Peroxides are involved in carotenoid losses and react with proteins causing losses of amino acids.

11. NITROGEN COMPONENTS

In fruits and vegetables, with the exception of pulses, the number of different nitrogen compounds is rather modest. Nevertheless modifications of these compounds can occur in relation to the different constituent characteristics.

Many phenomena have been identified:
a) Loss of amino acids due to dissolution
b) Formation of amino acids artefact
c) Maillard reaction at different stages:
   - Schiff bases
   - Degradation products and Amadori rearrangement
   - Pre-melanoidines
   - Melanoidines
d) Reactions between proteins
e) Reactions between proteins and oxidised lipids
f) Reactions between proteins and polyphenols

The importance of the biological value losses, being related to protein content, is different for fruits and vegetables and for legume seeds. With the exception of the latter, the lysine level is very low, also the serine and sulphurated amino acids involved in the formation of artefacts. On the other hand, these products, containing a good amount of glucides and carbonylic compounds, give rise to a high proportion of Schiff bases.

Legume proteins are badly damaged by heating; on the other hand, severe heating is needed to inactivate anti-tripsic factors, hemiglutinines and other anti-nutritional compounds.

11.1 Free amino acids losses

Trimming, peeling and blanching cause rather substantial losses (30% in peas and other vegetables) of free amino acids and soluble proteins (Thormann et al., 1977). On the other hand, with the loss of water and salt soluble proteins, an increase of alkali soluble proteins occurs. A modification of the albumin/globulin ratio at different denaturation levels occurs even in frozen products if they have not been previously blanched. Free amino acids in blanched and cooked potatoes, peas and asparagus show, after cooking, a 20 - 25% loss of basic amino acids and a 10% loss of asparagine.

Frying causes particularly high losses as Fitzpatrick (1976) demonstrated.

Even heavier losses are observed in juice making; only 23 - 29% of the total nitrogen of the fruit is available in the final product and only 30% of it is present as free amino acids (Baumann and Gierschner, 1973). These losses are attributed to lixiviation, to amino acids reactions with glucides and other carbonyl compounds, and to enzymatic clarification treatments.

11.2 Proteins reactions

Various reactions can take place as a consequence of thermic treatments; we can distinguish some different conditions

for heat damage, according to the suggestions of Mauron (1974):
  a) Mild heat treatments
     - in the presence of reducing glucides
     - in the presence of polyphenols
  b) Severe heat treatments
     - in the presence of reducing and non reducing glucides
     - in the presence of oxidised lipids
  c) Heating in alkaline conditions
  d) Oxidative treatments
  e) Treatments which disorganise the cell structure
     (deep freezing, mechanical treatments)

These conditions occur in different species and organs of vegetables and in different processes, such as:
- blanching
- canning
- home cooking
- extrusion
- alkaline blanching
- alkaline peeling (peaches and potatoes)
- cocoa alkalinisation
- alkaline extraction of proteins
- recovery, by coagulation, of proteins plasma of leaves
- freezing and deep-freezing
- juice hot extraction

The different problems regarding protein modification will be described in the lecture by Professor Bender. Therefore we will emphasise some particular aspects of vegetable materials.

The only case to be examined thoroughly is that of legumes whose proteins are extracted industrially. For fruits and vegetables we have a considerable lack of information. We can suppose that boiling and steaming on the one hand and home-style cooking on the other hand, cause very different effects. The loss of lysine as 1-deoxy-2-ketose is in any case very severe, as well as the melanine formation, which in many cases is needed. The amount of free and bonded amino acids is also important both for simple solubilisation and for the effect of mild alkalinisation and heating (frying, browning and roasting) in different cooking processes because of destruction, racemisation, cross-linking and aldolic condensations.

11.3 Proteins-polyphenols reactions

The reactions that mostly affect the availability of protein in vegetables are certainly the ones that involve phenolic compounds, ubiquitous in vegetables. We can distinguish phenolic acids and other low molecular weight compounds, flavans as simple and polymeric catechins whose metabolic and pharmacological activities are only partially demonstrated. Tannins react with proteins forming hydrogen bonds between phenolic groups and amide or peptic bonds of proteins, when heating or freezing causes cell disorganisation.

These reactions provoke a loss of digestibility and biological value of proteins, and therefore represent a problem, as yet unsolved, for formulations of good nutritional quality.

Many cases of loss of nutritional value of vegetable proteins must be attributed to reactions with polyphenols; among them we can mention proteins of cocoa and faba beans. For instance, the lack of methionin, characteristic of vegetable protein, can be ascribed to the coupling with quinones; methionin sulphoxydes and sulphones themselves can react and develop during polyphenol oxidation (Bosshard, 1972).

Horigome and Kandatsu (1968) demonstrated that the loss of protein biological value is due to phenolase in the substrate containing polyphenols. In the same way the loss of available lysine during leaf protein extraction has been ascribed to reactions with quinones. In sunflower, sorghum and cotton products, biological value losses are due to reactions of proteins with the different phenolic constituents of these species. A typical case is cotton in which the solvent extraction of gossypol gives a valuable protein material, while

the destruction of gossypol by heating causes a remarkable
loss of biological value.

A great number of reactions between amino acids and native
or oxidised polyphenols have been shown; these reactions produce,
among others, discolouration of juices and fermented beverages
(Cantarelli and Montedoro, 1974). The proper subtraction of
phenolic compounds, as well as the selection of phenotypes with
low phenolic content are interesting approaches from a
nutritional point of view.

This problem is connected with thermic treatments in at
least three different ways.

   a) Disorganisation of cytoplasm, caused by heating or
       freezing, promotes the formation of proteins-polyphenols
       adducts; this reaction is stressed if polyphenol
       oxidation occurs because of enzymatic or metal catalysis.
   b) Heat activates the phenolases (tyrosinase and laccase)
       which are highly diffused in fruits and vegetables.
   c) In order to avoid these enzymatic reactions, sulphites
       are widely used with considerable toxicological problems;
       therefore to improve nutritional quality it is necessary
       to select a heat treatment replacing the chemical
       inactivation.

11.4 Enzymatic proteins

Post-harvesting enzymatic anctivity is a major cause of
product deterioration; blanching is largely used to avoid this
process.

Many studies indicate different parameters of heat
inactivation for various enzymes, as well as different low
temperature levels for freeze blocking. Bolcato et al. (1973)
observed that, even at $-20^{\circ}C$, lipase, amylase and acid phosphatase keep their activities, unlike other oxidative enzymes. In
general peroxidase is considered the most stable enzyme, therefore its activity is used to test enzymatic stabilisation during

blanching (Burnette, 1977).

The LT-LT treatment (low temperature, long time) in blanching, recently proposed because of the firming effect related to the activation of pectin esterase, must be taken into consideration here. According to Steinbuch (1976) a previous mild blanching (20 min at $70^{\circ}C$ or 5 min at $75^{\circ}C$) followed by a water cooling and by a short high temperature treatment (eg, 4 min at $98^{\circ}C$) positively affects the texture of frozen vegetables after cooking. With this method, in fact, a leaching of soluble materials can occur due to the prolonged time in heated water; saturated water should be applied in this case.

12. VITAMINS

The fate of vitamins during the heat treatment of fruits and vegetables is well documented. It consists essentially of severe losses of vitamin C and thiamin; other B group vitamins and A vitamin being quite stable. Some data indicate an increase of riboflavin during canning.

It must be emphasised that the loss of vitamin C is related to the different characteristics of various vegetable organs, as well as to the operating conditions. The loss is due both to lixiviation and oxidation, this last form of degradation occurring even during freezing. A great improvement in the preservation of vitamin C is obtained by blanching with microwaves, as well as by de-aeration and by use of inert gases. Since the air content of vegetable tissues vary for different organs and species, these techniques can be utilised satisfactorily only for some materials.

13. MINERALS

During blanching there is a loss of potassium and phosphates while calcium, magnesium and iron do not diminish. We observe an increase in calcium using hard water. This increase is also a consequence of the activation of pectin esterase, which causes the release of available carboxyles. The most important losses are, in any case, due to the subsequent cooling and washing.

There is a lack of data concerning the essential trace elements whose role appears more and more important and whose number appears actually enlarged to 14 (or perhaps 16) elements*. Their presence in vegetables is known but there are only few data on the modifications occurring during technological treatments.

This review is far from being complete, many other aspects should be considered. The aim of this paper is to focus on those subjects that may have a nutritional interest.

*Fe, I, Cu, Mn, Mo, Zn, Se, Co, F, V, Ni, Sn, Si, As (Pb, Cd)

TABLE 1

NUTRIENTS IN DIFFERENT VEGETABLES

	Spinach	Brussels sprouts	Beans (green)	Peas	Asparagus	Cauliflower	Carrots	Potatoes	Lettuce
Glucides, g									
Simple sugars	0.3	2	2	4	2	2	3	2	1
polysacchar. available	0	1	14	10	0.5	0.5	0.3	18	0
unavailable	0.7	3	2.5	4	0.8	1.2	7	2	1.2
Lipids, g	0.3	0.5	0.8	0.4	0.2	0.5	0.4	0.1	0.2
Proteins, g	1.2	2.9	7.5	6.7	2.2	2.8	1.1	2	1.2
Vitamins									
Ascorbic a. mg	60	95	20	25	35	70	5	15-40	20
Thiamin mg	1	0.8	0.8	2.8	1.6	1.1	0.6	1	0.4
Riboflavin mg	2	1.6	1.2	1.2	1.9	1	0.4	0.3	0.8
Folic a. μg	100	70	30	30	150	30	10	6	20
Minerals, mg									
Calcium	80	35	57	25	20	22	35	8	25
Iron	3	1.3	0.8	0.7	0.9	1.1	0.8	0.7	0.5
Polyphenols	-	-	+-	-	-	-	-	+	-
Antinutritional factors	+	+	++	+	+	+	-	+	-
Caloric value, J	65	85	400	380	75	75	85	380	60
Rank for nutr. value*	2	3	4	5	6	8	10	14	26

All values per 100 g, except thiamin and riboflavin expressed per 1000 g. *Following Withe and Selvey, 1975

TABLE 2

NUTRIENTS IN DIFFERENT FRUITS

	Pomaceous	Berries	Stone f.	Citrus	Banana	Pineapple	Tomato	Olives
Glucides, g								
Simple sugars	10-14	7-18	10-12	7-10	10-16	7-18	3-4	3-6
Polysacchar. available	0.5	0	0-1	0	1-4	0	0-2	0
unavailable	2	2.5	1.2	1	2	1	2-3	2-4
Lipids, g	0.4-0.7	0.8	0.1	0.2-0.6	0.2	0.2	0.3	16-18
Proteins, g	0.3-0.4	1.4	0.5-1	1	1.2	0.4	1.1	1.5
Vitamins, mg								
Ascorbic acid	5	4-60	8	50	10	30	25	0
Thiamin	0.4	0.6	0.2	0.8	0.4	0.1	0.6	tr
Riboflavin	0.3	0.4	0.5	0.3	0.5	0.5	0.4	tr
Vitamin A, iu	90	80	800	170	200	100	700	40
Phenolics	++	+++	+	+	++	+-	+-	++
Antinutritional factors	-	-	+	-	++	-	+	+-
Caloric value, J	270	160-300	190-250	160-200	340-400	200	100	800
Minerals, mg								
Calcium	6	17-30	8	34	9	20	11	60
Iron	0.3	0.6-0.8	0.6	0.4	0.5	0.5	0.6	0.8

All values per 100 g, except for thiamin and riboflavin expressed as mg per kg

TABLE 3

PER CAPITA CONSUMPTION OF FRUITS AND VEGETABLES (Except starchy roots)

Country		Kg/year
AFRICA:		
	East and south	58
	North	86
	West and central	94
AMERICA:		
	Central	129
	North	188
	South	151
ASIA:		
	China	61
	Indonesia	49
	Japan	94
	Near east	145
	South east	53
EUROPE:		
	Western	145
USSR		82

FAO, 1970

TABLE 4

POSITION OF FRUITS AND VEGETABLES IN THE DIET (FRANCE)

Year	1966	1970	1972
Food considered as essential:			
Meat	26.5	30	28
Vegetables	21	14	19.5
Fruits	6	1	8
Preferred foods			
Meat	29		51.5
Vegetables	17		11.5
Fruits	4		2.5

From Claudian and Serville, 1975

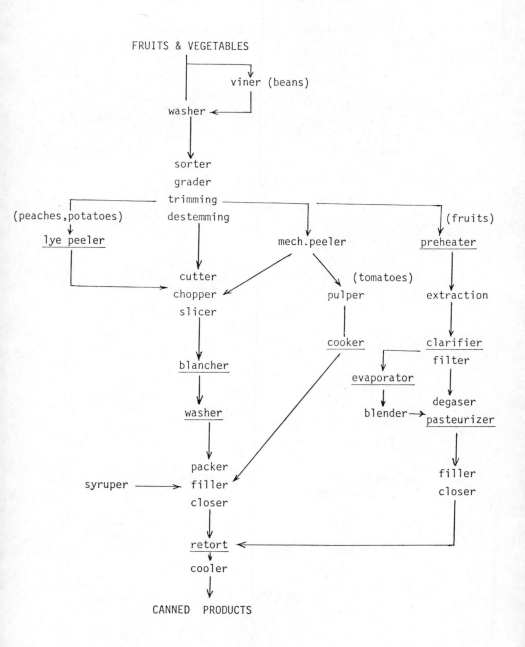

Fig. 1 Critical operations in canning

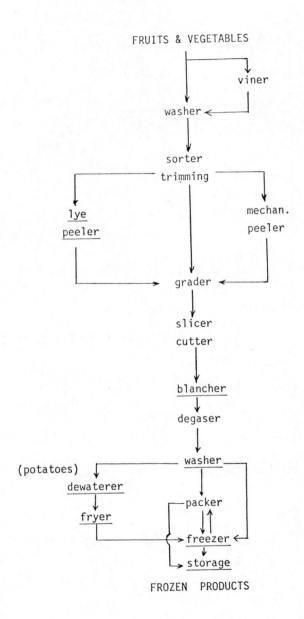

Fig. 2 Critical operations in freezing

REFERENCES

Articles

Baumann, G. and Gierschner, K. 1973. Rep. Int. Fed. Fruit Juices Prod. 13, 265.

Bolcato, V and Spettoli, P. 1973. Ind. Agr. 11, 76.

Bosshard, H. 1972. Helv. Chim. Acta 55, 32.

Brockmann, R., and Acker, L. 1977. Lebensm. Wiss. Technol. 10, 24.

Burnette, F.S. 1977. J. Food Sci. 42, 1.

Cain, R.F. 1975. Proc. Conf. Technology of fortification of foods, Nat. Acad. Sci. Washington.

Cantarelli, C. 1967. C.R. 2. Int. Symp. Oenol. Bordeaux 13-16 June, INRA Paris.

Cantarelli, C. and Montedoro, G. 1974. In: Natürliche u. synthetiche Zusatzstoffe in d. Nahrung d. Menschen, R. Amman and J. Hollò, Ed. Steinkoppf, Darmstadt.

Claudian, J. and Serville, Y. 1975. Chaiers Nutr. Diet. 10, 85.

Currier, A.B. 1957. Amer. J. Botan. 44, 478.

Dalal, K.B. and Salunkhe, D.K. 1964. Food Technol. 18 (8), 88.

Doesburg, J.J. 1966. Pectic substances in fresh and preserved fruit and vegetables - IBVT, Wageningen.

Fitzpatrick, T.J. and Porter, W.L. 1976. Am. Pot. J. 45, 138.

Gabor, M. 1975. The pharmacological properties of flavonoids, Proc. Int. Conf. Polyphenols, Groupe Polyphénols, Gargnano.

Haisman, D.R. and Clarke, M.W. 1975. J. Sci. Food Agr. 26, 111.

Horigome, T. and Kandatsu, M. 1968. Agr. Biol. Chem. 32, 1093

Jaffe, W.G., Moreno, R. and Wallis, V. 1973. Nutr. Rep. Int. 7, 169.

Labuza, T.P. 1972. CRC Crit. Rev. Food Techn. 3, 327.

Linehan, D.J. and Hughes, J.C. 1969. J. Sci. Food Agr. 20, 113.

Marchesini, A. 1971. Ann. Ist. Sper. Val. Tecn. Prod. Agr. 4, 271.

Mauron, J. 1974. Proc. 4. Int. Congr. Food Sci. Techn. 1, 564

Monzini, A., Crivelli, G., Bassi, M. and Buonocore, C. 1975. Structure of vegetables and modifications due to freezing. IVTPA, Milano.

Peguignot, G., Vinit, F., Chabert, C., Bodard, M. and Pesles, S. 1975. Ann. Nutr. Alim. 29, 439.

Rizek, R.L. and Snope, D.A. 1975. Trends in fruits and vegetables consumption and their nutritional implications. In: White, P.L. and Selvey, N., ed. l.c.

Thormann, M., Wolf, W., Spiess, W.E.L., Gierschner, K., Baumann, G. and Jung, G. 1977. Lebens. Wiss. Technol. 10, 28.

Ulrich, R. 1977. C.R. Jour. Alimentation Travail, Nancy.

Van Buren, J.P., Moyer, J.C., Robinson, W.B. and Hand, D.B. 1960. HSZ Phys. Chemie, 321, 107.

Weier, T.E. and Stocking, C.R. 1952. Adv. Food Res. 2, 312.

General Reviews

Aykroyd, W.R. and Doughty, J. 1964. Legumes in human nutrition. FAO Nutr. Studies, n.19, Roma.

Aylward, F. and Haisman, D.R. 1967. Adv. Food Res. 17, 1.

Brenner, S., Wodicka, V.O. and Dunlop, S.G. 1947. Food Techn. 1, 208.

Carpenter, K.J. 1973. Nutr. Abstr. and Rev. 43, 424.

Clifcorn, L.E. 1951. Adv. Food Res. 1, 39.

Jaffé, W.G. (ed.) 1973. Nutritional aspects of common beans and other legume seeds as animal and human food. Proc. Meet. Rivirao Preto.

Joslyn, M.A. and Ponting, J.D. 1953. Adv. Food Res. 3, 1.

Katz, S.H., Hediger, M.L. and Valleroy, L.A. 1974. Science, 184, 765.

Lee, F.A. 1958. Adv. Food Res. 8, 63.

Mathew, A.G. and Parpia, H.A.B. 1971. Adv. Food Res. 19, 75.

Montgomery, M.W. and Day, E.A. 1965. J. Food Sci. 30, 828.

PAG Symposium on nutritional improvement of food legumes by breeding - FAO, Rome, July 3 - 5.

Stadtman, E.R. 1951. Adv. Food Res. 1, 325.

Sternberg, M., Kim, C.Y. and Schwende, F.J. 1975. Science, 190, 992.

Synge, R.L.M. 1975. Qual. Plant. -Pl. Fds. Hum. Nutr. 24, 337.

Van Buren, J. 1970. C.R.C. Crit. Rev. Food Techn. 1, 5.

SESSION 2

PASTEURISATION/BLANCHING

Chairman: I.F. Vujicic

PASTEURISATION OF MEAT, FISH AND CONVENIENCE FOOD PRODUCTS

Thomas Ohlsson
SIK - The Swedish Food Institute
Fack, S-400 21 Göteborg, Sweden

ABSTRACT

Pasteurisation involves preservation of foods by heating to temperatures generally below 100°C. For long storage the products must also be preserved by means of some auxiliary procedure: low temperature, preservatives, acidity, etc. Product examples are semi- or ¾-preserved meat products like ham, wieners and other sausages, precooked convenience foods including ready-to-serve dishes and prepeeled potatoes.

Pasteurised products distributed and stored at refrigeration temperatures may have a great market potential as compared to, say, vacuum-packed products, since a substantial shelf life can be obtained, as also in comparison with frozen foods from a storage and handling point of view, and to sterilised foods from a sensory quality point of view.

A review of recent R and D work on different aspects of pasteurisation of meat, fish and convenience food products, shows that the investigations reported on are mainly concerned with the development of methods and/or packaging, or with comparisons of methods.

On the basis of the literature review and SIK original research work, an attempt is made to point out critical areas where continued and new R and D is needed. Work on heat inactivation and the occurrence of spore-forming psychrophilic or psychrotropic organisms (ie, Cl. botulinum type E) and of enzymes of animal origin is proposed, together with studies of their respective activities during refrigeration storage. Studies of sensory quality changes during heating and cooling and during storage are also suggested. Investigations of heat transfer and energy consumption, together with the development and optimisation of new and existing methods and equipment, are recommended. Finally, it is suggested that an attempt should be made to collect data for the establishment of a TTT-hypothesis for pasteurised food.

2. INTRODUCTION

Just over a century ago, Pasteur established that spoilage is caused by microbes and that heat will kill them. On that basis he also showed that the shelf life of beer and vinegar could be extended by heating them to close to boiling temperatures. Thus, for heat treatments - usually below 100°C - the term 'pasteurisation' was adopted. Today, pasteurisation is used for a variety of reasons (Ingram, 1971):

- to eliminate heat-sensitive pathogenic micro-organisms;
- to kill various organisms in order to clear the ground for an added starter organism;
- when a more severe heat treatment - sterilisation - would damage the product;
- to improve shelf life when the main spoilage organisms are not very heat-resistant;
- to improve shelf life when the surviving micro-organisms can be controlled by some auxiliary procedure.

The main objective of pasteurisation is the reduction and elimination of pathogenic and food-spoilage micro-organisms. Vegetative cells of bacteria are, with few exceptions, completely eliminated by the heat treatment. Spore-forming bacteria are substantially reduced during pasteurisation. Some of them survive, however, thus being the most resistant. The stability and safety of the pasteurised foods depend on the degree to which these residual micro-organisms can be controlled. The additional controlling factor needed may be low temperature, low pH, reduced water activity, the presence of inhibitory substances, or combinations of these factors. The complex interaction of these factors with the heat treatment gives a special interest to the subject, at the same time making it more difficult.

Pasteurisation will also lead to inactivation of many enzymes, owing to coagulation and hydrolysis of the enzyme proteins

During pasteurisation most vegetable enzymes are destroyed, the most important exception being peroxidase. Among the enzymes of animal origin, many remain intact after the heat treatment. Some enzymes are even activated, eg, in a partly inactivated biochemical system, where inhibitory substances or enzymes may be destroyed. Enzymes of microbial origin may also survive the heat treatment (even if the micro-organisms do not) and adversely affect storage life.

The inactivation of the abovementioned 'biological' constituents of foods is highly dependent on temperature. More chemically related changes, like degradation of nutrients and sensory quality loss, have a different, less pronounced temperature dependence. This is schematically illustrated in Table 1 (Lund, 1977), where the D-value stands for the time required, at a constant temperature, for a concentration or a count to change by one order of magnitude, and the z-value represents the temperature change necessary to change the reaction rate by one order of magnitude. These relationships form the background of the well-known HTST (High Temperature Short Time) principle, where the objective is to carry out the treatment at as high a temperature and in as short a time as possible.

TABLE 1
THERMAL CHANGES OF VARIOUS FOOD CONSTITUENTS (After Lund, 1977)

Constituent	z, °C	$D_{121°C}$
Vegetative cells	4 - 7	0.002 - 0.02
Spores	7 - 12	0.1 - 5.0
Enzymes	7 - 55	1 - 10
Vitamins	25 - 31	100 - 1000
Colour, texture, flavour	25 - 45	5 - 500

Although losses of nutrients owing to chemical degradation are small during pasteurisation, with the exception of vitamin C, important losses may occur owing to the juice lost during meat and fish protein coagulation and the loss of turgor in vegetables.

The changes in sensory quality parameters during pasteurisation can generally be regarded as initially favourable, but with severe heat treatment the sensory quality may be adversely affected, eg, in the form of textural changes.

Meat, fish and convenience food items to be pasteurised consist mainly of solid food particles, in which heat is transferred by conduction. Since foods have low thermal conductivity, heat penetration measurements and process calculations based on heat and mass transfer theory are important parameters in pasteurisation.

Process calculations based on the same principles as for sterilisation processes are sometimes employed. For the microbiological effect of a process, a pasteurisation F-value is used, often based on $65.5^{\circ}C$ (= $50^{\circ}F$) and a z-value of 6 - $6.5^{\circ}C$ (Andersen et al. 1965). Sometimes the F $150^{\circ}F$-value is renamed P-value to avoid confusion with the F $250^{\circ}F$-value (Shapton et al. 1971).

The time of storage is one of the most important factors with regard to pasteurised foods. Generally, attempts are made to increase it by combining one or more stabilising factors (low temperature, pH, etc.) Therefore, and also because the conditions and the performance both of the heat treatments and of storage, ie, the factors influencing storage life, are complex as well as interacting, accurate predictions of storage life for pasteurised food are very difficult to make. This is also illustrated by the fact that it has not yet been possible to construct a valid and reliable TTT-hypothesis for cool storage of foods.

On the market today you will find traditional pasteurised meat products such as canned ham, luncheon meat, canned sausage etc. Some pasteurised fish products, mainly shellfish, have been marketed, but only to a limited extent. During the past decade, convenience or ready-to-serve pasteurised foods have been introduced on the market in Europe. Many of these foods are packed in flexible packages to be reheated directly in the package to serving temperatures, eg, prepared dishes and pre-peeled potatoes.

The potential market for pasteurised, cool stored food is considered to be large, provided a sufficient storage life can be obtained. Compared to only vacuum-packed food, a substantially longer storage life at refrigeration temperatures is to be expected. This will facilitate more centralised production and distribution. Compared to frozen foods, the distribution will not be as complicated and capital-consuming. For the consumer the reheating is much quicker and more convenient. Compared to canned foods, much better retention of sensory and nutritional quality will be obtained, and a much wider product category can be processed. Finally, the overall energy consumption for processing - packaging - distribution - storage - reheating, is lower than for most other systems for food preservation.

3. REVIEW OF RECENT R AND D

Since many different heat treatment operations involve pasteurisation, the aim of this work is not to present a comprehensive review of all recent developments in this wide field. Instead, the purpose is to select a number of interesting lines of development, familiar to the authors, using them as a basis for recommending further R and D in the pasteurisation field.

3.1. Processing fundamentals
3.1.1. Microbial aspects

Pasteurisation is a food preservation technique used to eliminate some but not all of the vegetative microbes in foods. Most often heat is the microbiocidal agent of choice, but ionising radiation is likely to have a future as well. It should be noticed that pasteurisation processes have the dual purpose of producing foods that are free of pathogenic micro-organisms and also of reducing the risk of spoilage. Some heat resistance data are shown in Table 2 (Spencer, 1967).

TABLE 2

HEAT RESISTANCE OF MICRO-ORGANISMS-VEGETATIVE CELLS (After Spencer, 1967)

Micro-organism	Conditions	D_{140} value (min)	z value (°F)
Staph. aureus	Poultry stuffing	2.2	12.3
	custard	7.40-8.24	9.20-10.20
	chicken a la king	4.89-5.45	8.95-10.50
	skim milk	5.34	
Salmonellae	Poultry stuffing	2.75	10.1
	custard	2.39-2.49	13.0-15.8
	chicken a la king	0.39-0.40	8.7-9.2
	liquid egg pH 5.5	0.40-2.2	7.5-8.5
Salm. senftenberg 775 W	Custard	9.36-13.27	11.86-12.25
	chicken a la king	9.22-9.99	11.50-12.15
	liquid egg pH 5.5	9.5	12.3
Streptococcus faecalis	Saline	1-5	
	poultry stuffing	9.6	14.2
	various foods	10.5-15.3	12.3
Escherichia coli	Milk	12.5	10.2
Coliforms	Milk	0.3-1.3*	
Lactobacilli	Orange juice pH 3.7 1 x 10⁶ cells/ml	F_{150} = 0.28	7.
Bissochlamys nivea	Ascospores 10⁵ tube	F_{190} = 10	
Penicillium	Ascospores 16 x 10⁴/ml	F_{180} = 9.7	10.6

* D_{142} value

Table 2 provides information on the variation in survival times at one temperature for pathogens as well as for food spoiling organisms. There is, however, a lack of data especially regarding recently found pathogens like *Yersinia enterocolitica* and *Vibrio parahaemolyticus*. From a recent investigation (Hanna et al., 1977) a $D_{55°C}$-value of about 1.5 minutes could be calculated for *Y. enterocolitica* in skim-milk. It should be observed that the status of *Y. enterocolitica* as a food pathogen is uncertain, but the organism could be suspected of infecting humans via pork and some other meat products.

Corresponding resistance data for *Vibrio parahaemolyticus* are lacking. However, it is not until the last few years that this organism has been reported to occur in European sea waters and little work has been concerned with the heat resistance of this organism in sea foods. In Sweden, one outbreak of food poisoning might be attributed to the presence of *V. parahaemolyticus* in boiled crayfish. Most likely this was a result of post-process contamination, thus not caused by high heat resistance of the microbe.

There are also relatively few data on the presence and inactivation rates of different viruses in foods. Some data are shown in Table 3 (Rao, 1976) and Table 4 (La Rocca, 1971).

TABLE 3

VIRUSES ISOLATED FROM FOODS (After Rao, 1976)

Poliovirus 1 and 3	Raw milk (cow)
Para influenza 3	Raw milk (cow)
Bovine syncitial virus	Raw milk (cow)
Poliovirus, 1,2,3	Ground beef
Echovirus 6	Ground beef
Poliovirus 3	Mussels
Echovirus 3,5,6,8,9,12,13	Mussels
Coxsackie virus A-18	Mussels
Avian leukosis complex	Eggs
Infectious bronchitis virus	Eggs
Newcastle disease virus	Eggs

The heat resistance of a micro-organism is affected by the chemical and physical environment during heating. Thus factors like low pH-values and presence of inhibitory compounds will increase the inactivation rate, while a low water content will result in increased resistance. Simultaneous or alternating use of heat and radiation might also be used to increase the microbiocidal effect of a pasteurisation process. Thus, one way to extend shelf life of a pasteurised food is to make survival conditions worse for the microbes during the process.

TABLE 4

DESTRUCTION OF VIRUS (After La Rocca, 1971)

Temperature, °F	Time
Poliomyelitis	
143	15 min.
160	15 sec.
143	30 min.
160	15 sec.
131	15 min.
160	7.5 sec.
149	30 min.;
Coxsackie virus	
149	15 min.
160	7.5 sec.
143	30 min.
160	7.5 sec.
143	15 min.
160	7.5 sec.

Most often, however, altered conditions during pasteurisation are of minor importance and pasteurised foods have to be stored under conditions allowing a complementary preservation technique. Frequently, therefore, pasteurised foods are also fermented like pickles, refrigerated like milk and meat products, kept anaerobic like beer, or are depending on a low pH, or a water activity more or less reduced by sugar or salt, or relying on the presence of chemical preservatives.

Future microbiological work related to pasteurisation processes should continue to generate basic data on heat resistance under various conditions in the food during the process. Increased emphasis should be put on organisms like *Y. enterocolitica V. parahaemolyticus* and viruses.

Combination methods, like heat treatment combined with ionising radiation, might have a future, but more work should be done on the effect of post-pasteurisation storage conditions and the microflora surviving pasteurisation including non-proteolytic strains of *Cl. botulinum*.

3.1.2. Enzymatic aspects

Since, in the living organism, numerous enzymes are present and active in different metabolic processes, unprocessed foods contain enzymes that may drastically reduce shelf life. Some important enzymes which cause undesirable quality changes in foods are listed in Table 5. The enzymes are divided into four groups, related to changes in flavour, colour, texture/consistency, and nutritional value.

TABLE 5

ENZYMES RELATED TO FOOD QUALITY (After Svensson, 1976)

Enzyme	Catalysed reaction	Quality effect
Flavour		
Lipolytic acyl hydrolase (lipase, esterase etc).	Hydrolysis of lipids	Hydrolytic rancidity (soapy flavour)
Lipoxygenase	Oxidation of poly-unsaturated fatty acids	Oxidative rancidity ('green' flavour)
Peroxydase/catalase	?	'Off flavour' (?)
Protease	Hydrolysis of proteins	Bitterness
Colour		
Polyphenol oxidase	Oxidation of phenols	Dark colour
Texture/consistency		
Amylase	Hydrolysis of starch	Softness/loss in viscosity
Pectin methylesterase	Hydrolysis of pectin to pectic acid and methanol	Softness/loss in viscosity
Polygalacturonase	Hydrolysis of α-1.4 glycosidic linkages in pectic acid	Softness/loss in viscosity
Nutritional value		
Ascorbic acid oxidase	Oxidation of L-ascorbic acid	Loss of vitamin C content
Thiaminase	Hydrolysis of thiamine	Loss of vitamin B_1 content

Enzyme activity is destroyed by irreversible denaturation or by hydrolytic breakdown of the protein molecule. In the food industry heating is the most convenient method for enzyme inactivation. It is generally believed that the thermostability of an enzyme is determined principally by its amino acid sequence and the specific conformation derived from the sequence. Thermal inactivation of most enzymes follows first-order reaction kinetics, as shown for lipolytic acylhydrolase (lipase) from potato in Figure 1. The influence of temperature on enzyme inactivation is usually illustrated as the heat treatment time required to reduce the enzyme activity to 10% of the original value - D-value, as is shown for thermo-stable fractions of some potato enzymes ('lipase', lipoxygenase, polyphenol oxidase, and peroxidase) in Figure 2. The peroxidase is considerably more stable than the other enzymes. Besides, enzymes with high thermostability generally show far less temperature dependence in the relevant temperature range than do thermolabile enzymes. Thermostability data from the literature for some quality-related enzymes are summarised in Figure 3. Thermostability is expressed as the temperature required to reduce the enzyme activity to 10% of the original value in one minute. Inactivation data for enzymes originating from food-contaminating micro-organisms are included. Many investigations of the thermal inactivation of enzymes do not include the D-values or the one-minute heat treatment time. In those cases where plausible extrapolations could be made, the 'thermostability temperatures' have been estimated and are shown as brackets in the Figure. As regards enzymes with a 'thermostability temperature' of 100°C or more, only peroxidase originates from vegetables. All other enzymes are microbial ones. The peroxidase enzyme is the only _very_ thermostable plant enzyme, while several enzymes, such as lysozyme, ribonuclease, acid phosphatase, and phospholipase, which derive from animals, have been shown to be _extremely_ thermostable (Svensson, 1976).

3.1.3. Nutritional and sensory aspects

During pasteurisation losses of nutrients may for some products occur as a result of leakage of juices during coagulati of proteins, fat melting and breakage of cells in vegetables. Thermal degradation losses of nutrients and loss of digestibilit of protein are only slight (Hamm, 1977). Studies of the effect

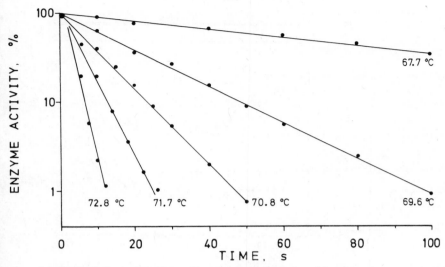

Fig. 1 Thermal inactivation of potato lipolytic acyl hydrolase in buffer extract of potato as a function of time and temperature (after Svensson, 1976)

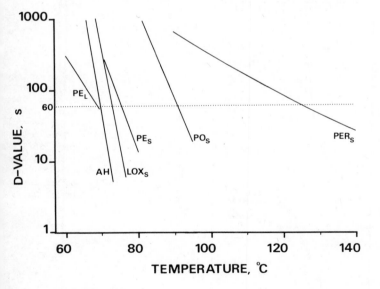

Fig. 2 Thermal inactivation of the thermo-stable fraction of potato lipolytic acyl hydrolase, lipoxygenase, polyphenol oxidase and peroxidase as a function of temperature (after Svensson, 1976)

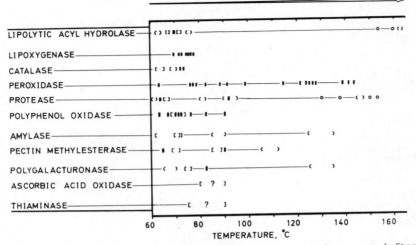

Fig. 3. Thermostability data for some quality-related enzymes (after Svensson, 1976).

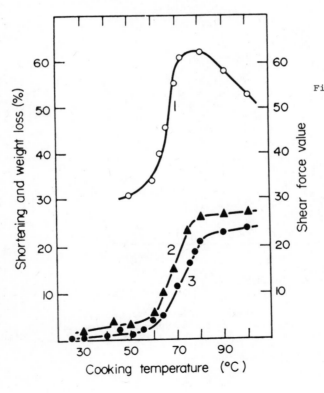

Fig. 4. The two-phase effect of cooking temperature on shear-force values. Standard deviations are given by vertical lines. Each point is the mean of 8 - 16 determinations from the muscle of four bulls (after Davey and Gilbert, 1974).

of nutrient loss during pasteurisation per se are very scarce, except for milk, where 10% losses were found in thiamine and vitamin C content during HTST-pasteurisation (Thompson, 1969). Recently, Poulsen (1977) reported 90% retention of vitamin C in vacuum-packed, pasteurised peeled potatoes, as compared to raw, and peeled ones.

In meat and fish products the heat treatment has an important and favourable effect on the sensory quality of the food, in the form of a tender texture and of a 'cooked' flavour.

Very important changes in the tenderness, firmness and water binding capacity of meat during heating are due to denaturing and more far-reaching reactions of myofibrillar proteins, and also to denaturing of connective tissue collagen. Investigations have shown that very drastic changes in myofibrillar proteins take place between $+30^{\circ}C$ and $+50^{\circ}C$, whilst at $+60^{\circ}C$ they are almost at an end. These changes are characterised by the development of protein molecules, accompanied by the association of the molecules leading to protein coagulation and to a loss in enzymatic activity. At temperatures above $+70^{\circ}C$ more radical processes take place, eg, oxidation of sulphhydryl to disulphide groups. Above $+80^{\circ}C$ rising temperatures cause increasing production of hydrogen sulphide, which is formed mainly from the sulphhydryl groups of myofibrillar protein (Hamm, 1977). These sulphurous components have an important effect on the flavour of the cooked meat products.

Changes in the tenderness, rigidity and water binding capacity of the meat during heating take place in two phases, the first being between $+30^{\circ}C$ and $+50^{\circ}C$, and the second between $+60^{\circ}C$ and $+90^{\circ}C$. There is little change in the temperature range from $+50^{\circ}C$ to $+60^{\circ}C$. Changes during the first phase are due to heat coagulation of the actomyosin system, whilst those during the second phase are ascribed to the denaturing of the collagen system and/or the formation of new cross-linkages within the coagulated actomyosin system. (Figure 4). Heat denaturing of the myoglobin begins at about $+65^{\circ}C$ and causes the meat to turn from red to greyish brown. Browning reactions on heating meat are the result of Maillard-type sugar/protein reactions and of caramelisation. These reactions also contribute to the aroma of the cooked meat. (Hamm, 1977).

Dagerskog (1976) reports on investigations of the textural changes of potatoes during cooling in water. The times necessary to reach a 'finish cooked' texture were determined at different heating temperatures, and a time-temperature relationship for the textural changes was established by means of the cook (C)-value concept, (Leonard et al. 1964), as illustrated in Figure 5. The C-value is calculated analogously to the sterilisation value (F-value) as follows:

$$C = \int_0^t 10^{\frac{T(t) - 100}{z_C}} dt$$

where z_C must be determined for the product and quality parameter under examination. Studies up to sterilisation temperatur have shown that z_C is generally between 25 and 35°C. (Ohlsson, 1977).

3.1.4. Heat transfer and process calculations

Studies of heat transfer in foods and packages to be pasteurised are important for the development and evaluation of processes and methods, as illustrated by Tändler (1972) in studies on pasteurisation of flexible packages of ready-to-serve foods.

With the increased availability of digital computers it is now also possible to combine temperature penetration measurement with calculations of the time-temperature development within th foods. By the use of a computer a very much larger number of parameters may be investigated than is feasible experimentally. It is also possible to combine heat transfer calculations with calculations of time-temperature dependent changes in enzymatic,

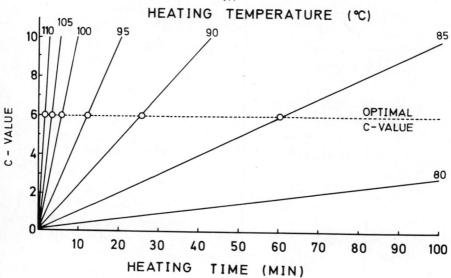
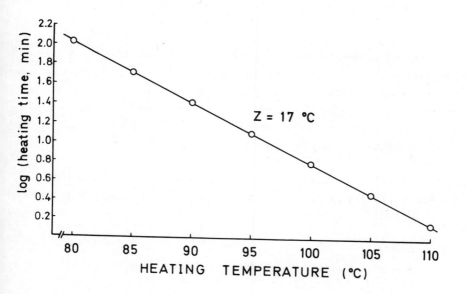

Fig. 5 Calculation of z-value for texture changes in potatoes, by heating of small potato cylinders (parenchyma tissue) in water of different temperatures. Optimal C-value corresponds to ready-cooking (6) (after Dagerskog, 1977)

nutritional, bacteriological, or sensory quality. This is illustrated for potatoes in Figure 6, (Dagerskog, 1977) where the temperature, cook-values, and remaining phenoloxidase activity during cooling after pasteurisation are presented.

3.2. Process methodology
3.2.1. Meat products

In Germany a proposal was made in 1970 to introduce the term 'three-quarter preserves' for products with heat treatment and a storage life somewhere between semi- and full preserves. (Leistner et al. 1970). They are to be heated to a F_s value of between 0.65 and 0.8, sufficient to inactivate mesophilic Bacillu spores, but not Clostridium spores. For this reason their storage life is limited to one year at a maximum temperature of +15°C The method is primarily intended for meat products that, with a complete sterilisation (Botulinum cook) value, have unacceptable sensory quality, like liver and black sausage, 'brühwurst' and others.

A way of decreasing the sensory changes induced by the high temperatures during sterilisation in liver sausage has been presented by Reichert and Stiebing (1977). By increasing the fat content by some 5%, the water activity was reduced enough for a pasteurisation to a core temperature of 95°C to be sufficient for bacteriological stability.

Promising results with dielectric pasteurisation of ham and precooked foods on a laboratory scale are known, but no experience from a large-scale plant has been reported. (Bengtsson, 1967).

The pasteurisation of fermented sausages was studied by Östlund (1970). For industrial applications 20 min. at 75°C and 15 min. at 80°C are recommended, based on experiments where no salmonellae or staphylococci survived these heat treatment combinations.

Stepwise heating has been investigated both for sausage (Reichert, 1976) and for pork (Bolshakov et al. 1971). Reichert

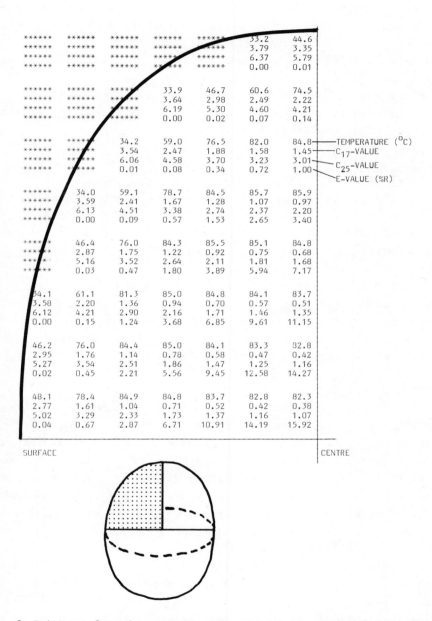

Fig. 6 Print-out from the computer programme with the simulated form of the potato tuber. C_{17}-values correspond to changes in texture and C_{25}-values to changes in sensory quality. Remaining enzyme activity corresponds to A-values (after Dagerskog, 1977)

found that a progressive increase of the processing temperature (to a three-quarter preserved product) resulted in more attractive products. The advantages obtained could also be demonstrated by improved surface cook-values. In contrast, Bolshakov et al. (1971) started with a higher temperature ($100^{\circ}C$) and finished with a lower ($75^{\circ}C$), with an overall cooling time of 65 minutes. Organoleptically the meat was juicier, tastier and more aromatic, as compared to conventional methods.

3.2.2. Poultry products

Klose et al. (1971) reported a new development for pasteurisation of the surface of whole ready-to-cook chickens by steam at $<72^{\circ}C$ for 4 minutes. The internal cavities could also be treated by means of a steam diffuser. A substantial reduction of the surface bacteria load was achieved.

For the production of pasteurised chicken rolls a scraped heat exchanger has been applied, using a temperature of $65^{\circ}C$. Higher product output, tighter product and process control have been achieved with this method, while still maintaining the quality of the product.(McGonigle, 1970).

3.2.3. Convenience foods

In the early sixties the so-called Nacka-method was developed in Sweden. It consisted in preparing and cooking the food conventionally to a minimum temperature of $80^{\circ}C$, hot packaging in plastic pouches, sealing (if possible under vacuum), followed by surface pasteurisation in boiling water for 3 minutes. After cooling to $+4^{\circ}C$ the packages were stored at a maximum temperature of $+4^{\circ}C$ for at most 3 weeks. It has been - and still is - used only to a very limited extent in Sweden.

Tändler (1973) has investigated the Nacka-method and modifications to it, concerning packaging methods and geometrics, heating and cooling techniques. He recommends that the core temperatures should be maintained for at least 10 minutes at at least $80^{\circ}C$ and that cooling must be done uninterruptedly to below $+3^{\circ}C$.

A similar American system has been presented by McGurkian (1969). In this system, however, storage life is stated to be 60 days with storage temperatures of between -2 and $0^{\circ}C$.

Recently, pasteurised peeled potato products have been introduced on the catering market in Denmark and Sweden. The peeled potatoes are packed in 3 - 5 kg pouches and are either pasteurised by steam heating of the packages for 40 minutes at 95 - $98^{\circ}C$ (Poulsen, 1977) or cooked in water and aseptically packed in steam. (Borg, 1977). Shelf life is 6 weeks at below $+4^{\circ}C$. The products are very convenient to use for institutional kitchens, especially on weekends, and have a very favourable sensory quality and price as compared to canned potatoes.

3.2.4. Packaging

For pasteurised foods packaging is required that can withstand the heat treatment temperatures and that have very low permeability of gases like water or oxygen during the storage period. Besides the traditional cans for meat products, new or improved flexible packaging materials are used, in some applications in connection with aluminium-foil, when zero permeability is required or desired. The problems of filling, evacuating and sealing the flexible packages are similar to those of sterilisation, which have been extensively investigated and discussed by Tändler (1976$_{a\ \&\ b}$). As in the case of sterilisation it is also important to ensure that the packaging material does not affect the aroma of the heat-treated food by migration of chemical compounds from the packaging material to the foods. An example of such interference is given by Jacobsson and Bengtsson (1972), where in-package-pasteurised beef had too high 'package material' off-flavour to be acceptable even directly after pasteurisation. Work in this field has been reported by Tändler (1973$_a$).

3.3. Processing comparisons
3.3.1. Meat products

Jacobsson and Bengtsson (1972) compared the quality of frozen and refrigerated-pasteurised cooked sliced beef. The beef slices were fried to 70 - 75°C centre temperature, packed hot in foil-laminates. Different headspace levels were used. The packs were pasteurised for about 10 minutes in a convection oven at 105°C to a centre temperature of 80°C, cooled to +10°C in 15 minutes and stored at +3°C and +8°C, for between 1 day and 3 weeks. The frozen samples were stored at -20°C for 2 months. For sensory evaluation on a 9-point preference scale the samples were reheated in a convection oven at 105 - 110°C.

The frozen samples were significantly better than the refrigerated ones even after a few days' storage. The vacuum-packed, refrigerated samples were, after two weeks, of the same quality as samples packed with 9 ml headspace and stored refrigerated for 4 days, (Figure 7). No differences in thiamine or riboflavine retention between the methods were found.

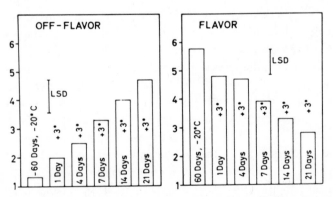

Fig. 7   Mean scores (average of 12 judgements) for off-flavour at 9 ml head space/75 g pouch. (Experiment 1) (After Jakobsson and Bengtsson, 1972).

A study of the headspace aroma compounds was also performed, showing good correlation between GLC-peak heights and sensory flavour scores (Persson and von Sydow, 1972), indicating advantages of using gas chromatography for quality control purposes as a complement to sensory evaluation.

Pasteurisation of wieners using different methods and temperatures was investigated by Ölmeskog (1972). Wieners are almost

sterile after smoking and cooling, but may be heavily reinfected
during cooling, mainly be lactobacillae. During storage this
will lead to the development of shelf-life limiting sour taste.
Among the other methods investigated, convection-oven heating
required too long times. IR-heating was not suitable for the
packaged wieners, whereas during microwave heating the packages
burst due to internal steam development. Surface pasteurisation
of individual wieners was much more successful than 3-pack pasteurisation, since the area between the wieners was difficult to
pasteurise.

In the main experiment the wieners were pasteurised to
three different $F150^{\circ}F$ levels in water at $85^{\circ}C$, giving the best
sensory quality retention. Times ranged from 50 to 130 seconds.
After 2 weeks' storage at $+3^{\circ}$ and $+8^{\circ}C$ the pasteurised wieners
still retained their high sensory quality. Thus, in conclusion,
shelf life can be markedly improved by pasteurisation of indi-
vidually packed sausages.

3.3.2. Convenience foods

A comparison between frozen and pasteurised-refrigerated
prepared foods has been reported on by Wadsworth, (1974). The
refrigeration temperature was recommended to be -1° to $2^{\circ}C$ for
a maximum of 30 days' storage. Comparable sensory quality was
found for several vegetable and entree dishes between frozen and
refrigerated, after 30 days' storage, except for one meat and
two vegetable items, where frozen samples were preferred.

3.4. Storage stability

For traditional types of heat-pasteurised foods, such as
canned hams, shelf life has been empirically established to be
in the order of more than 6 months in refrigerated storage. For
these products both heat treatment and chemical preservatives,
such as salt and nitrite, contribute to their storage stability.
In products such as precooked ready-to-serve foods, however,
heat treatment, raw material quality and packaging integrity are
alone responsible for giving the product sufficient storage
stability, recommended storage rarely exceeding 6 weeks at

refrigerated temperatures. In view of the great importance of
the stability and safety of pasteurised foods, it is surprising
that not more systematic work has been concerned with the combined effects of raw material and composition, processing
conditions, packaging and storage time and temperature on their
hygienic integrity in the products and on the retention of sensory quality and nutritional value. A number of papers have been
published over the years relating to empirical studies of a few
factors at a time.

Leistner et al. (1972) and Tändler (1972) both reviewed
work relating to the so-called Nacka system for factory-made
ready-to-serve meals, as previously summarised in this paper.
For the combination of hot filling, holding at $80^\circ C$ central temperature for at least 10 minutes after sealing, and rapid cooling,
they recommend a maximum storage time at -2° to $+2^\circ C$ of 2 - 3
weeks, both with regard to hygienic safety and sensory quality
retention.

Jakobsson and Bengtsson, (1972), pasteurised hot packed
meats under conditions applied commercially in Sweden (heating
in convection oven at $105^\circ C$ to a central temperature of $+80^\circ C$)
to products with a guaranteed shelf life of 3 - 4 weeks. Compared to a frozen reference, sensory quality deteriorated
noticeably in only 3 days when there was air in the sample headspace. Vacuum packaging reduced quality loss so that the equivalent level was reached in 2 weeks.

Several attempts have been made to develop accelerated
storage tests to predict the shelf life of heat-pasteurised hams
during refrigerated storage. Labots (1975) compared a range of
incubation temperatures and times and studied their effect on
the microflora and their relations to brine content, pH, etc.
He concluded that it is impossible to devise a high-temperature
accelerated storage test to predict shelf life at refrigerated
temperatures.

To predict and calculate accumulated quality loss during

frozen storage of foods, the so-called TTT-hypothesis (time-temperature-tolerance) was developed by van Arsdel and others, and has been applied with some success in practice. In principle, when the time required for a certain degree of quality change to develop is plotted against storage temperature in a semilogarithmic diagram, straight lines are obtained. The time required for a Just Noticeable Difference (JND) to develop is referred to as the High Quality Life (HQL), while the Practical Storage Life (PSL) is in the order of 2 - 5 times longer. Accumulated quality loss can be integrated from plots of 1/HQL or 1/PSL against time, for exposure to constant or fluctuating temperatures. The relations involved are influenced by product, processing method and packaging, the so-called TTT factors.

In recent years attempts have been made to establish the corresponding time-temperature relationship for foods stored at refrigerated temperatures. Jonsson (1974) reviewed the literature concerning the relevance of the TTT-hypothesis for refrigerated storage of foods. He concluded that conditions are much more complicated than for frozen storage, because of the microbial activity in this temperature range and the marked influence of water activity. In spite of this, it seems that the TTT theory can be applied to certain products, such as fish, meats and luncheon meats. The available data are very scarce, however, and a great deal of practical storage testing will be required for verification. Since microbiological activity is greatly reduced by pasteurisation, it appears likely that the TTT-theory would be more valid for pasteurised foods than for fresh foods.

Over the past five years several time-temperature-integrating devices (Kramer and Farquhar, 1976) have been developed for control purposes in the handling of frozen foods in distribution and storage. Such indicators could equally well be employed for monitoring the handling of refrigerated foods, provided a clear relationship can be established between quality change and time-temperature of cold storage. If so, this would be a significant improvement, considering that the frozen food chain is normally much better developed and controlled than the refrigerated food chain, and that exceeded time-temperature limits will normally, from the point of view of public health, be more hazardous for refrigerated than for frozen foods.

4. CRITICAL AREAS WHERE CONTINUED OR NEW R AND D IS NEEDED

4.1. Microbiological aspects

In pasteurisation most micro-organisms are inactivated. Some resistant organisms survive, however, mainly spore formers of the species Bacillae and Clostridiae. It is thus important to know their time and temperature requirements for heat inactivation in actual foods, and also the influence of different physical and chemical parameters in the foods. It is of special interest to study the psychrophilic or psychrotrophic spore-forming organisms, and their ability to sporulate and grow at refrigeration temperatures or close to them. The conditions for growth in actual foods and the influence of varying environmental factors, such as packaging in a vacuum or with protective gases, should be better clarified.

It is also essential to investigate the non-proteolytical species of the toxin producing *Cl. botulinum*, since they tend to grow at low temperatures. Such a study should also include meats.

4.2 Enzymatic aspects

During pasteurisation the enzyme proteins coagulate and hydrolyse to a great extent. Some survive, however, notably among enzymes of animal origin, but data on the time and temperature relationships for the inactivation of these enzymes in foods are very scarce. Their activity during refrigerated storage of the pasteurised product is also little known, especially the effect they might have on the sensory quality parameters of the foods. Another area that should be investigated is the possibility that partly inactivated enzymes, generally not active in foods, may affect the storage stability of the pasteurised food.

4.3 Nutritional and sensory aspects

The heat treatment causes more or less comprehensive changes in the nutritional and sensory quality of the foods, by deterioration of taste and aroma, or by melting out of fat or leakage of soluble proteins, which may give rise to a poorer appearance of the product. The leakage of juices from the foods may also lead to substantial losses of nutrients. It is thus important to investigate the changes in nutritional content and sensory quality during the heat treatment and the cooling operation in the production of a pasteurised product, and also to study it during the refrigerated storage.

4.4 Heat transfer aspects

For a successful pasteurisation process it is important that the technical possibilities for the heat treatment and the cooling should be well clarified. To obtain the best results it is necessary both to investigate and evaluate new heating methods, like infra-red heating and microwave heating, and to study further the older, traditional methods and try to optimise them for the pasteurisation process. For the study of the processing fundamentals, it is essential to take advantage of the well developed technologies for heat sterilisation processes, and also to use modern engineering aids, such as calculations of the heat transfer conditions by means of a digital computer.

Energy requirements for the heating and cooling processes are other important parameters that must be considered.

4.5 Storage stability

With a well-defined pasteurisation method it should be possible to collect sufficient storage-life data on a large number of foods processed in different countries and with different packages, etc. Preferably, collection of such storage-life data could be carried out on an international basis, and could serve as a foundation for attempts to formulate Just Noticeable Differences and High Quality Lives for different foods at different temperatures, in accordance with the TTT-hypothesis for frozen foods.

The establishment of a reliable TTT-rule for pasteurised refrigerated foods would be a great step forward in the pasteurisation procedure.

5. ACKNOWLEDGEMENTS

The author would like to thank Dr. Nils Bengtsson, Dr. Benkt Göran Snygg and Dr. Svante Svensson for their contributions to this paper and Professor Erik von Sydow for presenting it at the seminar.

REFERENCES

Andersen, E., Jul, M., and Riemann, H. 1965. In: Industriel levnedsmiddel-
	konservering. Bind 3. Teknisk forlag, København, Denmark.
Bengtsson, N.E., Green, W. and del Valle, F.R. 1970. J. Food Sci. 35,
	681-687.
Bolshakov, A., Oreshkin, E. and Bryanskaya, I. 1971. Myasnaya Industriya
	USSR 42 (2), 19-21.
Borg, A. 1977 Livsmedelsteknik 19, (2), 74-76.
Dagerskog, M. 1976. Physical, chemical and biological changes in food
	caused by thermal processing. p 77-100, Applied Science Publ. London.
Dagerskog, M. 1977. Mathematical modelling in food processing. Örenäs,
	Sweden. September 7-9, 1977. Proceedings p 269-288, Lund Institute
	of Technology.
Davey, C. and Gilbert, K.V. 1974. J. Sci. Food Agric. 25, 931.
Hamm, R. 1977. Fleischwirtschaft 10, 1846-1860.
Hanna, M.O., Stewart, J.C., Carpenter, Z.L. and Vanderzant, C. 1977. J. Food
	Sci. 42, (4), 1134-1136.
Ingram, M. 1971. SIK-Rapport No.292, A1-A44, SIK - Svenska Livsmedels-
	institutet, Göteborg, Sweden.
Jakobsson, B. and Bengtsson, N. 1972. J. Food Sci. 37, 230-233.
Jonsson, U. 1974. Scand. Refrigeration 5, 163-168.
Klose, A.A., Kaufman, V.F., Bayne, H.G. and Pool, M.F. 1971. Poultry
	Science, 50 (4), 1156-1160.
Kramer, A. and Farquhar, J.W. 1976. Food Technol. 30 (2), 50-56.
Labots, H. 1975. In: 21st. European meeting of meat research workers,
	Bern, Switzerland, August 31 - September 5, 1975, 67-69.
LaRocca, R. 1971. Food and bioengineering - fundamental and industrial
	aspects. 67 (108), 9-16.
Leistner, L., Wirth, F. and Takacs, J. 1970. Fleischwirtschaft 50, 216-217.
Leistner, L., Hechelmann, H.C. and Vukićević, Z. 1972. Fleischwirtschaft
	52, 993-996.
Leonard, S., Luh, B.S. and Simone, M. 1964. Food Technol. 18, 81-84.
Lund, D.B. 1977. Food Technol. 31 (2), 71-78.
McGonigle, T.P. 1975. Food Process. 36 (8) 88-89.
McGuckian, A.T. 1969. Cornell Hotel and Restaurant Administration
	Quarterly, 10 (1) 87-92, 99.

Ohlsson, T. 1977. In: Mathematical modelling in food processing, Örenäs, Sweden. September 7 - 9, 1977. Proceedings 77 -100, Lund Institute of Technology.

Persson, T. and von Sydow, E. 1972. J. Food Sci. 37, 234-239.

Poulsen, P. 1977. In: How ready are ready-to-serve foods? Karlsruhe, Germany. August 22 - 23, 1977. Symp. Summaries, 40.

Rao, V.C. 1976. J. Food Sci. & Technol. (Mysore) 13 (6), 287-293.

Reichert, J.E. 1976. Fleischwirtschaft 56 (5), 611-614.

Reichert, J.E. and Stiebing, A. 1977. Fleischwirtschaft 57, 910-921.

Shapton, D.A., Lovelock, D.W. and Laurita-Longo, R. 1971. J. of Appl. Bact. 34 (2), 491-500.

Spencer, R. 1967. Food Manufacture, 42 (6), 29-34.

Svensson, S. 1976. In: Physical, chemical and biological changes in food caused by thermal processing, 202-217. Applied Science Publ. London.

Thompson, S.Y. 1969. In: Ultra-high temperature processing of dairy products. Society of Dairy Technology, London, England.

Tändler, K. 1972. Fleischwirtschaft, 52, 845.

Tändler, K. 1973a. Fleischwirtschaft, 53, 1241-1244.

Tändler, K. 1976b. Fleischwirtschaft, 56, 539-547.

Tändler, K. 1976c. Fleischwirtschaft, 56, 1473-1484.

Wadsworth, C.K. 1974. Extended shelf life of pre-cooked refrigerated meals. Technical Report 75-15/US Army Natick Laboratories, 1-16.

Ölmeskog, S. 1972. M.Sc. Thesis, Chalmers University of Technology, Göteborg, Sweden.

Östlund, K. 1970. Nordisk Veterinaermidicin, 22 (12), 634-645.

PASTEURISATION AND THERMISATION OF MILK AND BLANCHING OF FRUIT AND VEGETABLES

J. Foley and J. Buckley
Department of Dairy and Food Technology,
University College, Cork, Ireland.

The word pasteurisation was coined from the name of Louis Pasteur who showed the value of heating in preserving wine and milk well over one hundred years ago. The term is now used to describe heat treatments which destroy certain undesirable micro-organisms and aid in product preservation. In the case of pasteurisation of milk, the minimum heat treatment is that required to destroy *Microbacterium tuberculosis*, the most heat resistant pathogenic organism likely to be present. Many time-temperature combinations will do this. Basically there are two methods which have fairly widespread legal recognition, the holder process and the high temperature short time method (HTST). Definitions of pasteurised milk may vary somewhat from one country to another. The holder method, which is very much declining in use, involves heating milk to 63 to 66°C for 30 minutes and subsequent cooling to 10°C. HTST pasteurisation requires heating milk to 72°C and holding for 15 seconds followed by cooling to 10°C.

In the case of liquid egg, the minimum pasteurisation treatment is that required to destory salmonella organisms. The heat sensitivity of egg albumin must also be taken into consideration and coagulation avoided. The time-temperature combination of 64°C x 2.5 minutes achieves these objectives.

In the brewing industry keg beer is pasteurised at 73°C x 28 seconds and cooled to 5°C. The trend with bottled beer is towards in-bottle pasteurisation at 60°C x 10 minutes. Great care is taken to avoid incorporation of air. Thus the bottles are filled to overflowing and vibrated before sealing to reduce the oxygen content and protect against oxidation. The efficiency of pasteurisation in the brewing industry is measured by incubation tests at 37°C for 3 to 7 days and plate counts using a membrane filter technique.

Pasteurisation is fundamental to the liquid milk industry and it is an integral stage in the manufacture of many dairy products. While the treatment conditions for pasteurised milk are precisely defined, the conditions for cream, ice cream mixes, egg nog and milk for manufacture are not specified in many countries.

INFLUENCE OF PASTEURISATION ON MILK CONSTITUENTS

Heating, depending on degree, alters the chemical, physical and biological nature of foods. Some proteins denature, starches gel in aqueous solution, proteins react with reducing sugars and changes in some minor constituents occur. Fortunately, pathogenic organisms are destroyed by relatively mild heat treatment and the changes brought about by pasteurisation do not significantly impair the flavour and natural goodness of milk.

Proteins

The effect of pasteurisation on milk proteins is small; casein is unaffected and the denaturation of serum proteins is barely evident. The immune globulins are most sensitive followed in order by blood serum albumin, β-lactoglobulin and α-lactalbumin. Euglobulin in milk promotes the clustering of fat and its denaturation leads to a diminution in creaming. High temperature short time pasteurisation gives a very slight loss in cream line. If the minimum pasteurising temperature is exceeded by a few degrees centigrade the depth of the cream line may be diminished by as much as 50%. Loss of cream line is, of course, not a consideration in pasteurised homogenised milk.

Fat

Milk fat is liquid at above 40°C. Small changes in emulsion distribution and stability may occur depending on the conditions and methods used. Plate heat exchangers are designed to give maximum turbulence to ensure uniform heat treatment and maximum heat transfer. Such treatment is not always conducive to preserving the emulsion stability of cream, high fat creams in particular. The destabilising effect is caused by the mechanical effects of pumping and the shearing effect on globules in the cooling and chilling of plate pasteurisers, rather than by temperature per se. (Foley et al., 1971).

Salts

The calcium phosphate in milk is less soluble at high than at low temperatures. Both soluble and colloidal calcium phosphate play an important role in rennet coagulation of milk for cheesemaking. The shift of soluble calcium phosphate to the colloidal state can be detected at minimum pasteurisation temperatures but the change is not sufficient to disturb normal cheesemaking. Hydrogen ions are liberated when calcium and phosphate become insoluble. The decreased pH is however counteracted by an increase due to removal of carbon dioxide on heating and no net change is observed.

Trace Elements

From the point of view of flavour deterioration in milk and dairy products, copper is by far the most important trace element. Copper migrates from the serum phase to the fat globule surfaces when milk or cream is heated and acts as an oxidation catalyst. Heat treatment within the range 60 to $90^\circ C$ promotes photocatalysed lipid oxidation of cream (Foley et al., 1977). Ripened cream butter manufactured from cream separated from pasteurised milk ($80^\circ C$ x 15 seconds) was more susceptible to oxidative changes than corresponding butter manufactured from cream which was separated from milk at $50^\circ C$ and the cream subsequently pasteurised ($80^\circ C$ x 15 seconds). The susceptibility to oxidation is related to changes in copper distribution between the serum and fat globule phases and to the ratio of serum to fat at the time of pasteurisation (Foley and King, 1977).

Enzymes

The enzymes of milk are heat labile. Pasteurisation inactivates aldolase, α-amalase, lipase and alkaline phosphatase; while catalase, xanthine oxidase, β-amylase, proteinase, peroxidase, acid phosphatase, superoxide dismutase and ribonuclease survive the treatment but are inactivated at higher temperatures. Enzymes which survive pasteurisation treatment are referred to as 'heat resistant' in this paper. The logarithm time for enzymes when plotted against time gives a straight line. Thus it is possible to predict from the position and slope of the destruction curves the time required at any given temperatures to achieve inactivation. Alkaline phosphatase, because of its close relationship with the destruction curve for *Microbacterium tuberculosis*, is used as an index of efficent pasteurisation of milk. Hill (1975) has recently reported the existence of the enzyme superoxide dismutase in bovine milk. The enzyme may have antioxidant activity.

It is considered that the enzymes of milk are constituents of the secretory epithelial cells which are released into milk during milking. Relatively little information is available on the physiological and nutritional significance of milk enzymes. The role of enzymes which survive pasteurisation in flavour deterioration of milk, cream and products made from pasteurised milk has not been sufficiently established and many aspects require further study. Is the keeping quality of milk associated with its lysozyme content? What are the roles if any of xanthine oxidase and superoxide dismutase in oxidation and photocatalysed oxidative changes, lipid oxidation in particular? Is the pro-oxidant influence of copper associated in any way with these enzymes? Does acid phosphatase, which has the ability to dephosphoralate casein (Bigham and Zittle, 1963) influence the quality of acid casein during long term storage? Does acid phosphatase dephosphoralate casein in cheese ripening?

Vitamins

The influence of pasteurisation on the vitamin content of milk has been the subject of many studies. In general, the effects of heat on the labile constituents of milk are less when using a high temperature for a short time than for a longer holding time at a lower temperature. It is further established that the factors which give rise to variations in the vitamin content of raw milks far exceed the effect of processing. Apart from losses in thiamine (3-10%) and Vitamin C (10-20%) HTST pasteurisation has very little influence on the nutritive value of milk. Small losses of vitamin B_{12} occur in the holder method of pasteurisation. Vitamins A, D, B_2, pantothenic acid, niacin, pyridoxine and biotin are unaffected. Little is reported on the effect of heat on vitamins E and K. They appear to be practically unaffected. The losses of vitamins C and B_2 and folic acid at higher temperatures are associated with the presence of oxygen. Milk is not regarded as a rich dietary source of vitamins B_2 and C and the losses during pasteurisation are of little nutritional significance.

Mild heat treatment does little if any damage to the nutritive value of milk proteins. Severe heat treatment may reduce the availability of lysine and of sulphur containing amino acids.

Quality after processing

Post-processing deterioration during storage and commercial distribution of pasteurised milk is related to the raw milk quality, the methods of packaging, the temperature and time of storage and the wholesale and retail system of distribution. The changing quality of raw milk is considered in detail in subsequent paragraphs. Pasteurisation, depending on the flora present, destroys 99% of the micro-organisms in milk. The surviving flora consist of thermoduric bacteria and spores. Most thermoduric bacteria remain relatively inert when proper storage temperatures are maintained. Psychrotrophic strains of the genus *Bacillus* have been identified (Witter, 1961). Their role in defects such as pin point flecking and bitty cream and their multiplication in pasteurised milk and cream held at $5°C$ needs clarification. Most problems in pasteurised milk arise from after contamination, from plant and packages and from failure to cool and hold at sufficiently low temperatures. Cartons and plastics used for product packaging are virtually sterile and the protection afforded against light by cartons is a decided advantage in protecting against oxidative changes of vitamins and lipids. The possibility of traces of volatile constituents from plastic packaging substances getting into milk is a topic which continues to require close scrutiny.

CHANGES IN MILK HANDLING

Changes both at factory and farm level over the past ten to fifteen years are to some extent reflected in the quality of raw and pasteurised milk. The swing towards pipeline milking systems, the use of refrigerated bulk holding tanks on the farms, every second day collection of milk, bulk haulage systems of milk assembly, have all contributed to changes in the general quality and biological quality in particular. Growth of psychrotrophic bacteria becomes a problem when raw milk is held for extended periods at low temperatures.

Over recent years the time lapse between milking and heat treatment of milk have increased. Alternate day pick-up has become standard practice, thus milk may be 36 to 48 hours old before it reaches the factory. It may be held in a milk silo for yet another 24 hours or possibly longer. Prolonged holding at 5 to $10°C$ allows the growth of psychrotrophic bacteria and the increased mechanical handling lowers the emulsion stability.

Microflora

The microflora of bulk refrigerated milk consist predominantly of the genera Pseudomonas, Achromobacter, Alcaligenes and Enterobacter (Thomas, 1974). Chapman et al. (1976) showed that the bacterial count of bulk milk stored at $5°C$ increased from 4.3×10^4 to 1.3×10^6 during a three-day storage period. The increase was accounted for by the psychrotrophic bacteria, many of which were lipolytic and caseolytic.

Bacterial lipases

Many recent studies indicate that while gram-negative psychrotrophic bacteria are inactivated by HTST pasteurisation the lipases and proteases produced by many of them survive this and much higher heat treatments. It appears that bacterial lipases do not cause significant fat hydrolysis in milk but because of their heat resistance may cause defects in some milk products such as cheese, UHT milk, and butter which may be held in storage for a long time. Overcast (1968) found that lipolytic enzymes produced by psychrotrophic bacteria could produce rancidity in pasteurised milk held for two days at $4^{\circ}C$. Law et al. (1977) found that Cheddar cheeses made from milks in which strains of lipolytic gram-negative rods had been allowed to multiply to $>10^7$ colonies/ml became rancid after four months even though the bacteria themselves had been killed by pasteurisation. Some strains of *Ps. fluorescens* and *Ps. fragi* produce lipases which retain 20-25% of their activity after heating to $100^{\circ}C$ for ten minutes. Driessen and Stadhouders (1974) found that Dutch cheese made from milk which had a count of 3.6×10^6/ml of *Alcaligenes viscolactis* before pasteurisation had increased FFA values.

Bacterial proteinases

The influence of heat stable proteinase got from *Ps. fluorescens* P26 on the shelf life of several dairy products was studied by White and Marshall (1973). The flavour of butter was not affected by the addition of eleven units of enzyme activity per ml in cream. Addition of the enzyme lowered the flavour of both Cheddar and Cottage cheese, but did not affect the flavour of milk when added twelve hours before pasteurisation. Adams et al. (1975) found that the addition of proteinase got from Pseudomonas M60 to UHT milk gave rise to a bitter flavour. Another study has shown that proteinases from a strain of *Ps. fluorescens* isolated from raw milk caused gelation of UHT sterilised milk (Law et al., 1977).

It appears that different psychrotrophic bacteria produce the lipase and proteinase enzymes at different stages of the

growth curve (Driessen and Stadhouders, 1974). With some
bacteria, the enzyme is only produced in the stationary phase
while others seem to produce heat resistant enzyme during the
logarithmic growth phase. In the case of some gram-negative
bacteria the production of a heat stable proteinase could be
detected at a count of 10^4 ogranisms/ml (Adams et al., 1975).

Milk lipases

The activity of milk lipase which manifests itself in the
form of free fatty acids in milk is a greater problem than
heretofore. Increased lipase activity is mainly associated
with increased mechanical handling and agitation of milk and
longer holding before heat treatment. This topic was recently
reviewed (IDF, 1975).

There are suggestions that in some areas the compositional
quality of milk has deteriorated over recent years. If this is
so, the decline may be associated with higher yielding cows or
perhaps in some cases with careless rinsing of pipeline milking
systems, farm tanks, tankers and milk silos and plant.

Aeration of milk

One of the consequences of the changes at farm and factory
outlined above is the increased agitation and aeration in pipeline milking machines, coolers and bulk farm tanks. After each
milking, the agitator is switched on until the milk is
sufficiently cooled and also before emptying. Agitation of
the milk continues during transport. On arrival at the factory
air agitation is often used to ensure a representative sample
for testing. The milk is kept agitated by air or mechanically
in the holding silos. Thus the milk reaching the pasteuriser
may contain significant levels of entrapped air. Aeration of
milk raises many questions both in relation to changes in
nutritional value before, during and after heat treatment and
in relation to other hydrolytic and oxidative changes. Aeration
of milk may cause fragmentation of liquid fat globules and
aggregation and partial churning of partially solidified
globules. The presence of air leads to cavitation of fat

globules during separation with a reduction in skimming efficiency. It also increases burning-on problems in pasteurisers and heat exchangers leading to poor efficiency of pasteurisation. Foaming is yet another problem associated with a mixture of air and milk. The problems brought about by incorporation of air in milk and the need to avoid deposits on the surfaces of heat exchangers and so achieve longer pasteurising runs has given rise to increased interest in pasteuriser installations incorporating vacuum deaeration units. Milk enters the vacuum deaerator at $63^{\circ}C$ and is flash cooled to $55^{\circ}C$. Deaeration is accompanied by concentration which is related in degree to the working vacuum. The influence of deaeration on the oxidative stability of milk lipids and vitamins of milk does not appear to have received much attention.

RESEARCH AND DEVELOPMENT IN MILK PASTEURISATION

At processing level, the amalgamation of factories and the closure of many has led to the growth of large centralised processing units. Milk is held in large silos before pasteurisation.

Conventional Pasteurisers

The trend is towards higher and higher capacity heat exchangers for pasteurisation with more sophisticated automated and sometimes computerised control systems. Rising energy costs have made regeneration up to 90% economic and the recent plants are designed to do this. Because of greater use of refrigeration at farm and factory level the chilled milk going to the pasteuriser obviates, in many cases, the need for a water cooling stage.

Pasteurisation by steam injection

Recent statistics show that there is a drop in the consumption of whole milk in the nine EEC countries. There is a significant increase in the market share of UHT aseptic milk in continental Europe and it is predicted that the trend will continue. This trend has created an interest in the development

of heat treatments which would give a midway product between
pasteurised and UHT milk. Encouraging results are reported for
one such process which involves indirect heating to $76°C$
followed by instant heating to $100°C$ by steam injection and
flash cooling to $76°C$ for 15 seconds before cooling to $5°C$
(Cox, 1973). Research on the development of high temperature
instantaneous pasteurisation by steam injection is also
reported from the USA (Dickerson et al., 1973). The method
is reported suitable for the pasteurisation of Grade A milk.
When milk is heated to temperatures of about $100°C$ for very
short holding times the term ultra-high temperature pasteurisation
is often used. It appears that UHT pasteurisation does not
give sufficient improvement in milk quality to justify the
extra cost of the process. The influence of these treatments
on the activity of bacterial enzymes and vitamin losses should
be of considerable interest.

Microwave pasteurisation

The application of microwave energy as a heat source for
continuous HTST pasteurisation of milk has been studied
(Hamid et al., 1969; Jaynes, 1975). When milk was pasteurised
at $72°C$ x 15 seconds using microwaves as a source of energy
there was no deleterious effect on flavour. Phosphatase tests
were negative and the bacterial counts were comparable with
milk pasteurised by conventional methods (Jaynes, 1975).
There appears to be no available information on the effects of
microwave pasteurisation on the nutritive value. The cost of
the method is a disadvantage.

Actinisation

A process for milk pasteurisation using infra-red radiation
and called Actinisation has been developed by Actini-France
(Stoutz, 1966). Milk is heated by passing it over horizontal
stainless steel plates under infra-red tubes or through a
quartz tube. A section for ultra-violet radiation for
enriching milk with vitamin D may be incorporated if required.
Here again the higher cost per unit of milk treated is a
definite disadvantage.

Standardisation and fortification

Other recent publications have related to the enrichment of pasteurised whole milk with iron (Demott, 1975; Edmondson et al., 1971), milk solids - not fat (Lang et al., 1976) and vitamins. Ferrous compounds tend to give an oxidised flavour in milk while ferric compounds resulted in lipolytic rancidity which was attributed to an increased heat resistance of lipase when ferric iron was present. The nutritional merits and de-merits of milk standardisation and fortification continues to be a topic of interest. Fluoridation of milk is recommended by some.

Pasteurisation of fermented dairy products such as yoghurts is also a controversial topic. The main problem associated with pasteurisation of fermented milk products is the contraction of casein on heating and separation of whey. This may be overcome by addition of stabilisers and suitable adjustment of the pH.

THERMISATION

The problems associated with psychrotrophic bacteria in refrigerated milk has given rise to a process called thermisation. In this process, milk is heated to about $63°C$ for 15 seconds or its equivalent before holding in silos. This treatment is sufficient to inactivate the lipolytic psychrotrophs and subsequent production of heat-resistant bacterial enzymes is thus prevented.

Studies in the Netherlands compared cheese made from fresh milk, milk which had been subjected to a thermisation treatment before holding for 48 hours, and milk held for 48 hours. The levels of free fatty acids were highest in cheese made from the milk held raw for 48 hours. The experiment indicates that thermisation inactivates the bacteria which produce thermoresistant lipases and therefore reduces fat hydrolysis during the ripening of cheese. Much of the milk delivered to cheese factories in the Netherlands is subjected to thermisation. Cheddar cheeses made from milks in which strains of a single species of lipolytic gram-negative rods had been allowed to multiply to over 10^7 colony/ml became rancid after four months despite the fact that the milk had been pasteurised before cheesemaking.

While thermisation is apparently beneficial in the manufacture of Dutch cheese, its more general use needs further study. The process involves extra capital and processing costs in terms of time, labour and energy. With the present lower returns from cheese than from intervention products, the trend in industry is to minimise costs consistent of course with the manufacture of a sound commercial grade cheese. It is the experience of many manufacturers that silo bulking of milk has led to greater day-to-day uniformity in Cheddar cheese making. There appears to be no evidence as yet to suggest that the change to longer holding of milk at low temperatures has led to any significant reduction in Cheddar cheese quality in Ireland. Despite this, all milk for cheese manufacture is thermised before holding at one large cheese factory.

When pasteurisation was first used in cheese manufacture, the main criticism of the process was that with the inactivation of milk lipase and the non-thermoduric flora the cheese tended to be milder in flavour. There is little doubt that some degree of lipolytic activity is necessary in the development of typical flavour of Cheddar cheese. The question appears to be what degree or intensity of lipolysis is desirable, and perhaps the answer will depend on the maturation time and subsequent required shelf life.

SUMMARY

Much of the published data of thermal inactivation of both natural and microbial enzymes in milk is expressed in terms of time at some defined temperature to achieve inactivation. The D values at different temperatures and the Z values would provide additional valuable information. Another aspect of heat-resistant milk enzymes is their significance in flavour stability of pasteurised milk and in possible changes in certain dairy products during long term storage. The deleterious effect of heat-resistant microbial enzymes on some dairy products has only come to be recognised fairly recently. The most appropriate means of overcoming the various problems caused by these enzymes is a topic requiring further study.

The processes of thermisation and deaeration of milk and the prolonged holding of milk before it is pasteurised have relevance not alone for the processor but also for the consumer. Is double heat treatment of milk acceptable for pasteurised drinking milk, or should it be confined to manufacturing milk? The influence of an extra heat treatment may have implications in oxidative changes.

Pasteurisation by infra-red radiation and by microwave radiations is unlikely to become of major significance in the milk industry and the main developments are likely to be related to further modifications in existing plate methods. Developments in UHT pasteurisation do not seem to have lived up to earlier expectations in extending the shelf life of pasteurised milk.

BLANCHING (Section 2)

Purpose

One of the more important processes in the preparation of fruit and vegetables for freezing, canning or dehydration is blanching. Heat, in the form of hot water or steam, is used to inactivate certain enzyme systems which cause undesirable flavour, colour and aroma changes in the finished product during storage. It is generally understood that blanching, as a preliminary treatment in the canning process, removes the tissue gases and effects a shrinking of the material so that adequate fill is obtained in the can. Enzyme inactivation per se is not so important, because it is likely that the heat used to remove the tissue gases will inactivate the enzymes. Any enzyme remaining in an active state will be inactivated by the cooking process. Blanching is used as a preliminary treatment before freezing to inactivate the enzymes in the tissues and shrink the material so that adequate fill is obtained in the container. Blanching is necessary in the dehydration process to inactivate the enzymes, because as in the case of preservation by freezing, no further cooking, previous to storage, is involved. Blanching has the added effect of improving cell permeability and thus aids moisture removal during drying.

While the process of blanching is necessary, it does lead to losses of nutrients and flavour. Losses (and gains in certain instances) of such inorganic substances as calcium, potassium, phosphates and iron, losses of such organic materials as sugars and nitrogenous substances as well as vitamins have been shown (Horner, 1936-1937a). Time and temperature blanching studies have revealed optimum conditions for the retention of maximum quantities of nutrients consistent with maintenance of desirable quality of the finished product (Kramer and Smith, 1947).

In very general terms, such vitamins as D, K, niacin, riboflavin, pantothenic acid and biotin, and other nutrients such as minerals, carbohydrates, lipids and essential amino acids, have good stability during heat processing and storage. About 85% retention of vitamins after process is suggested as a norm.

A lot of effort has been spent on (a) better understanding of the role of enzymes in products preserved by freezing and dehydration and (b) the most effective methods for the inactivation of these enzymes. The relative merits of water-blanching versus steam-blanching have been studied intensively. It seems that steam-blanching is the more effective of the two for the conservation of soluble nutrients. However, some authorities contend that under certain conditions, steam-blanching leaves some undesirable flavours (Lee, 1958).

Two conventional methods of blanching are commonly practised, namely, hot water and steam treatments. Less frequently used is the microwave process. A disadvantage of the first method is that water-soluble nutrients will pass into the blanching water, but an important advantage is that undesirable oxidation can be easily controlled by appropriate additions to the blanching bath. The other two methods, although they conserve the nutrients, are technically more difficult to control than the first, requiring specialised apparatus. Times and temperatures vary depending on the product which is to be processed. The 'holding time' after a temperature of 85 to $100°C$ has been attained is usually from three to five minutes.

EFFECT OF BLANCHING ON NUTRIENTS

Nutrients are sensitive in varying degrees, depending on the product, to pH, light, oxygen, trace metals and heat. Processing conditions which avoid extremes of these parameters give maximum nutrient retention.

Minerals

It is generally accepted that loss of minerals in process water is not significant enough to be of serious nutritional concern, even though considerable losses of potassium and phosphates occur during the blanching of most vegetables (Horner, 1936-1937a). Kramer and Smith (1947) made a survey of the effect of time, temperature and type of blanching on the mineral composition of peas and beans. Steam blanching was found to cause no significant changes in composition. The effect of time was more important than temperature in the case of water blanching.

Sugars and proteins

Losses were found to increase with time of blanching in water. The smaller the pieces (eg peas) the greater the losses. Steam blanching reduces the amount of sugar losses considerably (Horner, 1936-1937b). Again the effect of time was found to be more important than temperature for water blanching. Under most processing conditions the protein content of foods remains fairly stable. Certain processing procedures can result in the combination of some amino acids with other food components, making them non-available for nutritional purposes.

Vitamins

The major vitamin losses in processing result from the leaching out of water-soluble vitamins in such procedures as washing and blanching. The greater the amount of product surface area exposed during these processes (diced products, for example) the greater becomes the potential for soluble vitamin losses. Amount of loss also relates to the time and temperature of the water treatment. High temperature short time water blanching has been shown to be superior to low temperature long time water blanching, and steam blanching superior to both of the former methods. Under extremely poor conditions loss of vitamins during blanching has been reported as high as 70%. Under proper conditions losses can be held to as low as 5 to 10%.

Fat soluble vitamins

Both vitamin A and provitamin A and carotenoids demonstrate good stability in food processing. However, they can be destroyed at high temperatures in the presence of oxygen. Vitamin D is extremely stable and no appreciable losses have been noted as a result of good processing and storage methods. Vitamin D is sensitive to oxygen and light.

Water soluble vitamins

Vitamin B_1 can suffer considerable damage during food processing. Primarily, it is lost as a result of leaching. It is also unstable at neutral and alkaline pH and therefore alkaline natural waters may destroy this vitamin. Blanched and frozen peas retain as high as 94% of the thiamine content of the fresh washed vegetable, and 68% in the case of lima beans. In canning these same products, the retention is 52% in the case of peas and 42% for lima beans. Vitamin B_2 stands up better to both brine grading and blanching in vegetables than does vitamin B_1. Studies on lima beans indicate 87 to 76% retention respectively. In the case of blanched cabbage 80% retention was noted.

Niacin is possibly the most stable of the B vitamins. It is unaffected by light, heat, oxygen, acid or alkali. Therefore, the only losses that can occur are from leaching into process water. Blanching of green peas for three minutes in 99 to 100°C water can cause losses of 15%. Similar processes in brine could result in losses in the range of 27 to 39%.

No significant folic acid losses have been reported during the blanching of vegetables.

Vitamin C is the most easily destroyed of all the vitamins. The greatest losses of this vitamin are due to leaching into process water. Steam-blanched peas showed greater retention of ascorbic acid than those blanched in the conventional water blancher (Holmquist et al., 1954).

Other substances

Chlorophyll is insoluble in water and therefore does not leave the cells unless the cell walls are ruptured or destroyed. Blanching does not bring the colour to the surface of the green vegetables, and, since subsequent cooling merely hardens the cell walls, no changes in the chlorophyll content occur (Lee, 1958).

Blanching does not noticeably affect the carotene content of vegetables. Greater carotene retention can be expected when vegetables to be frozen are blanched prior to storage (Zscheile et al., 1943).

EFFECT OF BLANCHING ON ENZYMES

Blanching time should be long enough to inactivate the enzymes responsible for deterioration, yet not long enough to soften the texture of the vegetables for freezing. The undesirable effects due to enzymes are: discolouration, off-flavours, loss of viscosity, as well as tissue disintegration.

Both peroxidase and catalase enzymes have been shown to produce off-flavours in peas and beans (Wagenknecht and Lec, 1958).

Discolouration in the form of browning of plant tissues, particularly after injury, is a familiar problem. The oxidases, such as the phenol oxidases, are the primary browning agents and they are also responsible for the oxidation of vitamin C. Phenolic oxidation products may also contribute to flavour changes.

The browning of sliced potatoes and apples can be prevented by the addition of inhibitors or antioxidants such as sulphur dioxide or vitamin C. The effect of the latter in turn can be reduced by ascorbic acid oxidase. Edible organic acids can be included in the blanching medium to reduce enzyme activity.

TESTS FOR THE EFFICIENCY OF BLANCHING

Blanching prior to freezing is used to inactivate all enzymes which may cause deterioration during subsequent frozen storage. The qualitative test for catalase can be deceptive in determining proper blanching times for some products. The peroxidase test may be used as an additional index of blanching. Peroxidase, being a more heat-resistant enzyme, allows a greater margin of safety in blanching than does catalase for most vegetables. However, a positive peroxidase result on some vegetables such as snap beans and asparagus does not indicate underblanching as these vegetables retain their quality during storage, provided the catalase test is negative. A further disadvantage of the peroxidase test is the regeneration of peroxidase activity during storage. The phenolase test, for vegetables and fruits which discolour, is the most definite index of adequate blanching (Lee, 1958). A satisfactory blanching time for vegetables to be frozen is the time necessary to inactivate catalase plus an additional 50% of the inactivation time as a safety factor.

NEW BLANCHING TECHNIQUES

Hot water blanching is one of the most effective washing operations possible. It gives not only a final cleansing but also washes away certain raw flavours and reduces the bacterial load. With many products it helps to preserve the colour.

In the past, most research work on the blanching process was concentreated on the problems associated with conventional hot water and steam-blanching such as textural damage, leaching or loss of water soluble constituents, including flavour, vitamins and minerals. There was little economic or legislative motivation for pollution control. However, during the last five years many in-plant modification projects for pollution control in the food industry have been activated. Surveys of individual unit operations in the canning process indicate that blanching contributes significantly to overall plant effluent. In most cases over 50% of the plant biochemical oxygen demand (BOD) is due to blanching and cooling. Consequently many investigations have been undertaken recently to design blanching operations that would significantly reduce the generation of waste water (Lee, 1975).

1. Fluidised-bed blanching

Fluidised-bed blanching was developed by Mitchell et al. (1968) in Australia for use on green peas to achieve uniform short-time blanching at precisely controlled times and temperatures. The fluidising medium used was a mixture of saturated steam and air. In the continuous-type fluidised-bed blancher the fluidising mixture is pumped around a system of ducts and steam is injected continuously into the duct attached to the suction side of the fan. The fluidising mixture is blown through a perforated gas distributor plate and then through the bed and returned to the suction side of the fan. The slope of the bed is adjustable to control the depth of peas on the bed.

2. In-can blanching

This method replaces a traditional blancher with an overflow hot water (96°C) brining unit to bring green peas to blanching temperature. A jet of hot water directed into a can filled with peas stirs the contents and supplies sufficient heat to blanch the peas in 20 to 40 seconds. When blanching is completed the flow of water to the can is stopped, the can head-space is adjusted, salt is added, and the can is closed. It is suggested that this method of blanching would reduce the incidence of damage since the peas could be placed in the cans immediately after washing. It is calculated that approximately 1 600 gallons of hot water per ton of peas are required, therefore limiting the usage of in-can blanching as a pollution reducing system. However, it may be improved by recirculating the blanch water or recovering the heat in a heat exchanger (Mitchell, 1972).

3. Thermocyle blancher (venturi effect)

Venturi recycling of heat under a sealed blancher dome provides uniform temperature and increases efficiency and capacity. Incoming steam passes through the venturi to create a vacuum in the recycling tube. This draws available heat from the dome of the blancher's hood via the recycling tube. Although there is no detailed scientific information available up to now, this steam recycling blancher appears to be promising for energy savings (up to 50% saving in steam requirements), low liquid waste discharge volumes and elimination of steam clouds in working area.

4. Hydrostatically sealed steam blancher

This unit was designed to combine washing, blanching and cooling in a single unit that is about half the size of a standard steam blancher of the same width, but with a blanching capacity of approximately twice as much. Steam consumption is

only about .077 kg of steam per kg of product - one half that of
the conventional exhaust steam blancher (.155 kg of steam per
kg of product). Overall it is reported that this steam blancher
uses 50% less energy and 50% less water for cooling than is
normally required (Ray, 1975).

5. Individual quick blanch (IQB)

This process was developed as a modified steam blanch by
Lazar et al., 1971. IQB is a two stage process: (1) each
piece of product is exposed to a heat source until temperature
is in the range desired, and (2) the piece is held adiabatically
until the temperature has equilibrated. In this system, pieces
of vegetable are spread in a single layer on a mesh belt moving
through a steam chest. Here, maximum heating rates result from
complete exposure of each piece to live steam. After the
relatively short exposure to live steam, and before the interior
of the pieces become too hot, the product is discharged as a
deep bed onto another belt moving through an insulated chamber
to equilibrate the product temperature at a temperature high
enough to stop enzyme activity and to achieve a desired texture.
This method reduces leaching from the product and thereby
reduces effluent BOD. Further reduction in leaching and waste
effluent was achieved by 'preconditioning', ie partially
warming and drying the food surface with air before the steam
heating step. Brown et al. (1974) modified the original IQB
system into a compact assembly of two stacked circular vibratory
trays and a vertical insulated tube for adiabatic holding.

6. Hot-gas blanching

Combusted gases from a natural gas furnace are blown down
through the product which is held and conveyed by belts.
Steam is used to reduce dehydration losses and to increase heat
transfer of the gas medium. Hot-gas blanching produces less
yield and is more expensive than conventional blanching (Klinker,
1975).

7. Microwave Blanching

The first reported studies of microwave blanching were

carried out by Proctor and Goldblith (1948). In-package blanching on a continuous basis operating at 915 MHz was reported in 1963 (Anon. 1963). Jeppson (1964) recommended the use of saturated steam in combination with microwave energy as a more economical approach. It is questionable if microwaves alone or in combination with steam would be significantly better than conventional steam blanching of vegetables.

RESEARCH AND DEVELOPMENT

In the past most research on the blanching process was concentrated on problems such as leaching of the water-soluble constituents and was associated with the conventional hot water and steam type blanchers. In recent times, for economic and legislative reasons, pollution control as well as maximisation of nutrient retention have been the priority areas.

Blanching has been shown to contribute significantly, in some cases up to 50%, to total plant biochemical oxygen demand. As a result, the design of blanching operations has been and will be a priority area for some time to come.

The major development in blanching in the past few years has been to minimise water wastage and thus decrease the disposal problem. There have been two approaches to this. First, cooling of the blanched product with chilled air instead of water sprays or fluming. The difficulty with this technique is that evaporation losses decrease yields and this is not commercially acceptable. However, for some vegetables this has been overcome by saturating them with a water spray prior to entry into the air cooler. Under controlled conditions evaporation losses equal the sprayed-on water.

The second approach has been the use of the Individual Quick Blanch (IQB) process, which is primarily used for steam blanching of vegetables as already outlined. The blancher effluent is reduced by 50% over conventional steam blanching with less loss of soluble solids. Product texture generally is better and less overcooking of surfaces occurs. There is no data available on nutritional changes and further studies are needed in this area.

The optimisation of thermal processes for nutrient retention, in the case of blanching, will probably be based on considerations other than thermal losses, such as oxidative degradation, damage to product and leaching losses. It must be stated, however, that considerations other than nutrient retention, such as pollution control, will govern to a greater extent the blanching processes used by the food industry.

More research is needed to improve tests for adequacy of blanching. The specific enzyme reactions that cause off-flavour, off-colour, etc could be more positively identified in the various products. This would then dictate the enzyme(s) to be tested for in the different products. On the basis of this, improved time-temperature relationships for specific enzyme inactivation could be developed. This in turn would lead to improved overall product quality.

To conclude, it is clear that as more fruit and vegetables are processed, efforts must be made to reduce the amount of effluent as well as to cut nutrient losses. It is a double expense to be destroying or losing naturally present nutrients and then adding them back again by way of fortification.

REFERENCES (Section 1)

Adams, D.M., Barach, J.T. and Speck, M.L. 1975. J. Dairy Sci. 58, 828.
Bigham, E.W. and Zittle, C.A. 1963. Arch. Biochem. Biophys. 101, 471.
Chapman, Helen R., Sharpe, Elisabeth M. and Law, B.A. 1976. Dairy Inds. 2, 42.
Cox, G.F. 1973. The Milk Industry, 5, 9.
Demott, B.J. 1975. J. Milk Fd. Technol., 38, 7: 4-6.
Dickerson, R.W., Stroup, W.H., Thompson, H.E. and Read, A.B. 1973. J. Milk Fd. Technol., 36, 8: 417.
Driessen, F.M. and Stadhouders, J. 1974. Neth. Milk Dairy J., 28, 10.
Edmondson, L.F., Douglas, F.W. and Avants, J.K. 1971. J. Dairy Sci., 54, 10: 1422.
Foley, J., Brady, J. and Reynolds, P.J. 1971. Soc. of Dairy Technol 24, 1: 54.
Foley, J., Gleeson, J.J. and King, J.J. 1977. J. Food Prot. 40, 1: 25.
Foley, J. and King, J.J. 1977. J. Food Prot. 40, 7: 480.
Hamid, M.A.K., Boulanger, S.C., Tong, S.C., Gallop, R.A., Pereira, R.R. 1969. J. Microwave Pr. 4, 4: 272.
Hill, R.D. 1975. Aust. J. Dairy Technol. 21, 74.
IDF Document No. 86, 101, 1975.
Jaynes, H.O. 1975. J. Milk Food Technol. 38, 7: 386.
Lang, M., Forment, R. and Duckley, W.L. 1976. J. Dairy Sci. 59, 9: 1560.
Law, B.A., Andrews, A.T. and Sharpe, Elisabeth M. 1977. J. Dairy Res. 44, 145.
Law, B.A., Sharpe, Elisabeth M. and Chapman, Helen R. 1976. J. Dairy Res. 43, 459.
Overcast, W.W. 1968. J. Dairy Sci. 51, 1336.
Stoutz, W.P. 1966. XVII Int. Dairy Congr. 567.
Thomas, S.B. 1974. Dairy Inds., 39, 279.
White C.H. and Marshall, R.T. 1973. J. Dairy Sci., 56, 849.
Witter, L.D. 1961. J. Dairy Sci. 44, 983.

REFERENCES (Section 2)

Anon. 1963. The utilisation of microwave ovens for blanching prior to freezing. Quick Frozen Foods, 26, 273-275.

Barratt, B. 1973. Nutrition: 2. Effects of Processing Food in Canada, Feb. 28-31.

Brown et al., 1974. A reduced effluent blanch cooling method using a vibratory conveyor, Food Technology, 36, 696.

Haurghorst, C.R. 1973. Venturi tubes recycle heat in blancher. Food Engineering, 45 (1) 89.

Holmquist et al., 1954. Steam blanching of peas, Food Technol. 8, 437-445.

Horner, G. 1936-1937a. The losses of soluble solids in the blanching of vegetables. Ann. Rept. Fruit Vegetable Preserv. Research Sta. Campden, Univ. Bristol, 37-40.

Horner, G. 1936-1937b. Progress report on the mineral content of canned vegetables. 11. Ann. Rept. Fruit Vegetable Preserv. Research Sta. Campden, Univ. Bristol, 51-56.

Jeppson, M.R. 1964. Techniques of continuous microwave food processing. Cornell H. and R.A. Quarterly 5 (1) 60-64.

Klinker, W.J. 1975. Hot gas blanching studies. Jamesville Wisconsin Report No. 8063.

Kramer, A. and Smith, M.H. 1947. Effect of duration and temperature of blanch on proximate and mineral composition of certain vegetables. Ind. Eng. Chem. 39, 1007-1009.

Lazer et al., 1971. A new concept in blanching. Food Technology, 25, 684.

Lee, C.Y. 1975. New blanching techniques, Korean J. Food Sci. Technol. 7, 2, 100-106.

Lee, F.A. 1958. The Blanching Process, Advances in Food Research 8, 93.

Mitchell et al., 1968. Fluidised-bed blanching of green peas for processing. Food Technol. 22, 717.

Mitchell, R.S. 1972. In-can blanching of green peas. J. Food Technol. 7, 409.

Proctor, B.E. and Goldblith, S.A. 1948. Radar energy for rapid food cooking and blanching and its effects on vitamin content, J. Food Technol. 2, 95-104.

Ray, A. 1975. Steam blancher uses 50% less energy. Food Processing 36, (1) 64.

Wagenknecht and Lec. 1958. Enzyme action and off-flavour in frozen peas. Food Research, 23, 25.

Zscheile, E.P. et al., 1943. Carotene content of fresh and frozen green vegetables. Food Research, 8, 299-313.

SESSION 3

STERILISATION

Chairman: J. McCarthy

MODERN HEAT PRESERVATION OF CANNED MEAT AND MEAT PRODUCTS

F. Wirth

Federal Meat Research Institute, Institute of Technology,
Kulmbach, Federal Republic of Germany.

INTRODUCTION

After the last world war, canned meat stabilised by heat-treatment, had to overcome a lot of prejudices in many countries of Europe. The content of the canned meat products was less responsible for this aversion than the way of treatment. With the aim of better protection against spoilage and food poisoning, and also the lack of good technical methods, the products were often 'over-heated'. The results were deviations in appearance, texture, colour and flavour, as well as losses of nutrients.

Since that time, comsumption practice has been affected by tourism, 'convenience foods' for the housewife, and private and public reserves of cans.

First of all it must be mentioned that during the last twenty years conditions have been created for great improvements in the quality of many kinds of canned meat, by the development technology for the preparation of food prior to filling in cans, by modern technical equipment, and by the use of new methods of heat treatment.

There is no doubt that canning will also have great importance for the next decade. The heat treatment will certainly continue to be important for the storage of meat, as it is hardly to be expected that any cheaper method of preservation will be able to replace heat-treatment in the near future. Therefore it would seem to be profitable to purchase modern types of retorts for meat factories as an investment.

BACTERIOLOGICAL AIM OF CANNING MEAT

The bacteriological aim for producing canned meat and meat products is characterised by two demands. First those micro-organisms and their spores have to be destroyed, which can multiply during storage of the can, and which can form toxins and so endanger human health (problem of health). For a can to be safe from the sanitary point of view, the destruction of *Clostridium botulinum* must be primarily guaranteed, as this is both the most dangerous and also the most heat-resistant toxin-producer. If *Clostridium botulinum* is destroyed, all other toxin-formers - that means food-poisoners - will be eliminated.

On the other hand, micro-organisms which spoil the cans, should also be destroyed (economical problem). A can, in order to be stable from the economic point of view, has to guarantee the elimination of *Clostridium sporogenes*, which is a putridity causer, and also a very heat-stable spore-former. Indeed, there are several heat-stable types of germs among the causes of spoilage, for instance *Clostridium thermosaccharolyticum*. But this organism and also the other thermophiles, mostly multiply only at temperatures above $40^{\circ}C$ to $50^{\circ}C$. That means temperatures which appear in our climatic zone only under very extreme conditions of storage, or by virtue of a very retarded cooling of the cans after the heating-process (incorrect cooling). Heat-treatments which lead to destruction of *Clostridium botulinum* and *Clostridium sporogenes*, result in cans stable during storage without any special conditions of cooling.

But for many kinds of canned meat such an intense heat-treatment is still today connected with important changes of appearance, flavour, texture etc, even when using modern technical methods. Careful heating leads to lower stability, that means it requires limitations in storage-temperatures and storage-time. On this basis - the heat-treatment and the resulting storage-time - Leistner et al.(1970) proposed a classification of canned meat (Table 1).

TABLE 1

RECOMMENDED CLASSIFICATION OF CANNED MEATS   (Leistner et al., 1970)

Type	Preservation	Treatment	Eliminated	Maximal shelflife
1	Semi	> 65-75°C in centre	vegetative organisms	6 months < + 5°C
11	Three-quarter	Fo 0.6-0.8	same as 1, but also mesophilic *Bacillus* spores	6-12 months < + 15°C
111	Full	Fo 4.0-5.0	same as 11, but also mesophilic *Clostridium* spores	4 years < + 25°C
1V	Tropical	Fo 12.0-15.0	same as 111, but also thermophilic *Bacillus* and *Clostridium* spores	1 year > + 40°C

Semi-preserved canned meat (Type 1) is only moderately heated (mainly temperatures between 65°C to 75°C at the centre). Higher temperatures would change the typical character of the product too much (for instance flavour and texture of ham). In order to prevent the surviving micro-organisms from quickly multiplying, the semi-preserves must be stored under cool conditions. The possible storage-time depends on the storage-temperature (for instance maximum 6 months at 5°C or maximum 3 months at 10°C). Generally speaking, semi-preserves can only be stored for a short time because of progressive bacteriological alteration.

With respect to heat-treatment and stability three-quarter-preserved canned meat (Type 11) is placed between the semi and the full-preserves. This type is only permitted to include spores of mesophilic clostridies as well as heat-stable spores of thermophils. But - besides all vegetative kinds of microbes - the spores from the mesophilic types of micro-organisms of the family Bacillus, which often are responsible for the spoilage of meat and meat products, are extensively destroyed. According to our experiments, heat-treatment with F -values from 0.6 to 0.8 are sufficient. These can be achieved by heating up to between 108°C and 112°C in the centre of the food, which are acceptable from the economical point of view and which are also non-deleterious to quality. Considering the arrestation of germination, we think it necessary that a three-quarter-preserve be stored at a maximum temperature of 15°C and for not longer than 12 months. There are many kinds of so treated cans on the market today, mostly meat products to which an intensive heating to 120°C causes important sensory changes (for instance luncheon meat, liver sausages, blood sausages, jellies).

Fully-preserved canned meat (Type lll) has to be safe from a health point of view and storage stable without refrigeration. Therefore the elimination of *Clostridium botulinum* and *Clostridium sporogenes* is necessary. This can be achieved by heating with F-values of 4.0 to 5.0 in which case mostly autoclave-temperatures between 117°C and 130°C are used. So the fully-preserved canned meat can be stored for a long time. A limit to storage-time - eg after four years at 25°C - is not caused by microbes but by chemical spoilage. Heating used for full-preserves does not eliminate the thermophilic sporeformers, which often are extremely heat-resistant. That is why storage at temperatures above 40°C is not possible.

For the tropical-preserved canned meat (Type lV) the destruction of thermophilic spore-formers is also considered. The strong heating which is necessary for tropicals (F = 12.0 to 15.0) is until now only possible for some kinds of canned meat without impairment of palatibility - and then only under special conditions. These cans, which may be stored in tropical zones, are stable from the bacteriological point of view even at temperatures above 40°C. But the quicker chemical alterations at extremely high storage-temperatures only allow a relatively short storage time.

For many years there has existed a system of classifying cans according to pH into low-acids foods (pH > 4.5) acid foods (pH 4.0 - 4.5) and high-acid foods (pH < 4.0). This classification is based upon the growth limit of *Clostridium botulinum* at pH 4.5. Apart from some exceptions, meat products belong to low acid foods with pH-values of about 6.0. These low-acid pH-values call for an intensive heat-treatment in the production of canned meat (Table 2).

TABLE 2
pH VALUES OF CANNED MEATS

Product	pH after processing
Blood sausage	6.8
Chopped pork	6.3
Pasteurised ham	6.3
Luncheon meat	6.2
Canned frankfurters	6.2
Liver sausage	6.2
Corned beef	6.1
Chopped beef	6.0
Gulasch	5.5
Jelly sausage	5.0
Jelly sausage, adjusted	4.5

HEAT INFLUENCE ON CANNED MEAT

Having a high content of protein, meat is a food which is sensitive to heat. In addition, muscle- and fat tissue are bad conductors of heat, especially the latter. Depending on ingredients and consistency, the penetration of heat in some meat products occurs mainly by the unsatisfactory method of conduction heating. Therefore the main problem of heat sterilisation of meat concerns the kind of heat penetration of the can. The answer depends on the method of sterilisation, for the heat penetration from the wall to the centre of the can either takes place by quick convection of liquid or by slow conduction through compact pieces. Combinations of both kinds of heat penetration are usually present. If a lot of liquid ingredients are present we find a free circulation of liquids between the compact meat pieces, for example in meat dishes with gravy and frankfurters in brine, or if sufficient liquid arises from the heating processes, for instance when the fat or liver sausage liquefies, or through the release of tissue liquid of raw meat. This convection heating produces a relatively rapid increase of the temperature within the centre of the can and renders possible the application of favourable heating methods. On the other hand there are products which are compact, like ham, or those which coagulate to a compact mass during the heating process at relatively low temperatures - about 65°C - for instance blood sausage or Bologna type sausage. In these cases heat penetration takes place by direct contact from particle to particle. This conduction heating needs a very long time to achieve the necessary high temperature in the middle of the can.

HEATING METHODS

1. Rotation sterilisation

The most important progress in canning technology within the last decades was the introduction of movement sterilisation, when the cans were moved during heating in the autoclave. Under these conditions, we are able to heat several kinds of canned meat products for a shorter time and therefore more carefully. The result is a better product with regard to flavour and nutrition for the same microbiological effect. The application of movement sterilisation on canned meat products is only convenient for fillings already containing a high percentage of liquid or for those in which the liquid arises during the heating by liquefying of the fat. The mechanical movement, largely supported by a moving gas bubble from the head space of the can, ensures a permanent mixing of the contents. Therefore we get a rapid heat penetration to the thermal centre of the can. So it is possible to reduce the sterilisation time by more than 50% compared with conventional methods (Figure 1).

By rotation sterilisation one can influence consistency, appearance and flavour as the influence of high temperature on the filled product is essentially shorter and, in addition, -

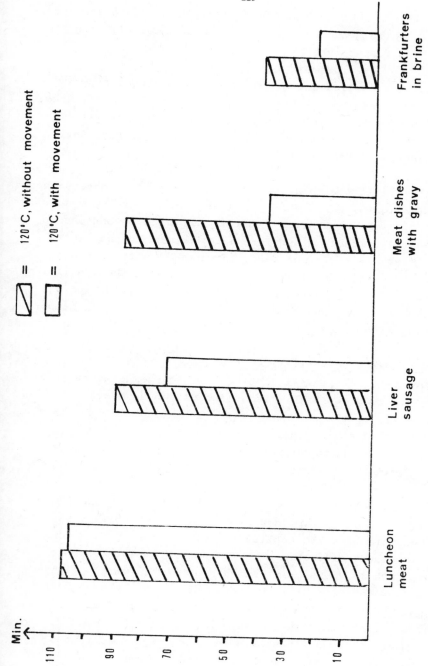

Fig. 1. Influence of rotation-sterilisation on heat penetration of some canned meat products.

because of the continual mixing - there is no fixed zone near the wall, which would be exposed for a particularly long time to the high temperature.

From the application of 30 rotations of the can per minute there results the best effect for a range of fillings. For products with a higher proportion of liquid, or with especially watery consistency, 10 - 20 rotations per minute often have the same effect.

With fillings with a greater proportion of slightly viscous or watery liquid, for instance canned frankfurters, the actual acceleration of penetration is mainly achieved by setting up an intensive current. Therefore a head space in the can is dispensible. On the other hand, in products with liquid of greater viscosity, for instance gravy in meat dishes, a suitable head space in the can has to exist, so that the gas bubble, which moves during the rotation in the can, can support the mixing effect.

2. High-temperature-short-time-method (HTST-method)

During recent years the method of high-temperature-short-time was introduced in the production of canned meats. This method relies on the fact that a short heating at high temperatures is often less deleterious to the food than a longer heating at lower temperatures, both in terms of nutritional value as well as flavour.

Considerable improvements in quality may be reached by HTST, which is often combined with rotation-sterilisation. A prior condition for the application of the method is a considerable amount of liquid with low content of proteins in the can, such as we have in canned meat dishes with a high content of water or meat products in gravy or brine. On the one hand, the liquid can prevent a long contact of any single part of the contents with the very highly heated can wall, and on the other it renders possible an extremely rapid rise of temperature in all zones of the can. So you can reach

temperatures which quickly eliminate heat resistent spores in
a short time. After reaching the desired sterilisation effect,
the high temperatures, which would diminish the quality, have to
be reduced immediately by refrigeration. The application of the
high-temperature-short-time method has no advantage for sausage
conserves like luncheon meat and liver sausage-type products.
In the absence of sufficient liquid, parts of the contents,
especially the parts near the walls, may be excessively exposed
to high temperatures resulting in considerable heat damage.

3. Optimal ranges of temperature

Sterilisation of canned meat and meat products which can
be stored for a long time, need not necessarily occur at about
$120^\circ C$, but can be heated at temperatures above and below to
obtain the same effect of sterilisation. Each kind of meat
product depends largely on the composition, consistency and
heat-sensitivity. Contents with relatively high amounts of
liquid and low amounts of protein are very well suited for
temperatures of sterilisation between $+125^\circ C$ and $+140^\circ C$ by the
high-temperature-short-time process. A heat treatment of $+120^\circ C$
should essentially be reserved for those products which either
have lower proportions of liquid, or where liquid in the form
of juice with rich proteins and liquid fat is separated either
too slowly or in insufficient quantity. Such fillings must not
be very heat-sensitive. In particular, products which are very
sensitive to high temperatures, for instance liver sausage, are
sterilised best between $110^\circ C$ and $115^\circ C$. With the same
effectiveness of sterilisation a difference in temperature of
about $5^\circ C$ may often produce enormous improvements in palatibility
(Table 3).

4. Vacuum application in cans

Just like many other foods, meat products show undesired
reactions and changes due to the influence of oxygen. Air in
canned meat influences colour, flavour and consistency of the
products. Therefore, in the treatment of meat products today
modern techniques have developed ways to remove air - and with

it oxygen - from the material. During heating, a more or less large proportion of the muscle pigment reacts with oxygen, included in the chopped meat and the can, to form brown metmyochromogen. Thus the colour of the meat product as an overall result of several possible combinations of colours of myoglobin either desired or not, is unfavourably influenced by metmyochromogen. More important than that are the changes of the flavour in many kinds of canned meat by the effect of enclosed air. Of particular importance is the ageing of fats, caused by oxidation. This applies to the oils present in the spices as well as the animal fats (Table 4).

TABLE 3

OPTIMAL RANGES OF HEATING FOR CANNED MEAT PRODUCTS

Range of heating temperatures	Products
130°C (140°C)	Goulash Meat dishes with gravy Meat soups
117° - 120°C	Corned beef Frankfurters in brine Chopped beef with juice Chopped pork with juice
110° - 115°C	Liver sausage Luncheon meat Blood sausage

TABLE 4

INFLUENCE OF AIR (OXYGEN) ON CHOPPED AND/OR EMULSIFIED MEAT PRODUCTS

Chemical changes	Faults of colour (pale, grey, bad colour preservation) Faults of flavour (rancidity of fats, flat aroma, old aroma)
Physical changes	Faults of consistence (too soft) Faults of structure (too spongy)

Possibilities of evacuation of air during the production of canned meat exist at the following stages:

a) Mincing and mixing of the raw material
   (Vacuum-cutter, Vacuum-mixer)
b) Filling and closing of the cans.
   (Vacuum-filler, Vacuum-closing machine).

The expected effect of the method depends mainly on the type of meat product and its composition. For products with large pieces and a spongy structure, for example meat in its own juice, an evacuation of air in the filling machine and during the closing of the can is enough. Vacuum chopping and mixing of such products would have little benefit because the air might stream into the spongy structured mass during subsequent treatment.

However, other conditions of evacuation are used in meat products which have a rather homogeneous or emulsion-like consistency through an intensive chopping and mixing process (luncheon meat, Bologna-type sausages, liver sausage). Here small and very small air bubbles in large numbers are mixed into the product during cutting in rapidly driven machines. A subsequent vacuum treatment is only able to remove a little of this air from the tough emulsion. The heating process leads to a connected coagulation of the protein, the air remains enclosed in this fixed emulsion. Therefore chopping and mixing under vacuum is the best method to prevent mixing large amounts of air into the machine from the very beginning.

FULLY-PRESERVED CANNED MEAT

The aim for all canned meat products is fully-preserved products. The use of the above mentioned methods allows this aim to be achieved by the canning of many meat products. These cans may be stored for a long time and their contents have high palatabilities and nutritive values.

The use of modern technologies, which leads to better quality (palatability, nutritive value) of long stored cans, can best be shown by two examples:

1. Canned Goulash

The industrial production of canned meat dishes is often very similar to their preparation in the kitchen, but only industry has the possibility to turn away from the largely wasteful methods of production of former days. Technical innovations in machinery in addition to developments of chemical admixtures are able to achieve the necessary condition. The availability of concentrated roast-flavour and improved binders (special starches) for sauces made it possible to develop a new technology. This method is based mainly on the idea of producing the gravy during the process of sterilisation in the can. All ingredients of the gravy - binders, salt, spices, and aromatic essences - are put together as a dry mixture and then filled into the can. The necessary liquid is made up by water. The fat for the gravy is added to the raw meat. The formation of the gravy - its binding and emulsifying - from the dry mixture, fat and water takes place during the process of sterilisation in the autoclave by movement of the can. So you have to fill only three ingredients into the can:

   a) Raw meat, together with fat (liquified fat or oil)

   b) Dry mixture of binders, salt, spices etc.

   c) Water.

A condition for the use of this technology is a movement-sterilisation (rotation-effect), because the emulsification of the gravy is brought about by the moving contents, mainly the meat-pieces.

The method leads to many advantages for the producer: the wasteful procedure of gravy production is not necessary; production can always be stopped without losses; the product shows a quick heat-penetration during the sterilisation, so that a good product with a high sterilisation effect can be

achieved; the method is much more hygienic than common ways of
production and it is also more economical; the product can
easily be standardised.

2. Canned Frankfurters

Only a few years ago canned frankfurters were only produced
as semi-preserved. They were often sterilised at temperatures
of about 100°C or a little higher. Bad consistency of the
sausages and the frequent bursting of natural casings were
the main reasons that necessitated this low heating temperature.
During the last years, however, fundamental developments have
been made that allow the heating of canned frankfurters to
three-quarter-preserved and full-preserved products and still
retain the desired qualities. Essential for this is a modern
autoclave with the possibility of making very quick changes of
high temperature, as well as a safe regulation of pressure
during heating and cooling.

These technical conditions make use of high-temperature-
short-time-methods (HTST) possible, which are applicable due
to the high proportion of liquid in the can. Movement
sterilisation also has some more advantages (Figure 2).

Figure 2 shows three heating methods and the resulting
values of lethality. Whilst at a temperature of 115°C 24
minutes heating in the autoclave were necessary, we only had
to heat 9 minutes at 125°C in order to achieve the same
sterilisation-effect. Besides the economical advantage of the
more intensive use of machines, there is a better quality of
sausages. For these heating-methods the frankfurters have to
be prepared with a special technology during their production:

The composition, the raw material, techniques of cutting
and filling, forming nitrosomyoglobin and smoking have to be
adapted to the end product, the canned frankfurters. These
technologies are therefore often different from those of
producing fresh sausages. There are many possible measures
for improvement of texture and consistency: the use of meat with

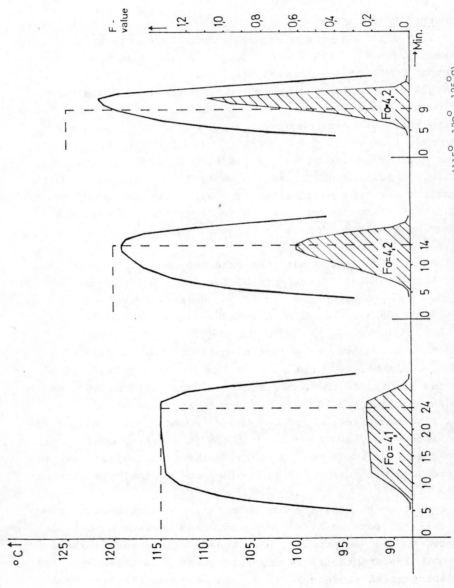

Fig. 2. Fully preserved frankfurters in brine. Different heating ranges (115°, 120°, 125°C).

high content of ATP, added phosphates, protein-emulsifiers high-salting of the meat, vacuum-cutting, improved techniques of cutting, and so on.

A hot smoking technique is used to produce a good smoked colour, a good smoked flavour, and the hardening of the casings. In addition it supports colouring and bacteriological stability of the sausages. It is a time saving method - and that means it is more economical. If the sausages are dried before they are filled in cans and sterilised, the heating will be less risky (bursting of casings).

THREE-QUARTER-PRESERVED CANNED MEAT

For some kinds of canned meat, the heating to full-preservation is today still only possible with certain changes of quality (texture, appearance, flavour). These are mainly products, in which the mass coagulates early and so a predominantly conduction method of heating results (Bologna-type-sausages, liver sausages). Here it is very important to use tins which are favourable for quicker heat-penetration, ie they have to be small in diameter. It is always surprising how seldom this simple method for improvement of quality is noted in practice.

The relatively low water activity of canned liver sausage probably offers the possibility to achieve enough storage stability of this product with careful heating between 90° and 100°C. An a_w-value of 0.95, which is the limitation for growth of Bacillus and Clostridium - the important micro-organisms for canned meat - can be reached by the composition of liver sausage. A more intensive heating of jellies is largely limited by the instability of the gelatines. It is possible to achieve stability against *Clostridium botulinum* in jellies by lowering the pH-value below 4.5 so that storage-stable cans of high quality can also be produced with careful heating.

REDOX POTENTIAL

Based on results of measurements in many kinds of canned meat and meat products we believe that there is an optimal range of redox-potential for canned meat with regard to flavour. This zone is somewhere between $Eh' = 20$ and -150 mV and is obviously influenced by the raw material, intensity of heating and storage-time. Modifications within this range apply to different foods, depending on the nature of the product, luncheon meat-type-products (for instance corned beef) with values of about -100 mV, a little lower are liver sausages and blood sausages (about -120 mV), pork usually has values of about -120 to -150 mV.

The measurements show that the redox-range which is favourable for the flavour of the food, can be achieved by adding either reducing or oxidising substances, dependent on the product. In products with high Eh' -values (for instance luncheon meat-types) we were able to lower the redox potential and so improve flavour by withdrawal of oxygen or adding reducing substances (for instance ascorbic acid). In products with naturally low potentials we could increase the Eh' -value and also improve flavour by adding oxidising substances (for instance nitrite in cured meat products). So influencing the redox potential could be of great importance in the future for can production.

RAW MATERIAL

It should be mentioned that the production of high-quality canned meat starts with the selection of the raw material. The effectiveness of the bacterial enzymes as well as the enzymes of meat have a decisive influence on the quality of the final product. The bacteriological influence seems to be more important than the effect of meat enzymes, because it is generally more intensive. A hygienic production and treatment and a sufficiently cool storage are very important until the meat is canned. The optimal times for canning meat are one day for pork after slaughter and about three days for beef.

NUTRITIVE VALUE OF CANNED MEAT

The value of meat for human nutrition is based on its relatively high (20%) content of proteins, as well as on the high biological value and the good digestibility of these proteins. In addition, the contents of B-vitamins, mineral substances and trace elements are of great importance. So the main part of the B-vitamin requirement can be fulfilled by eating meat, mostly pork. The human need of iron also is considerably covered by eating meat, organs and blood.

The digestibility of the muscle proteins is already excellent in the raw condition. It is only with the proteins of connective tissue that a better usefulness can be reached by heating. Alterations of the composition of aminoacids are not expected at temperatures beneath $100^{\circ}C$ (semi-preserved). At temperatures of about $120^{\circ}C$ and higher (fully-preserved) losses of sulphur containing aminoacids (Cystein and Methionin) will be found. But if we critically appraise our present knowledge and consider modern heating methods, we need not accept losses of the biological value of meat-proteins in excess of 10%.

With regard to the alteration of vitamins by heat, the vitamins of the B-group are of main interest (B1, B2, Niacin). Vitamin B1 is relatively heat sensitive. When producing cans you have to expect losses of about 30 to 50%. Vitamin B2 is unfavourably influenced by higher temperatures too; losses of about 20 to 30% have been reported in published literature. Niacin is less heat-sensitive, but more oxygen-sensitive.

It seems to be remarkable, that wet heating processes with temperatures of about $100^{\circ}C - 120^{\circ}C$ - such as canning - show much smaller losses of B-vitamins than dry cooking-methods - like roasting, grilling and frying - where temperatures between 180° and $350^{\circ}C$ act on the surface. For preservation of vitamins it is also important to exclude oxygen from the can. Vacuum treatment during canning can help to preserve the value.

If cans are stored for long times, alterations of proteins will only be expected at higher storage-temperatures. Loss of vitamins refers mainly to vitamin B 1 (up to 30%), while vitamin B 2 and Niacin are relatively stable during storage. The storage-temperature is of greatest influence.

Summarising, we can state that meat can be canned so carefully that the loss of proteins and vitamins will not be higher, or may even be lower, than the common kinds of cooking in the kitchen. But even today these possibilities of modern canning are not used everywhere, although all technical conditions are on hand.

SUMMARY

Being a high-value source of protein, meat is a food sensitive to heat. Muscle and fat tissue, especially the latter, are poor conductors of heat, and, depending on ingredient and consistency, heat penetrates into many meat-preserves partly or even largely by conduction, which is unfavourable. Moreover, almost all meat-preserves come into the category of low-acid foods, with pH-values about 6.0. Thus to produce canned meat which may be stored without special temperature conditions it is necessary to inactivate the formers of mesophil aerobic and anaerobic spores; particular attention must be paid to the heat-resistant spores of *Clostridium botulinum* and *Clostridium sporogenes*.

The intensive heating necessary for high sterilisation values demands a modern technical and technological procedure if the changes caused by the heat are to be restricted and the products to remain satisfactory both in palatability and from the point of view of nutritive value. Notable progress was achieved when movement sterilisation was introduced for canning meat heated mainly by convection, ie for fillings already containing a high percentage of liquid (eg goulash) or in which the liquid arises during the heating process as the fat liquefies (eg liver sausage) or through the release of tissue liquid (eg beef in its own juice, corned beef). It is possible to reduce the time required for sterilisation by more than 50% compared with conventional methods. With some fillings further advantages in quality are obtained through short, intensive heating (HTST); the precondition being a circulating liquid with low protein contents, guaranteeing a speedy distribution of the heat while protecting the heat-sensitive solid material (eg frankfurters in brine). The optimum temperatures for sterilising the different kinds of meat products should be noted; they lie between $110°$ and $130°C$, and a difference of even $5°C$, while giving the same sterilisation effect, may have a distinct influence on the flavour of the product. For most fillings the removal of oxygen by vacuum processing, both during production and in closing the containers, can bring real improvements in colour and flavour and, by structural compression, in consistency too. For all kinds of canned meat products there is also a range of optimum redox-potential, varying between $Eh' = -20$ and -150 mV and which, depending on the product, may be positively influenced by the removal of oxygen or by the addition of reducing substances (eg ascorbic acid) or even by oxidising additives (eg $NaNO_2$ as curing salt).

For a number of kinds of canned meat it is not at present possible to heat them sufficiently to make them fully preserved canned meat without greatly detracting from their palatability. In such cases the use of suitably shaped containers allowing quick heat penetration is particularly important. For canned jellies a pH-value about 4.5 and for liver sausage preserves an a_w-value of 0.95 or less, will guarantee adequate security for storage over a fairly long period, after careful heating ($100°$ to $110°C$) with its resultant high flavour value.

The value of meat for human nutrition is based on its relatively high content of proteins with high biological value and good digestibility and the content of B-vitamins, mineral substances and trace elements. Meat can be canned so carefully under modern conditions that the loss of proteins and vitamins will not be higher, or will even be lower, than with the common kinds of cooking in the kitchen.

QUALITY ASPECTS OF THERMAL STERILISATION PROCESSES

H. Burton
National Institute for Research in Dairying
Reading RG2 9AT, England

ABSTRACT

1. Quality is normally improved by increasing processing temperature and reducing the time.

2. Thermal death data for microorganisms need to be extended to higher temperatures.

3. Discrepancies between laboratory thermal death data and results obtained for practical plant at high processing temperatures need investigation.

4. The inactivation of heat resistant enzymes by long holding at relatively low temperatures should be studied with a view to practical application.

5. Vitamin losses during processing and storage, and the effects of interaction between different food constituents, need definition.

6. More emphasis should be given to change of quality on storage, as affected by temperature and container characteristics such as oxygen permeability, headspace and light transmission.

7. Methods of container sterilisation during aseptic filling should be investigated and given a scientific basis.

8. Techniques for the ultra-high-temperature processing of liquids containing particles should be studied.

INTRODUCTION

Thermal sterilisation processes as applied to meat and meat products have been considered by the previous speaker, and I must attempt to consider sterilisation processes applied to all other foods. My own area of work has been with milk and milk products, so what I say will inevitably be biased by that fact, and my comments on the sterilisation of fruits, vegetables etc. are not authoritative. Furthermore, this is a very personal survey of a wide field.

The main purpose of sterilisation is to allow a product to keep for a suitable period, usually months or years, without refrigeration. This requires 'commercial sterility', in which the proportion of packs containing a microorganism capable of multiplying in the sterilised material, whether the organisms have survived the heat treatment or entered as post-treatment contaminants, is so small as to be negligible in public health or commercial terms.

Although the bacteriological effects of the sterilisation process are of primary importance, the parallel chemical and biochemical effects of the heat treatment are also of great significance. They are usually adverse effects, such as those which affect organoleptic and nutritional quality of the sterilised products. Some of the chemical and biochemical effects are in fact beneficial, such as those which give an element of cooking, or which improve the nutritive value of the product by the inactivation of naturally occurring toxins such as enzyme inhibitor and haemagglutinins.

In considering the quality of sterilised products, and the influence of processing methods on quality, we must therefore deal with these aspects: the effectiveness of the sterilisation process as determining bacterial quality; the effect of the process and subsequent storage on those properties such as colour, flavour and texture which can be recognised by the consumer; and the effects of processing and storage on nutritive value, which are not usually recognisable by the consumer.

STERILISING EQUIPMENT

The conventional sterilisation process for foodstuffs is an in-container process. After pretreatment, the product is filled into a suitable container and sealed, and then given a suitable time-temperature treatment.

The can has been the most common type of container, of tinplate with double-seamed lids, but glass jars with suitable closures have been used for a proportion of the market. In recent years there have been moves away from the traditional types of container, for example to deep-drawn aluminium or composite cans with easy-open lids, and to sachets formed of laminated plastics.

The batch steam autoclave was for many years the standard steriliser. It was progressively modified, for example by incorporating rotating reels to improve heat transfer to the can contents and cooling water sprays to improve cooling, and by the development of air counter-pressure systems to prevent can or sachet distortion through internal pressure.

Continuous steam sterilisers have become more common for all types of container. The first of these, the spiral pressure cooker, is still widely used but the hydrostatic steriliser, first used on a large scale for the processing of in-bottle sterilised milk, is now used extensively for all types of foodstuffs in all forms of container. Other novel continuous sterilisers are now available, although not used to such an extent as yet. Continuous sterilisers impart a small amount of agitation to the product in the container, and can be pressurised with air. Their great advantage, however, is in uniformity of heating, and economy of steam and labour.

Flame sterilisers, in which a rapidly-rotating can is passed over a series of gas flames, are claimed to have advantages for certain types of product but they are not yet widely used.

It is found, as will be discussed later, that higher processing temperatures applied for shorter times normally give less chemical change for a given sterilising effect than lower temperatures. For this reason, there has been a tendency towards higher processing temperatures. To reduce the processing times correspondingly, higher heat transfer rates to the product are required. This has led to high agitation rates to reduce surface heat transfer coefficients with cans and bottles, and in part to the use of evacuated flexible packs to give minimum intervening material between steam and product. However, these advantages cannot be obtained if thermal conduction into relatively large volumes of solid product is the factor controlling processing time

Ultimately, increase of processing temperature and decrease of time leads to the so-called 'ultra-high-temperature' processes using a temperature of $135^\circ - 150^\circ C$ with holding times of a few seconds. These conditions cannot be obtained with a product in a container: the product is processed in bulk in continuous flow in a suitable heat exchanger, and is then aseptically filled into sterile containers and sealed without contamination. Ultra-high-temperature processes involve the satisfactory performance of two separate and sophisticated stages, the sterilisation of the produ in continuous flow, and aseptic filling.

BACTERIOLOGICAL ASPECTS OF THERMAL STERILISATION

The effects of thermal sterilisation depend on the temperatur and time relationships reached in the product during the process used, and the kinetics of the beneficial and adverse reactions that occur.

It is usually assumed that the effects of time-temperature treatments on the thermal inactivation of bacterial spores are well known. For more than 50 years, sterilisation processes have been compared using the classical methods of Bigelow, Ball and their successors. It has been assumed that thermal inactivation follows a first order reaction, giving a linear relation between logarithm of number or proportion of surviving spores and time at a constant temperature. It has further been assumed that the variation of rate of thermal inactivation with temperature can be described by a simple relationship such as the z-value (which is almost universally used in the food processing industry) or the mathematically equivalent Q_{10} value. Both of these criteria assume that for constant increments of temperature, the rate of thermal inactivation changes in a constant ratio. It seems more likely that rate of spore inactivation varies with temperature according to an Arrhenius relationship, which would imply a constant activation energy and an increasing z-value or falling Q_{10} value with rising temperature, but this concept is not yet accepted industrially.

The food processing industry has traditionally based its design and comparison of sterilisation processes on data for the thermal inactivation of *Clostridium botulinum* spores, and a z-value of $18^{\circ}F$ (corresponding to a Q_{10} of 10) has been accepted almost without question.

Corresponding kinetic data for the other chemical and biochemical reactions taking place during sterilisation is very sparse (Lund, 1975). Such evidence as there is, which relates mainly to thiamine loss, enzyme inactivation, colour change in milk, and arbitrary measure of flavour and texture, shows that these reactions are significantly less temperature-dependent than is spore inactivation. It follows that increased processing temperature, with correspondingly reduced time, will lead to smaller amounts of chemical change for equivalent sterilising processes.

The realisation of this fact has caused the trend towards higher sterilisation temperatures and shorter processing times, referred to earlier, and in the limit has led to continuous ultra-high temperature (UHT) sterilisation processes followed by aseptic filling. Attempts to predict the sterilising effects of ultra-high-temperature processes from data obtained at lower temperatures have caused difficulties, and have led to studies which have an important bearing on sterilisation processes as a whole.

THERMAL DEATH DATA

Thermal death curves themselves need critical re-examination. Classically they have always been assumed to be first order or exponential, converting to straight lines on semi-logarithmic co-ordinates. The strength of this assumption is shown by these quotations from an authoritative textbook (Ball and Olsen, 1957):

"...We feel that it is not unreasonable to believe that perfect experimentation will eventually show this curve to be truly exponential - without the sigmoid characteristic that has appeared at times in experimental results" and, "...we believe that careful workers who take the factor of heat penetration into account will have less difficulty in demonstrating logarithmic death rates than those who neglect heat penetration".

This contention can no longer be sustained. For example, careful experimentation at our Institute has regularly shown the deviation from linearity of thermal death curves for spores, through the development of 'tails' to the curves at low proportions of survivors. This has been found with spores of *B. subtilis* *B. cereus* and, most recently, *B. stearothermophilus* (Davies et al., 1977) - (Figure 1).

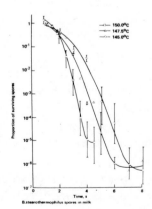

Fig. 1   Thermal death curves for
         B. Stearothermophilus spores

Studies at the Food Research Association in England have now shown a similar effect with *Cl. botulinum* spores (B. Jarvis and P. Neaves, personal communication) - (Figure 2). Similar 'tails' to death curves have been found for death by γ-irradiation (Grecz et al., 1973) - (Figure 3) and by Cerf and Hermier (1972) - (Figure 4) for organisms in H_2O_2. Although the French workers have attempted to test the interesting theory that the 'tails' are artefacts caused by statistical errors at low number of survivors, their results are inconclusive (C. Jacob, J-P. Ley and O. Cerf, personal communication). Most workers who have found these 'tails' would probably consider them a genuine occurrence of a small number of highly resistant surviving organisms in a normal spore population.

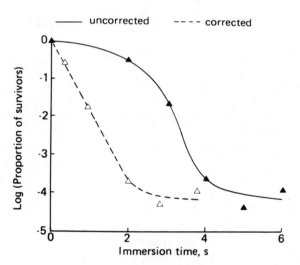

Fig. 2 Thermal death curve for *Cl. botulinum* (NCTC 7242) spores in H_2O at $130°C$ (derived from Jarvis and Neaves, pers. comm.)

The assumption that thermal death rates vary with temperature according to a constant z-value or Q_{10} relationship must also now be questioned, as the result of experiments which extend classical capillary tube studies of thermal death rates to higher temperatures. It now appears that, for *B. stearothermophilus* spores, the Arrhenius relationship best describes the variation of death

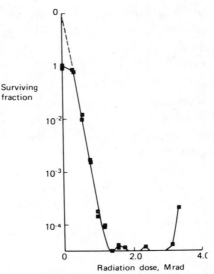

Fig.3 Survival curve for *Clostridium botulinum* spores subjected to gamma radiation. From Grecz et al., (1973).

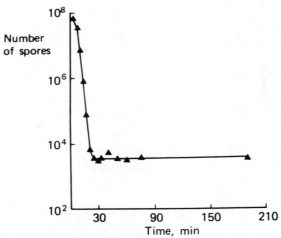

Fig.4 Death curves for *B. subtilis* SJ2 spores in H_2O_2 (230 g/l, 26°C, pH 7.7) From Cerf and Hermier, 1972.

rate with temperature over the range 120°- 150°C (Davies et al., 1977; Perkin et al., 1977) - (Figure 5). A constant z or Q_{10} value is a satisfactory approximation if the temperature range in the process is small, as in the conventional processes used previously, but extrapolations based on a constant z or Q_{10} will lead to significant over-estimates of sterilising effects at the higher temperatures which are being increasingly used.

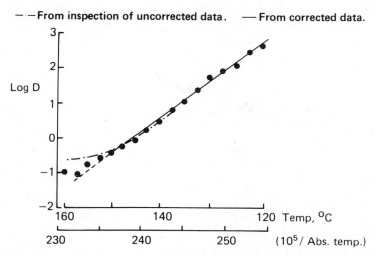

Fig.5 Variation of decimal reduction time with temperature for *B. stearothermophilus* spores suspended in milk.

More recent work by Jarvis and Neaves (private communication) suggests that thermal death rates for *Cl. botulinum* spores deviate from the constant z or Q_{10} pattern even more than would be accounted for by an Arrhenius relationship (Figure 6). They have also found it difficult to reproduce the classical data for *Cl. botulinum* spore destruction on which the analysis of sterilisation processes is traditionally based.

These results have great significance for the interpretation and comparison of sterilisation processes, particularly at high temperatures, and need further study and confirmation.

Additional difficulties have arisen when attempts have been made to compare capillary tube data with sterilising effects obtained in practical ultra-high-temperature sterilisers.

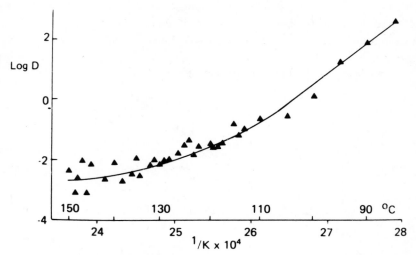

Fig. 6 Variation of death rate with temperature for *Cl. botulinum* (NCTC 7242) spores in H_2O (derived from Jarvis and Neaves, pers, comm.)

Burton et al. (1977) have found a much greater change of sterilising effect with temperature in practical sterilisers than would be expected from capillary tube results obtained at the same temperatures (Figure 7, Table 1).

TABLE 1

COMPARISON OF KINETIC DATA FOR THERMAL DEATH OF *B. stearothermophilus* SPORES IN CAPILLARY TUBES AND IN PRACTICAL UHT PLANT

Activation energies calculated from Arrhenius plots		
	Activation energy, kJ/mol	
	Water	Milk
Capillaries	375	342
Batch UHT	447	580
Continuous UHT	492	546
Corresponding Q_{10} values for temperatures of $134°-148°C$		
	Water	Milk
Capillaries	15-13	12-10
Batch UHT	26-21	68-51
Continuous UHT	36-28	53-41

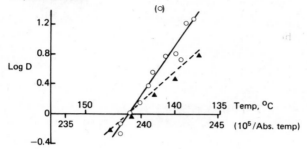

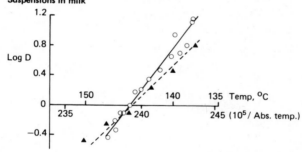

Fig. 7 Comparison of variations of thermal death rate with temperature for capillary tube experiments and experiments with practical UHT plant.

Similar unexpectedly large variations of sterilising effect with temperature obtained in experimental or laboratory sterilisers operating in the UHT range have been found previously by Burton (1970) in Shinfield, Oquendo et al. (1975) in Karlsruhe (although their published results are open to question) and Snygg in Göteborg (personal communication).

Daudin and Cerf at Jouy-en-Josas (personal communication) claim that very rapid temperature changes have a lethal effect on spores which increases with the imposed temperature rise. This effect is additional to the integrated effect of time and temperature. Such a thermal shock would lead to an unexpectedly large variation of sterilising effect with temperature as reported in the above work, but it would not entirely explain the results of Burton et al. (1977).

This discrepancy between thermal death results from capillary tube techniques and practical sterilisers needs further investigation, as does the claim of Daudin and Cerf of a significant thermal shock effect which could contribute to the sterilising effect of UHT processes.

The influence of spore environment on destruction, and on germination and outgrowth of survivors, is a problem which has always been recognised. It may be impracticable to hope to understand all the factors which are important over the whole range of foodstuffs but efforts should be made to understand general principles, as in the studies of the survival of micro-organisms in oil-water systems by Senhaji and Loncin (1977) and Senhaji (1977) at Karlsruhe.

DETERMINATION OF SPOILAGE LEVELS

Whether a sterilised product is produced by an in-container process or by a UHT process followed by aseptic filling, bacteriological quality in terms of overall potential spoilage levels can only be determined by an incubation test of a relatively small proportion of that total output. Interpretation depends on a very small number of positive samples.

With some products, the characteristic form of spoilage is blowing caused by gas production: in such a situation, a relatively simple partial check is possible, but only if the distribution chain can be organised so as to include an effective incubation period. With UHT milk, on the other hand, spoilage by blowing is extremely rare. The more typical forms of spoilage need the destructive examination of packs for their identification.

No doubt food manufacturers on a large scale have sampling schemes which, combined with accumulated experience, give satisfactory control of batch sterility, but it is difficult to determine what these are, and on what they are based. A recent book (ICMSF, 1974) deals with this problem, but does not seem to be of great help to the practical processor. Frequently the problem is never confronted. For example, International Dairy Federation standards give methods of test for sterilised and UHT milk (IDF, 1969) through which it can be determined precisely whether a single container of milk is commercially sterile. However, the standards say nothing about the extension of the methods to a whole batch of sterilised or UHT milk beyond saying that sampling 'should be based on statistical principles' and that batch quality 'should be based on a statistical interpretation of the sample results!'

Perhaps there should be more discussion of this subject, to encourage the adoption of standardised sampling plans which are acceptable to, and provide useful information for, the practical processor.

CHEMICAL AND ORGANOLEPTIC ASPECTS OF THERMAL STERILISATION

Although increasing processing temperature in general improves the quality of the product, there are exceptions, of which perhaps the most important concern the possible survival of enzymes. Enzymes cause quality deterioration during storage, but are inactivated by blanching and sterilisation. However, certain enzymes are extremely heat stable, notably some of the enzymes of vegetable origin (eg horseradish peroxidase) and proteases and lipases of bacterial origin (eg from *Pseudomonas fluorescens*). This inherent stability, combined with the low activation energy which is characteristic of chemical reactions, means that at high processing temperatures the enzymes may not be completely inactivated, and may even regenerate during storage (Bengtsson et al., 1973; Speck and

Adams, 1976; Law et al., 1977; Adams, 1977). The quality of the processed product consequently suffers.

To deal with this type of enzyme, the processing temperature must be lowered and the time increased, possibly to the extent that other chemical changes now limit product quality. However, Barach et al. (1976) for proteases from pseudomonads and Adams (1977) for horseradish peroxidase have independently shown that the enzymes can be inactivated by prolonged holding at relatively very low temperatures, eg 60 min. at $55^{\circ}C$, apparently by a mechanism other than or additional to heat. This effect needs further study, to examine the possibility of including it into processes in which incomplete inactivation of enzymes is a problem.

Flavour of a processed product is normally improved by increasing the processing temperatures and decreasing the time, although it is recognised that there are some products in which a somewhat artificial flavour developed by processing is expected and demanded by the consumer. Change of flavour through processing rarely seems to be a problem, except perhaps with milk. Here we have a bland product in which the development of even a slight flavour is a disadvantage, certainly when the consumer can compare sterilised milk with a good pasteurised milk as is still possible in some countries. Furthermore, the flavour of UHT-sterilised milk can be shown to deteriorate slowly during storage (Figure 8), so that the useful life of the milk is often set by flavour development. There have been claims that certain procedures, generally aimed at oxidation of -SH groups (Samuelsson and Borgström, 1973; Swaisgood et al., 1976), will prevent the development of the characteristic UHT flavour. Investigation of these claims, and a study of the development of UHT flavour, might lead to an increased acceptance of UHT milk in a wider market.

It is less easy to generalise about texture changes and the effect of processing. Rate of change of texture seems to vary with temperature as do other chemical changes, but a desirable texture frequently represents an optimum between an uncooked and an overcooked state. In this situation an increase in processing temperature could cause either an improvement or a deterioration in product texture, depending on the desirable characteristics.

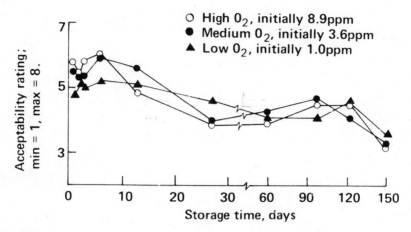

Fig. 8 Effect of initial dissolved oxygen content in acceptability of UHT processed milk during storage in the dark at 20°C. Taste panel 6.7 persons. Each sample judged in duplicate or triplicate.

NUTRITIONAL ASPECTS OF THERMAL STERILISATION

Thermally sterilised products may show significant losses in nutritional quality when compared with the corresponding fresh product, as for example in the data given by Schroeder (1971). However, the overall process comprises many stages additional to the thermal process, and it is probable that much of the recorded loss is due to such incidental processes as peeling or blanching, causing a selective loss by mechanical separation or by leaching of soluble constituents. This section of the paper will only attempt to deal with changes brought about by the thermal process.

Thermal sterilisation seems to have comparatively little effect on protein quality, although there can be a significant loss of lysine in the presence of reducing sugars, which is made worse at the optimum moisture content of 10-14% for the reaction. Loss of protein quality is reduced by higher-temperature, shorter-time processes, but there may be in certain foodstuffs a need for compromise between reduction of the nutritive value of protein and the destruction of, for example, toxic factors and the development of suitable texture.

Vitamin destruction seems to be generally characterised by a low energy of activation, although kinetic data is scarce for vitamins other than thiamine: even with thiamine, there is a recent report that destruction rates can be very markedly affected by components often present in foods (McBee and Marshall, 1977). With a low activation energy for thermal destruction, vitamin loss is minimised by the use of UHT processes, as has been clearly shown for milk (Rolls and Porter, 1973). There are, however, complicating factors which make general predictions difficult. For example, reduced ascorbic acid, which is heat stable, may be oxidised at an early stage of handling or processing to dehydroascorbic acid, which is heat labile: the thermal sterilisation process is then blamed for the loss of the vitamin. There may be interactions between vitamins which make the effect of the process alone difficult to interpret: loss of vitamin B_{12} seems to be associated with the oxidative degradation of ascorbic acid (Ford, 1957).

Folic acid may be an important vitamin, and it is one for which little basic information is available. Published figures for canned foods (Schroeder, 1971) suggest about 50% loss during processing, but it seems probable that most of this represents the loss of water soluble vitamin during pre-processing rather than during thermal treatment.

More information on the effects of thermal processing on vitamins, and on the effects of other related compositional factors, seem to be desirable.

Although the effect of thermal processing as such on nutritive value seems to be comparatively small, particularly with the newer and higher temperature processes, it must be remembered that thermal processing is followed by storage and the adverse effects of storage may be greater than those of processing (Lund, 1975). This can easily be observed with sterilised milk, where very extended storage at room temperature leads to a brown colour and strong flavour typical of a much more severe sterilisation process. In this example, it appears that the rate of chemical change at room temperature can be related to that at sterilisation temperatures by the normal kinetic constants. It is doubtful whether it is possible to make a similar comparison for other storage changes of nutritional significance.

Other factors than storage temperature are of importance in controlling the loss of some nutrients. For example, oxygen is very important for the loss of ascorbic and folic acids during storage of sterile milk. This loss can only be prevented by the removal and subsequent exclusion of oxygen. Folic acid is potentially an important vitamin, and the young and elderly may be particularly dependent on its presence in processed foods. Nevertheless UHT milk is often commercially produced with high levels of oxygen so that the vitamin disappears within a few days, and large sums have been spent on developing aseptic filling systems for UHT milk in bottles blow-moulded from polyolefines which are very permeable to oxygen so that even de-aeration, which is practically possible, could not prevent vitamin loss.

Closer examination of storage factors which influence nutritional quality, and the relationships between the different constituents of the food and the content of, for example, oxygen should be studied more closely. It seems very probable that the storage of a sterile product has more effect on its nutritional quality than did the heat treatment that was responsible for the sterility.

PRACTICAL PROBLEMS IN THERMAL STERILISATION

The problem in obtaining commercial sterility is that of consistently obtaining suitable time-temperature conditions within the product. Prevention of post-sterilisation contamination is an ancillary problem, which becomes a problem of equal importance when an ultra-high-temperature process in a continuous heat exchanger is followed by aseptic filling.

Many of the problems of in-container sterilisation have been solved, but on a practical level the solutions have not always been applied. Perhaps this is because thermo-bacteriology has become a playground for applied mathematicians, and this has obscured some of the practical problems and hindered the adoption of solutions.

For example, the interrelations between air distribution and content, heat transfer rates and levels of spoilage within a single load in a batch steriliser were demonstrated many years ago (cf Burton et al., 1953), as were the effect of different steam supply systems and rates, and air venting systems. Yet these subjects still seem to be under discussion and not to be well understood at the factory level. With the increased use of counter-pressure sterilisers, both batch and continuous, a closer study of the effect of air on heat transfer rates and the effect of air distribution systems might be useful. Such a study has now become more practicable with the availability of more sophisticated instruments for the accurate recording of temperatures within containers in batch and continuous sterilisers.

Heat recovery systems are little used. Batch sterilisers have notoriously high steam consumptions, and continuous sterilisers use only about half as much steam per unit of output (Singh, 1977). However, in both cases there is potential for the recovery of waste heat. For example a new and sophisticated continuous steriliser has been observed to reject 0.75MW of heat as hot water at $85^{\circ}C$. Avoidance of this sort of waste, and the solution of other industrial problems, often requires no research but only the application of existing knowledge.

Most quality factors improve with the use of higher processing temperatures for shorter times. The principal exception to this is the survival of heat-resistant enzymes. The possibility that these can be inactivated at low temperatures, if found to be a technical possibility, removes a major objection to the wider use of high temperature processes.

Problems of heat penetration restrict the advantages to be obtained from high temperature processing of conduction heating packs, although it is theoretically possible to apply a temperature 'wave' passing through the pack, excited by successive heating and cooling.

Ultra-high-temperature systems based on continuous-flow processing in heat exchangers are most simply applied to liquids, Newtonian or non-Newtonian, and of course they are already in wide use for milk and milk-based products. The fundamental problem posed by recent research is whether sterilisation under these conditions is in fact similar to that under laboratory conditions. If the sterilisation processes are found to be similar, then it must follow that recent experiments with continuous heat exchangers have been defective in characterising time and temperature conditions within them, in spite of careful experimentation. This would imply the need for closer study of different types of practical equipment, so that the factors controlling their sterilisation performance can be better understood.

UHT processing in a heat exchanger cannot be separated from aseptic filling, so that problems associated with aseptic filling must be considered although they are not problems of the thermal sterilisation of foodstuffs. The majority of aseptic filling systems which are now available rely for container and closure sterilisation on combinations of hydrogen peroxide and heat. There are, however, no published data which allow such a process to be designed satisfactorily. The published data on sterilisation by hot hydrogen peroxide refer to conditions quite different from those existing in practical equipment (Bockelmann and Bockelmann, 1972. The combinations used by different manufacturers differ so widely that it is not conceivable that they can all be equally effective. Further work on this subject is needed.

Problems of pack characteristics as influencing organoleptic and nutritional quality changes during storage need constant attention. In the past the can, which is opaque and hermetic, has been the most usual container. New types of container are now being advanced for commercial application, and the effect of permeability, headspace and light transmission need to be kept in mind.

The application of UHT methods to foods comprising solid particles suspended in liquid raises many questions. Some of these are fundamental, for example those concerning the sterilisation of the particles: can they be assumed to be sterile in the interior so that only the outer layers need active sterilisation, or must the whole particle be given a sterilising process? If the whole particle is to be sterilised, is it possible to derive the heat from the suspension liquid without that liquid being over-processed, or is it preferable to sterilise the particles and liquid separately and then mix them aseptically before or at filling? Other problems are technological, such as the design of pumps, valves and filling mechanisms to operate with particulate materials in suspension.

A decision has ultimately to be made on whether it is worth while to adopt such a process, or whether equally satisfactory results in terms of product quality and processing costs could not be obtained by a refinement of in-container methods.

Microwave sterilisation processes have been considered for this purpose and much work has been done by, for example, Bengtsson, and his colleagues in Sweden, but the technological problems are considerable. Non-uniformity of heating caused by the finite penetration of microwaves into lossy materials, and the distortion of field patterns by high-permittivity foods, cause problems at least as difficult as those presented by conventional conduction heating. There is also the additional factor that the product temperature does not rise to a limiting value set by the temperature of a heating medium as in most other processing methods. It is not clear why radio-frequency heating at frequencies of the order of 100MHz has been abandoned in favour of microwave methods, since temperature variations caused by lack of penetration into the food would be much reduced.

CONCLUSIONS

Thermal sterilisation in industry has an enviable record for quality over many years. However, the use of higher processing temperatures has introduced new unknowns which need resolution. The aim should always be to preserve nutritional quality within the limits of bacteriological safety, and this may mean a critical examination of safety factors so that processes are not more severe than they need to be. Nutritional problems with sterilised foodstuffs do not seem to have arisen so far, but such foods have always formed part of a mixed diet. As a higher proportion of foodstuffs in developed countries is processed on the way to the consumer, the balancing effect of a mixed diet is in danger of being lost, and the need to ensure that processing does not cause excessive loss of important nutrients becomes more urgent.

REFERENCES

Adams, J.B. 1977. The inactivation and regeneration of peroxidase in relation to the high-temperature short-time processing of vegetables. Campden Food Preservation Research Association Technical Bulletin No. 34 (in preparation).

Ball, C.O. and Olson, F.C.W. 1957. Sterilisation in Food Technology. New York: McGraw Hill.

Barach, J.T., Adams, D.M. and Speck, M.L. 1976. J. Dairy Sci. 59, 391-395.

Bengtsson, K., Gardhage, L. and Isaksson, B. 1973. Milchwissenschaft 28, 495-499.

Bockelmann, I. von and Bockelmann, B. von. 1972. Lebensm. -Wiss. u. Technol. 5, 221-225.

Burton, H. 1970. J. Dairy Res. 37, 227-231.

Burton, H., Akam, D.N., Thiel, C.C., Grinsted, E. and Clegg, L.F.L. 1953. J. Soc. Dairy Technol. 6, 98-113.

Burton, H., Perkin, A.G., Davies, F.L. and Underwood, H. 1977. J. Fd. Technol. 12, 149-161.

Cerf, O. and Hermier, J. 1972. Lait, 52, (511-512) 1-20

Davies, F.L., Underwood, H., Perkin, A.G. and Burton, H. 1977. J. Fd Technol. 12, 115-129.

Ford, J.E. 1957. J. Dairy Res. 24, 360-364.

Grecz, N., Lo, H., Kennedy, E.J. and Durban, E. 1973. In: Radiation preservation of food. International Atomic Energy Agency Symposium Series. Vienna: IAEA.

ICMSF (International Commission on Microbiological Specifications for Foods) 1974. Micro-organisms in foods. 2. Sampling for microbiological analysis: Principles and specific applications. Toronto. University of Toronto Press.

IDF (International Dairy Federation) 1969. Control methods for sterilised milk. IDF 48. Brussels: IDF.

Law, B.A., Andrews, A.T. and Sharpe, M.E. 1977. J. Dairy Res. 59, 786-789.

Lund, D.B. 1975. In: Nutritional evaluation of food processing. Eds. R.S. Harris and E. Karmas. pp. 205-240. Westport: AVI.

McBee, L.E. and Marshall, R.T. 1977. Abstract Paper No. D35. USDA Annual Meeting, Ames, Iowa.

Oquendo, R., Valdivieso, L., Stahl, R. and Loncin, M. 1975. Lebensm. -Wiss. u. Technol. 8, 181-182.

Perkin, A.G., Burton, H., Underwood, H.M. and Davies, F.L. 1977. J. Fd Technol. 12, 131-148.

Rolls, B.A. and Porter, J.W.G. 1973. Proc. Nutr. Soc. 32, 9-15.

Samuelsson, E.-G. and Borgström, S. 1973. Milchwissenschaft 28, 25-26.

Schroeder, H.A. 1971. Am. J. clin. Nutr. 24, 562-573.

Senhaji, A.F. 1977. J. Fd Technol. 12, 217-230.

Senhaji, A.F. and Loncin, M. 1977. J. Fd Technol. 12, 203-216.

Singh, R.P. 1977. Food Technology 31, 57-60.

Speck, M.L. and Adams, D.M. 1976. J. Dairy Sci. 59, 786-789.

Swaisgood, H.E., Janolino, V.G. and Horton, H.R. 1976. Cited in Swaisgood, H.E., Dairy Ice Cr. Field (1977) 160, (1) 48, 50, 60.

SESSION 4

DEHYDRATION

Chairman: G. Fabriani

THE STATE OF THE ART OF FOOD DEHYDRATION

H.A. Leniger and S. Bruin
Food Science Department, Process Engineering Group,
Agricultural University, Wageningen, The Netherlands.

ABSTRACT

1. *Drying is an extremely important preservation method.*
2. *Knowledge of the drying process is still very limited.*
3. *Only recently there has been a break-through in the knowledge of the physics of the process, but much theoretical and experimental work remains to be done.*
4. *During drying many changes in the product may occur. Up till now the relation between 'quality' and drying conditions is obscure.*
5. *The knowledge of reaction mechanisms and of the influence of temperature and water activity on the conversion rates is inadequate.*
6. *The complicated relation between the physical aspects and the changes occurring during drying has to be studied thoroughly.*
7. *Although a lot of work has been done in the field of storage of dehydrated foodstuffs, in this field too much work remains to be done.*

1. INTRODUCTION

The programme of this seminar lists two papers on food dehydration. Fortunately we had the opportunity to see Dr. Escher's paper before writing the final text of our own contribution so that we were able to take into account the contents of the second paper and to avoid too much overlap. Whereas Dr. Escher in his paper deals in particular with practical aspects of the dehydration process and experimental results described in literature, our principal aim is to give a systematic survey of the more theoretical or fundamental aspects of the process. A further objective is to point out where we stand nowadays, what we can do with the available knowledge and insight, what the chances are of making good the deficiencies in our knowledge by further work and which research themes can be most recommended.

Unfortunately there was no time to make a detailed study of the literature; only a few references are given. We are of the opinion that there is a need for a clear and up to date survey and our first recommendation is to promote the publication of such a survey. This seems to be an activity excellently suited for international co-operation.

With food dehydration we encounter three groups of problems:

a. Physical aspects of the dehydration process

In principle knowledge of the physical phenomena makes it possible to calculate temperature and moisture profiles in a drying material as a function of time and therefore of rates of drying and drying times.

b. Changes in foodstuffs during dehydration

The dehydration process influences the quality of foodstuffs in the widest sense of the word. One can distinguish between:

b_1. Influence of drying on micro-organisms and their metabolic processes.

b_2. Influence on enzymes and enzymatic conversions.
b_3. Chemical conversions during dehydration.
b_4. Physical changes in foodstuffs during dehydration.

All changes depend on temperature, water activity and time, therefore on temperature and moisture profiles as a function of time. Furthermore such factors as composition and structure of the foodstuff, the pH, and the presence of oxygen play a part. Many conversions also occur during the period preceding the dehydration process and during the storage after drying. Therefore one always has to consider the entire history of a product from the time of harvesting or winning until consumption. The changes observed during storage before and after drying are influenced by the same factors as those which occur during the drying process. As a rule the storage time is very long compared with the drying time, whereas the storage temperature usually is much lower than the product temperature during dehydration. Some reactions do not occur at a perceptible rate unless the temperature is much higher than the usual storage temperatures.

c. Optimisation of the drying process

The design of drying equipment has to be done in such a way that a favourable combination of cost and quality will be attained. With the cost of drying the energy consumption is the main factor. This is influenced primarily by the choice of drying conditions and final moisture content. Starting with a certain quality of the raw material, the same factors determine the quality of the end-product. The moisture content of a dried foodstuff also influences greatly the keepability, therefore the quality after a certain shelf-life.

The dehydration process may be considered as a heat treatment. Therefore it is self-evident, that there are analogies with other heat treatments such as cooking, baking, roasting, sterilisation and evaporation. However, an important characteristic of the drying process is that the heat treatment takes place as the water activity decreases markedly. Under these circumstances the physical aspects as well as the quality aspects of the drying process are much more complicated than with other treatments.

We intend to analyse successively the three categories of problems in more detail, but we restrict ourselves to drying in air (convection drying) because it is impossible to discuss other methods in the available time.

2. PHYSICAL ASPECTS

Knowledge of the mechanism of drying, that is to say of the heat and mass transfer phenomena during drying, enables us to understand the course of the process and in simple cases even to calculate it. Such calculations result in temperature and moisture profiles, drying rates and drying times as a function of type, shape and dimensions of the product and as a function of the drying conditions.

This knowledge is of special importance for the optimisation of the drying process and the design of driers.

When temperatures - and moisture content distributions inside drying materials or drying particles are known a clear comprehension of chemical or physico chemical changes during drying is possible. If one knows enough about the influence of temperature and water activity on the rate of conversions, it is possible to derive the conversion rates of chemical reactions from the calculated profiles. In other words it then is possible to relate the quality of the products to the drying conditions.

Of course rates of drying as a function of drying conditions can also be determined empirically. In complex cases this may take less time and may be more accurate than making calculations. From the results of experiments conclusions can be drawn about the mechanism of drying. However, the measurement of temperature and moisture profiles is very difficult, except in simple models. It goes without saying that an increase in our knowledge of the physical aspects of drying would very much contribute to a better understanding, assist in organising experimental programmes and facilitate the interpretation of experimental results.

We therefore suggest that further work in the area of developing calculation schemes for drying processes be supported. We also advocate expansion of the work regarding physical properties involved in drying.

In the first place knowledge of the binding of water in foodstuffs and of the influence of the temperature on the water binding is required, in other words, of the water activity as a function of moisture contents and temperatures. Much is known about sorption isotherms and there are a few useful literature surveys. However data on the influence of the temperature are scarce and it is self-evident that such data are important because product temperatures during drying may vary widely.

In principle the influence of the temperature on the water activity may be expressed by the simple equation:

$$\frac{d \ln a_w}{d (\frac{1}{T})} = \frac{\Delta H}{R}$$

where a_w = water activity (dimensionless), T = abs. temp. ($^\circ$K), R = gas constant for water vapour (J/kg. $^\circ$C) and ΔH = differential heat of sorption (J/kg). If the water is not bound, $\Delta H = 0$, so that temperature has no effect on the water activity. One can say that ΔH may be used to measure the non-ideal behaviour of water in the system. ΔH varies strongly for different foods depending on the moisture content of each particular foodstuff. Above a certain moisture content the heat of sorption no longer differs markedly from the heat of evaporation; ΔH is then about zero. Hence, with higher moisture contents, the sorption isotherms for a range of temperatures get closer and closer together and ultimately blend into a single line. For knowledge of ΔH, one depends on experimental work. Few reliable figures are available.

Furthermore knowledge of properties involved in heat and mass transfer inside and outside a drying material is required. In order to calculate temperature equilibration, data regarding heat transfer coefficients, thermal conductivity and thermal diffusivity are necessary. These data are amply available, also as a function of moisture content. Only in cases of complicated structures, such as anisotropic and porous materials the knowledge is still inadequate.

Calculation of moisture equalisation requires knowledge of the mass transfer coefficients and the water diffusivities or apparent diffusion coefficients. Data concerning mass transfer coefficients can be derived from direct correlations or from better known heat transfer coefficients. However, data about water diffusivities are extremely scarce.

It is essential to know the influence of the moisture content on the transport properties and helpful to know their temperature dependence. It is known that water diffusivity depends considerably on temperature and very strongly indeed on the moisture content, but there are very few reliable figures. This is caused by the fact that experimental determination of this property is extremely difficult and that internal moisture tranfer can take place in different ways.

In liquid solutions and gel-like materials transport of water only takes place by molecular diffusion, a relatively simple phenomenon. In many other materials gasfilled cavities, capillaries, cell walls, inter- and intracellular spaces occur. In such cases water can be transported by water diffusion, vapour diffusion, internal evaporation/condensation effects and capillary flow. Often there is a mixture of various transport mechanisms and the share of the different mechanisms varies from place to place and changes as drying progresses. In such complicated situations one speaks of an apparent diffusion coefficient.

As an illustration of the strong influence of the water content on the diffusion coefficients we refer to Figure 1 for data on cellophane, gelatine solutions, starch gel, maltodextrin solutions, coffee extract and amylopectin solutions.

Calculation of temperature and moisture distributions during drying requires the solution of the local equations of change of energy and mass (water), pertaining to an infinitesimally small volume element in the drying material. These equations are partial differential equations with coefficients depending on moisture content and temperature.

The solution of these equations in the most general form is only possible with numerical techniques and in the case of simple geometric shapes of the drying material. As the coefficients vary markedly, the programming is quite complicated and a great number of mathematical operations is needed.

Fortunately there are important practical situations where simplifications are possible. Table 1 presents a classification of calculation schemes in the order of increasing difficulties.

The most simple type of material is a liquid solution or a gel; in such a material transport of water only takes place by molecular diffusion (type 1). It is characterised by a continuous water phase with a relatively high thermal diffusivity a (m^2/s). On the other hand, as we have shown in Figure 1, the water diffusivity D (m^2/s) is low, especially at lower moisture contents. Consequently, the dimensionless ratio a/D has a very high value, or the so called Luikov-number D/a is very low. One can also say that in this case the Biot-number for heat transfer $Bi_h = \alpha R/\lambda$ is very small (α = heat transfer coefficient $(W/m^2.°C)$, R is characteristic dimension (m) and λ = thermal conductivity $(W/m.°C)$), whereas the Biot-number for mass transfer $Bi_m = \kappa R/D$ is very high (κ = mass transfer coefficient (m/s)).

It is easily understood that under these circumstances the temperature profiles inside the drying material may be

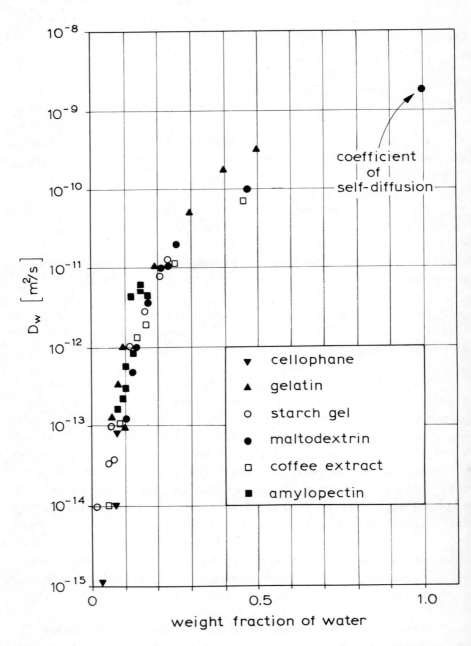

Fig. 1.

considered to be flat, whereas the moisture profiles show steep
gradients. This has been illustrated qualitatively in Figure 2.
As the temperature in the material does not vary with the position
an overall energy balance can be used to calculate the course
of the product temperature as a function of time. In the Table
this has been stated as a macroscopic balance.

Spray drying of solutions and fluid bed drying of gel-like
particles like starch are important examples of processes which
can be calculated relatively simply.

In recent years there have been significant developments in
this type of calculation by the work of Thijssen and co-workers
in the Netherlands (Thijssen and Rulkens, 1968; Rulkens and
Thijssen, 1972; Rulkens, 1973; Kerkhof, 1975; Kerkhof and
Schoeber, 1974; Schoeber, 1976; Schoeber and Thijssen, 1975;
Lijn, 1976a; Lijn et al., 1972; Lijn, 1976b). Eventually this
has led to short-cut calculation methods, simplified calculation
procedures, which yield excellent approximations and can be
used for engineering calculations on simple computing systems.
Moreover the data needed for these calculations can be obtained
directly from simple dehydration experiments.

The short-cut calculation methods originated from
regularities in drying processes, observed by Schoeber (1976)
when he analysed a large number of computer simulations. On
these he founded his 'regular regime theory' and he derived
calculation methods for the drying of materials in which
diffusion is the governing mechanism. It is impossible to deal
with these methods in more detail within the scope of this paper
and we have to refer to the original literature. In order to
illustrate that there is close agreement between the short-cut
method and the numerical solution of the diffusion equation,
we present in Figure 3 drying curves (average water content as
a function of time) at two different holding temperatures for
spray drying of a maltose-water solution (Kerkhof, 1977).

TABLE 1

CLASSIFICATION OF VARIOUS CALCULATION SCHEMES FOR DRYING PROCESSES

Type of drying material	Energy equation	Water diffusion equation	Heat- and mass flux equations
I: liquid solutions (non shrinking)	macroscopic balance $$\rho C_p \frac{dT}{dt} = \frac{A}{V} q_s$$	microscopic balance (liquid diffusion) $$\frac{\partial X_w}{\partial t} = -\nabla \cdot j_w$$	$j_w = -D_w \rho_l \nabla X_w \quad D_w = f(X_w)$ q = not needed
II: liquid solutions (shrinking)	macroscopic balance $$\rho C_p \frac{dT}{dt} = \frac{A}{V} q_s$$	microscopic balance in solid-fixed co-ordinates (liquid diffusion) $$\frac{\partial X_w}{\partial t} = -\nabla \cdot j_w$$	$j_w = -D_w \rho_l \nabla X_w \quad D_w = f(X_w)$ q = not needed
III: porous bodies (small shrinking effects)	microscopic balance $$\rho C_p \frac{\partial T}{\partial t} = -\nabla \cdot q + \phi_q$$	microscopic balance (vapor diffusion + capillary movement of water) $$\frac{\partial X_w}{\partial t} = -\nabla \cdot j_w$$	$j_w = -\rho_l (D_{XZ} \nabla X_w = D_{TZ} \nabla T)$ $q = -\lambda \nabla T + j_{wv} \Delta H$
IV: capillary-porous bodies (shrinking)	microscopic balance $$\rho C_p \frac{\partial T}{\partial t} = -\nabla \cdot q + \phi_q$$	microscopic balance (vapor diffusion + liquid diffusion + capillary movement of water). $$\frac{\partial X_w}{\partial t} = -\nabla \cdot j_w$$	$j_w = -\rho_l (D'_X \nabla X_w + D'_T \nabla T)$ $q_w = -\lambda \nabla T + j_{wv} \Delta H$

Difficulties increase ⟶

LEGEND TO TABLE 1

T	=	abs. temperature
t	=	time
ρ	=	density of drying particle
C_p	=	specific heat of drying particle
A	=	surface of drying particle
V	=	volume of drying particle
q	=	heat flux vector
q_s	=	heat flux at surface of drying material
ϕ_q	=	heat generation inside particle
X_w	=	water content
j_w	=	mass flux of water (total)
$j_{w,i}$	=	mass flux of water in the vapor phase
D_w	=	water diffusivity
D_{Xl}	=	apparent diffusion coefficient for capillary water movement
D_{Tl}	=	thermal diffusion coefficient for capillary water movement
D_X'	=	apparent diffusion coefficient
D_T'	=	apparent thermal diffusion coefficient
ρ_l	=	density of water
λ	=	thermal conductivity
ΔH	=	enthalpy
∇	=	Nabla operator
$\cdot$	=	vector in product

*

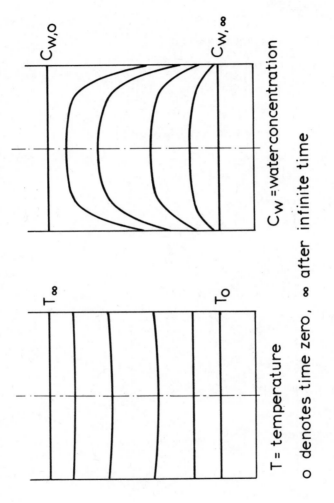

Fig. 2. Temperature and moisture equilibration.

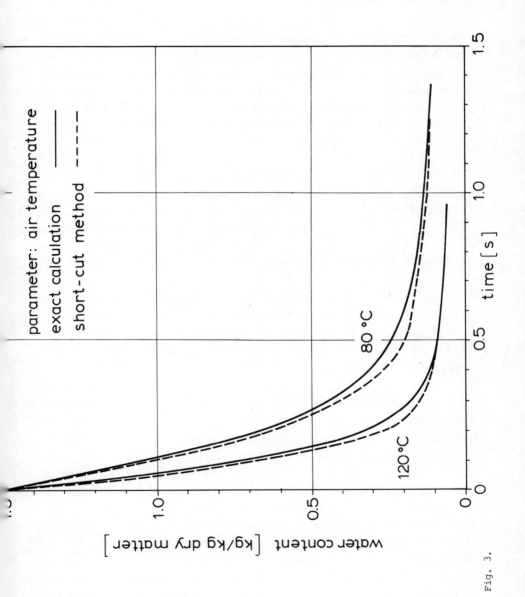

Fig. 3.

There is no doubt that good progress has been made, but much remains to be done.

In the meantime the calculations have also been applied to shrinking liquid solutions (group II of the Table) under the assumption that the excess partial molar volume of mixing of water and the solids in solution is zero. In this case a simple co-ordinate transformation suffices; instead of co-ordinates fixed in space co-ordinates fixed with reference to increments in solid content are used.

All studies mentioned so far have been carried out with boundary conditions to a particle (sphere), cylinder and slab that imply a constant temperature and moisture content in the bulk of the drying medium. Some calculations have already been done (Kerkhof and Schoeber, 1974; Lijn, 1976a) for a dispersion of droplets in cocurrence with the drying air.

In future calculations have to be extended to other types of boundary conditions, eg pertaining to fluid bed drying, drying of packed beds of granular materials with through draft of air or drying of layers of granules with overdraft of the drying medium.

We mentioned already that materials with pores, capillaries etc, belonging to groups III and IV of the Table are much more complicated. In such materials the transport of water or water vapour can be extremely complex. Little is known about the thermal diffusivity and very little regarding the water diffusivity in such complicated drying materials. In the literature on mathematical modelling of drying one frequently encounters studies based on equations derived by Luikov, but these equations (ie a microscopic energy balance and a water diffusion equation) are usually solved analytically, which is only possible when assuming that the physical properties are constant. We would like to stress that such studies at best are of theoretical value only because, as we have seen before the water diffusivity in particular varies strongly with the moisture content. Therefore completely wrong conclusions can be drawn from calculations neglecting this fact.

It follows that the work described above has to be continued and extended. Drying of more complicated materials has to be studied, also drying of irregularly shaped particles. With this work knowledge of physical properties is essential.

Besides transfer of moisture and moisture vapour during drying, transport of other volatile components (flavours) and of non-volatile components of low-molecular weight, such as sugars and salt, takes place; this feature is referred to as migration. These phenomena will be mentioned briefly in Chapter 3.

The same physical aspects play a part during the storage of dried products. Much theoretical and experimental work has been done regarding the rate of heating/cooling and drying/humidifying of stored materials, either packed or not, at changing external temperatures and humidities. From changes in temperature and moisture content of products several research workers concluded which conversions will take place. In other words they studied the influence of external storage conditions and packaging on the keepability and quality of the products. However also in this field the knowledge is still very limited, eg little is known about heat and mass transfer phenomena in big bulks of unpacked or packed commodities.

3 CHANGES IN FOODSTUFFS DURING DEHYDRATION

This subject will be dealt with at some length by the next speaker. In order to minimise overlap we will confine ourselves to some introductory remarks.

In this field the knowledge varies rather strongly. During the drying process or during part of it, micro-organisms may multiply rapidly and they may cause undesirable conversions. However the conditions may also cause destruction of micro-organisms. Besides the environment, the combination of temperature and water activity determines what happens. Inadequate knowledge is available regarding the influence of this combination on the growth rate and the rate of metabolic processes. The influence on the rate of reversible inactivation by desiccation and of destruction by heat is still less clear. Taking into consideration that temperatures and water activities in a drying product depend on place and time this proves to be a very complicated problem.

Still more complicated and less well known is the behaviour of enzymes during drying. One does not know enough about the limiting combinations of temperature and water activity above which enzymatic conversions take place and below which enzymes are inactive. The rate of reactions as a function of temperature and water activity in the range in which conversions are observed is insufficiently known. Outside this range a reversible inactivation by desiccation may occur, but also an irreversible inactivation by high temperatures. What actually occurs with various combinations of temperature and water activity has been studied insufficiently.

Often the drying conditions cause rapid enzymatic conversions. If these changes are undesirable either the enzymes have to be inactivated prior to the drying or such drying conditions have to be selected that a rapid inactivation occurs. The rest activity of enzymes after drying influences of course the changes in quality during storage. Sowing seeds and brewing barley have to be dried in such a way that the enzymes are hardly damaged, or at least are not irreversibly inactivated. During drying of some products desirable enzymatic reactions take place; in such cases the drying conditions have to follow a certain programme. It follows that it is highly desirable to know more about the behaviour of enzymes during drying processes.

The problem of chemical conversions during drying is extremely complicated. First there are desirable and undesirable

chemical reactions. When one wants to preserve the initial
quality as much as possible, all chemical changes have to be
considered as undesirable, but this is an extreme case which can
at best be approximated. In other extreme cases certain chemical
conversions are so much desired that it is preferable to speak
of a heat treatment instead of a drying process. Examples are
roasting of coffee and cocoabeans and baking processes. In
between these two limits there is a whole spectrum of dehydration
processes with a complex of desirable and undesirable conversions.
Such drying conditions have then to be chosen that a certain
selectivity with respect to desired reactions will be attained.
Of course this is closely bound up with the traditional
character of many foodstuffs.

With chemical conversions the reaction mechanism and the
influence of concentration, temperature and water activity on
the rates of reactions are important. It is self-evident that
the number of possible reactions in many foodstuffs is very
large and that in principle several reaction mechanisms may be involved. Simple, first order degradation reactions have been
studied many times. An example is the destruction of heat labile
components, such as vitamins and chlorophyll. The hydrolysis
of sugars and esters can also be considered as first order
reactions. As a matter of fact in such a complicated reaction
mixture as a foodstuff many reactions with a much more complicated
mechanism will occur. We think of second order reactions and
consecutive reactions, which may be responsible for causing off-
flavours, discolouration and the like. Little is known about
these reactions and one cannot expect that our knowledge will
soon increase.

The rate constants k of the reactions depend on temperature,
concentration of the reactants and concentration of water (water
activity). The question arises in which cases the temperature
dependence may be described by the simple Arrhenius equation

$$k = k_o \exp.(-\Delta E/RT)$$

where k_o is the so called frequency factor, ΔE = activation energy (kJ/mol)°, R = gasconstant and T = abs. temperature. Strictly speaking the frequency factor depends on the temperature and can be considered only as a constant when ΔE is sufficiently high. The value of ΔE depends on the reaction mechanism and varies widely. Reactions which are catalysed by enzymes have a lower activation energy then chemical reactions, but in both groups there are considerable differences. When discussing conversion rates it is perhaps an inadmissible simplification to use average values for both groups of reactions. Moreover data on activation energies are scarce. The importance of the value of ΔE is clearly illustrated in Figure 4, which presents the ratio of the rate constant k_t at temperature $t°C$ and the rate constant at $0°C$ as a function of temperature. For ΔE = 50 kcal/mole the conversion rate increases over a temperature range from $0°C$ to $100°C$ by more than ten orders of magnitude. This means that if at $100°C$ product temperature a certain thermal degradation reaction is still acceptable at a residence time of the product at that temperature of say one second the residence time at $0°C$ may not exceed 1700 years.

It is not quite clear for which types of reaction mechanisms the Arrhenius equation cannot be applied and what errors are the result when using the simple equation in such cases. Anyhow, one has to be very careful when the mass transfer of the reactants may be a limiting factor. With enzymatic reactions the Arrhenius equation may describe the temperature dependence accurately at low temperatures, but at higher temperatures the rate of reaction decreases because inactivation of the enzymes becomes more and more important.

During the drying process concentrations change also and it is not clear to what extent thereby the chemical changes are influenced. The influence of the water activity is important but insufficiently understood. In general chemical reactions are slower as the water activity decreases. As during drying the product temperature increases whereas the water activity decreases, k may first increase when the temperature effect is dominating whereas k may decrease later on when the influence of

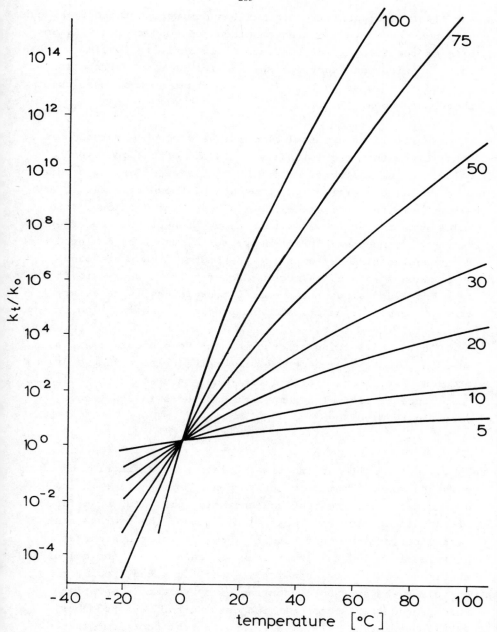

Fig. 4. Effect of temperature on relative value of rate constant, with activation energy kcal/mole as parameter.

the lower water activity becomes the dominating factor. However
with non-enzymatic browning-reactions one observes a maximum
rate of reaction at intermediate water activity levels, whereas
various oxidation reactions show a minimum rate of reaction at
a certain water activity. It is important to know more about
these phenomena.

The field of chemical conversions during dehydration is so
complicated that provisionally it is out of the question to
calculate the influence of drying at different drying conditions
on chemical conversions with the aid of known temperature and
water activity profiles, except in very simple cases. This
holds true in particular with sensory properties. For the time
being it is much simpler to carry out a number of drying experiments and to find out with the assistance of a taste panel
what changes in quality occur. However it is self-evident that
more knowledge about the chemical conversions is necessary in
order to explain, understand and predict changes during drying.

It is interesting to observe that some progress has been
made in calculating the conversion rate during a simple drying
process. In the case of a first order rate equation the local
rest concentration of a component somewhere in a drying
particle is given by:

$$\frac{C}{C_o} = \exp.\left(-\int_0^{t_e} k\,dt\right)$$

in which t_e is the drying time. Since k is a function of
moisture content which in turn depends on the place in the
drying material, the rest concentration will also be a function
of the place. For an exact calculation of the average rest
concentration one would have to know the water concentration
distribution and the temperature at each moment and calculate
the local conversion as a function of time, and finally
calculate the average concentration by volume integration.
This procedure has been carried out (Kerkhof and Schoeber,
1974), but Kerkhof (1977) also applied the short cut method,
described in the previous chapter. Then, only the average
water concentration and temperature in the drying material as

funtions of time need to be known. The simplest approximation
is to assume a uniform value of k over the material, as
corresponding with the average moisture content $\bar{m}$. This leads
to

$$\frac{\bar{c}}{c_o} = \exp.\left(-\int_0^{t_e} \bar{k}\, dt\right)$$

with $\bar{k}$ the value of k at $\bar{m}$.

Figure 5 shows that inactivation of phosphatase could be
calculated reasonably well during drying of droplets at 120°C,
but that the short cut method yielded unsatisfactory results
at 80°C. This could be explained and some improvements could
be introduced. This is represented by the dotted line. For
details we refer to the literature.

In Section 2 it has been mentioned already that much
research work has been done with regard to chemical changes
which occur under various conditions during storage of packed
or unpacked foodstuffs. It is understandable that this problem
receives so much attention because slow changes during a very
long storage can influence the quality more strongly than fast
reactions during the short drying process.

There is also an influence on the quality because of
physical changes during drying. One may think in particular
of diffusion and vaporisation. Soluble components like sugar
and salt migrate by diffusion, whereas volatile components
such as flavours may be lost by vaporisation. Much is known
about loss of flavours, less regarding migration. Many physical
properties also change as a result of chemical conversions.
Clear examples are changes in colour and in texture. The latter
can be caused eg by changes in starch, in cell walls and in
proteins.

4. OPTIMISATION OF THE DRYING PROCESS

Optimisation of a dehydration process may be defined as
the selection from a number of alternatives of that process

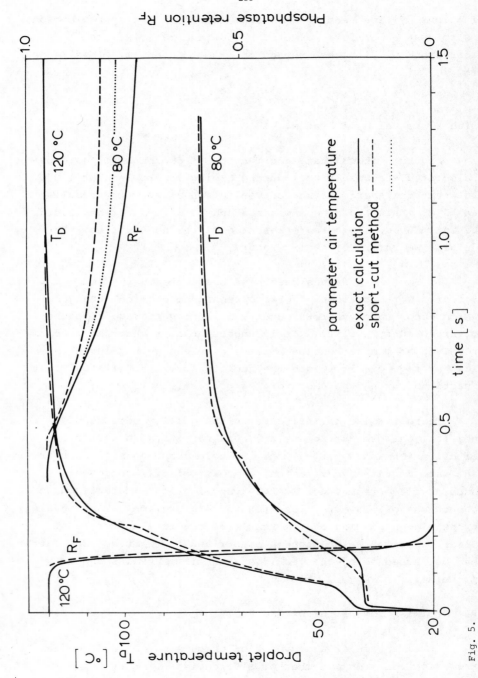

Fig. 5.

which yields the maximum rate of return on invested capital. Obviously there are many possibilities in between two extreme cases:

1. In the most simple case we consider the drying only and leave aside the rest of the flow sheet; furthermore we assume that the quantity and the quality of the product to be dried is fixed and the required final moisture content is known.

2. On the other hand, one can imagine that the only starting point is the type of product; quality of raw materials, flow sheet, formulation, scale of production, drying conditions, final moisture content, packaging, storage conditions etc are variables. Sometimes it is essential to consider the entire flow sheet eg when drying is preceded by another concentration process, such as evaporation. Then, in the optimisation procedure the concentration at the end of the evaporation and at the beginning of the drying step is of the utmost importance.

In the first case the optimisation procedure boils down to the following steps:

a) To study the influence of the drying conditions on the quality;

b) To relate the quality to the selling price;

c) To calculate the drying costs as a function of the drying conditions;

d) To calculate the profits and rates of return as a function of the drying conditions.

It may be remarked that drying conditions and types of driers are more or less interrelated, but if desirable these aspects may be considered as independent variables. We will show below that the optimisation procedure, even in this most simple case, meets with various obstacles.

With the more general case the optimisation seems to be hardly feasible; anyhow it would be a formidable task. It presupposes knowledge of the influence of several factors on the quality of the end product and of the relations between product quality, selling price, sales volume (scale of production) and the cost price. The problem is the more complicated because many factors are interrelated.

Still one has to keep in mind that in practice one meets a great variety of drying processes. Sometimes the quality of the raw materials strongly influences the result, in other cases this factor may be negligible. Some commodities are highly heat sensitive so that drying conditions play a dominant role in optimisation; the quality of other products hardly depends on the drying conditions. Often there is a very limited choice of drying conditions. If one wants to preserve the quality of sowing seeds and brewing barley one is committed to certain maximum product temperatures. In accomplishing desirable reactions when drying malt and cocoabeans it is also necessary to operate at narrow temperature and time ranges. In some cases the scale of operation has a significant influence on the cost price, but this certainly is not always true. The expected storage conditions and shelf life, the projected way of packaging, all these factors have to be taken into account.

And finally the cost of drying, when calculated per kg of final product varies tremendously, mainly because the amount of water to be evaporated per kg endproduct may differ by a factor of one hundred.

It follows that in each case one has to determine carefully which factors have to be taken into consideration and which may be ignored.

Now returning to the simple case, the first difficulty concerns the relation between the quality and the selling price. In the case of bulk products and semi-manufactured products, intended for further processing, as well as in the case of end-products for consumers, it is feasible to give an indication of the total quality by obtaining a weighted average of all quality factors involved. However as a rule this is far from simple.

With final products this can hardly ever be done in an objective way. Some properties can be evaluated only sensorially with the assistance of a consumer's panel and the importance of quality factors often is a matter of opinion. Nevertheless a quantification of the total quality concept is possible. Recently this has been discussed amply by the first author (Leniger, 1977). In many cases, in particular with bulk and semi-manufactured goods the quality concept is simpler. With the technological quality sometimes a few or even one quality factor dominates so strongly that other properties may be left out of consideration. It also occurs that a product has only to meet some or even one requirement.

As far as we know there is no general method to determine the correlation between quality and selling price. We assumed a constant quality of dried product and one can think of cases in which the quality of this product has a strong influence on the selling price, whereas on the other hand, one can imagine that there is no influence at all. Marketing experts have to find out in some way or another the price of a certain quality product.

The most interesting aspect of the optimisation process is the relation between quality and drying conditions. Assuming that quality deterioration during drying results from first order enzymatic and/or chemical reactions, one can derive from the Arrhenius equation that changes in quality will be the same if the following condition is satisfied:

$$\ln t = C + \frac{\Delta E}{RT}$$

with t = residence time, C = a constant of the product, ΔE = activation energy of the reactions, T = the <u>constant</u> process temperature. Consequently, when the product residence time in the dryer is plotted against the reciprocal of the absolute process temperature, straight lines with slope ΔE indicate conditions yielding equal thermal degradation for those values of ΔE. This is illustrated in Figure 6. Each line represents combinations of temperature and time yielding the same quality.

For several reasons the relation between quality and drying conditions is complex. Quality deterioration may be caused by reactions which cannot be described by simple equations, chemical and enzymatic reactions have different activation energies, and in a dryer one cannot speak of a certain process temperature. The product temperature changes from the entrance of the dryer to the outlet and so does the water activity. Therefore one has to find out an 'average' influence of the drying conditions on the quality. Nevertheless, in principle there are combinations of product temperatures, moisture contents and time which yield the same quality and other combinations resulting in a better or worse quality.

The situation is still more complicated because there are only limited possibilities to vary the drying conditions. The rate of drying strongly depends on the size of the product and in practice the choice of drying conditions may be very limited indeed. Sometimes it is possible to reduce the size of the drying material and in this way increase considerably the rate of drying. Taking into account that the activation energy of enzymatic and chemical reactions is higher than of physical processes, one arrives at the conclusion that conversions of an enzymatic and chemical nature during drying can be prevented to a greater extent as the drying process is carried out at lower temperatures and with shorter residence times. This means that one has to find out first of all which are the mildest possible combinations of air conditions and times. This has to be done experimentally. Next, more severe drying conditions, yielding inferior qualities have to applied. The drying costs have to be calculated for combinations yielding equal quality (situated on the same line in Figure 6) as well as for combinatio resulting in more quality degradation.

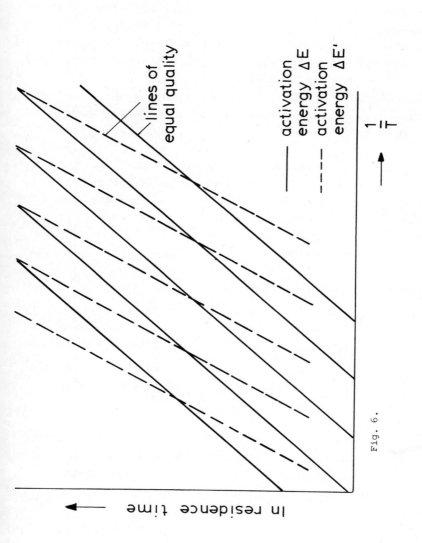

Fig. 6.

Finally, making use of the relation between quality and selling price, the profitability can be calculated.

The drying costs proper are, when starting with a certain material, determined by the final moisture content, the type of drier, the drying conditions and the scale of the process. In the above example we assumed the final moisture content and the scale of the process constant, but it is not difficult to incorporate these factors as variables in the optimisation procedure. As a matter of fact one has to consider various types of driers.

In order to simplify the calculation of the drying costs, one can restrict oneself to considering energy costs and capital costs. All other elements are relatively unimportant. Today and in the future the cost of energy is by far the most important; this factor is governed especially by the drying conditions and the final moisture content.

Therefore it is not difficult to predict that:

- More severe drying conditions are compatible with an increasing difference between initial and final moisture content and with decreasing heat sensitivity of the product;

- Milder conditions are obvious when products are more sensitive and when less water has to be evaporated.

In conclusion, it is a matter of course that the optimisation procedure will benefit considerably from improved methods for calculation of the course of the drying under various drying conditions and from more insight in the influence of the process on the product quality.

5. IMPORTANCE OF RESEARCH IN THE FIELD OF DEHYDRATION

With raw materials, semi-manufactured products as well as end-products, drying is a very important preservation method.

The influence of the drying process proper on the quality of the dried products is quite small; losses of heat labile components are minute, off-flavours and discolouration can be prevented to a large extent; however some loss of water binding properties can hardly be avoided. The quality depends on the drying conditions, a better quality corresponding with higher costs. It follows that an interesting compromise between quality and costs arises. Dried products have excellent keeping properties; under favourable conditions of water activity and temperature they can be stored practically indefinitely. Because of the low weight (and sometimes volume) transport and storage costs are low. The history of the product before drying, including the quality of the raw materials, is important and has to be taken into account.

In many cases drying is the most efficient and cheapest way of preservation. However, when a large volume of water has to be evaporated per kg dry product, the energy consumption is high. This is a serious disadvantage, forcing us to study energy saving measures or even to consider alternative methods of concentration and preservation. With many products there are hardly any alternatives.

Although drying is a very old operation, from the foregoing it can be concluded that our knowledge is very limited indeed. This is caused by the fact that drying is an extremely complicated process. This holds true for the physical as well as the chemical and biochemical aspects.

As far as the physical aspects are concerned, only recently there has been a break-through, made possible by the development of computing systems. Much progress has been made in the field of calculating temperature and moisture profiles in drying materials, drying rates and drying times. But much remains to be done. Theories and calculation procedures have to be developed for more complicated materials being dried under more complex conditions. This work has to be supported heavily by experimental work, in particular concerning physical properties.

More knowledge and insight is essential for a clear comprehension of the process and for the optimisation of the process with regard to quality and costs.

Knowledge of all kinds of changes which occur during dehydration and of the influence of temperature, water activity and other factors on these changes is very limited indeed. In particular, there is a strong need for more data regarding the relation between drying conditions and quality. It is true that such a relation can be determined experimentally, but more knowledge of the reaction mechanisms etc will be of great help in understanding the observed phenomena, in predicting the influence of all kinds of variations and in designing experimental work.

A lot of work has been carried out in regard to the storage of dried, packed and unpacked, products. In this field the information seems to be a little less inadequate, but many problems still have to be solved.

In view of the tremendous significance of dehydration of foodstuffs further research work in this field must be recommended.

REFERENCES

Kerkhof, P.J.A.M. 1975. Ph.D. -Thesis, Eindhoven University of Technology.

Kerkhof, P.J.A.M. and Schoeber, W.J.A.H. 1974. In: 'Advances in preconcentration and drying of foods', Ed. A. Spicer, Appl. Sci. Publ. (London).

Kerkhof, P.J.A.M. 1977. Paper presented at EFChE-symposium, Sweden, September 7 - 9, 1977 (in press).

Leniger, H.A. 1977. Proceedings of the 7th European Symposium Food, Eindhoven September 1977, pag. 146-162.

Lijn, J.V.D. 1976. Ph.D. -Thesis, Agricultural University, Wageningen.

Lijn, J.V.D., Rulkens, W.H. and Kerkhof, P.J.A.M. 1972. Int. Symp. on Heat and Mass Transfer Problems in Food Engineering, Wageningen.

Lijn, J.V.D. 1976. Chem. Engng. Sci., $\underline{31}$, 929.

Rulkens, W.H. and Thijssen, H.A.C. 1972. J. Food Technol. $\underline{7}$, 95.

Rulkens, W.H. 1973. Ph.D.-Thesis, Eindhoven University of Technology.

Schoeber, W.J.A.H. 1976. Ph.D.;Thesis, Eindhoven University of Technology.

Schoeber, W.J.A.H. and Thijssen, H.A.C. 1975. 69th Ann. Meeting of A.I. Ch. E., Los Angeles. November 16 - 20.

Thijssen, H.A.C. and Rulkens, W.H. 1968. De Ingenieur $\underline{80}$, Ch. 45.

QUALITY AND NUTRITIONAL ASPECTS OF FOOD DEHYDRATION

F. Escher* and B. Blanc**

*Department of Food Science, Swiss Federal Institute of
Technology, Universitätstr. 2, CH-8092 Zurich,
**Swiss Federal Dairy Research Institute,
CH-3097 Liebefeld-Bern, Switzerland.

ABSTRACT

Food dehydration comprises a series of unit operations (eg blanching, drying, packaging, etc), all of which contribute to the final quality of the reconstituted food. In judging quality of dehydrated foods, priority has to be given to wholesomeness and nutritive value, although subjective quality factors (colour, texture, flavour) should be considered as much as possible. In this paper, the general relationship between dehydration and quality changes is discussed, and effects of dehydration on nutritional quality of plant foods and dairy products are summarised. The nutritive value of most products is more affected by pretreatment of raw material and storage of dried products than by the drying operation itself. Research efforts in the area of quality aspects of dehydration should be concentrated on theoretical description of relations between dehydration process and quality, on procuring a valid data base for optimising processes in point of view of quality, energy requirements and costs, and on evaluating the relative importance of dehydration among other food preservation methods.

INTRODUCTION

Removal of water by dehydration is one of the oldest and most effective ways of preservation of agricultural raw materials, of ingredients for further processing and of consumer products. The preservation principle, which dehydration has in common with many other processing methods (eg salting, sugaring, freezing), is the lowering of water activity, ie the reduction of availability of water for microbial growth and chemical and biochemical reactions. According to the extent of moisture removal, distinction is usually made between concentration (partial water removal from liquid and semi-liquid foods) and dehydration (near full moisture removal).

In dealing with quality aspects of food dehydration, it would be advisable not to consider dehydration as an isolated unit operation. The production of dried foods normally comprises a series of unit operations (Figure 1), all of which are interdependent in their contribution to the final quality of the consumed product. Quite frequently, cases are encountered in which the decisive losses of overall or nutritive quality is not induced by the drying operation itself, but by either unfavourable design of the preparatory steps or by improper storage and reconstitution conditions. Furthermore, final product quality is influenced by the selection of suitable raw material, as was discussed in the Commodity Session of this Seminar.

Two examples may illustrate these points (1) The loss of ascorbic acid in producing drum-dried apple sauce is mainly due to oxidation of ascorbic acid during the cooking process and to a far lesser extent due to destruction during the drying operation (Figure 2). (2) The retention of volatile flavour components in spray drying liquid foods greatly increases with increasing total solids in the feed, which shows that the preparatory step (preconcentration, addition of solids) has a significant influence on the final product quality (Figure 3).

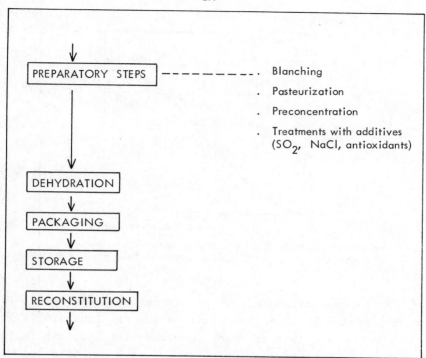

Fig. 1. Food dehydration process.

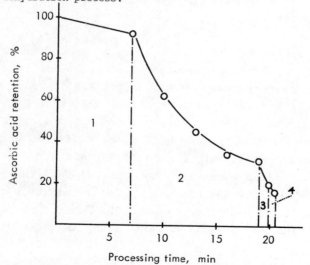

Fig. 2. Ascorbic acid retention during the different steps in manufacturing drum-dried apple flakes (Escher and Neukom, 1969).
1. Cutting.   2. Blanching.   3. Pulping.   4. Drum-drying

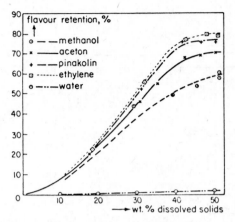

Fig. 3. Retention of various volatile compounds during spray drying of coffee extract vs. initial dissolved-solids content (Thijssen and Rulkens, 1968).

If the quality aspects of the various unit operations in the dehydration processes are still discussed separately, it is because of the lack of exact knowledge relating the influence of each step into the final product quality, and also because data on quality changes in some steps are still scarce to-date (Labuza, 1975).

The present state of technology of food dehydration has been the subject of Professor Leniger's contribution in this seminar and it was also reviewed by Holdsworth (1971), King (1971), Karel (1973) and van Arsdel et al. (1973a, b). Specific reference of the effect of moisture removal and storage at low moisture on the nutritional quality of foods was extensively made by Duckworth (1975), Labuza (1972) and Bluestein and Labuza (1975). This paper summarises the general laws that govern quality and nutritional changes during dehydration and during storage of low moisture food products. Nutritional aspects of drying and storing plant foods and dairy products will be given priority.

DESCRIBING OPTIMAL QUALITY OF DEHYDRATED FOODS

For any successful approach to analyse relations between a processing method and food quality, it will be important to define those quality attributes that are most important to the respective process. Table 1 shows a series of possible quality descriptors for dehydrated foods, whereby it is assumed that overall quality can be described, (1) by the characteristics of the dry product, (2) by the characteristics of reconstitution, and (3) by the characteristics of the reconstituted product.

TABLE 1
QUALITY CHARACTERISTICS OF DEHYDRATED FOODS

Stage	Descriptor
Dry, dehydrated	Bulk density
	Susceptibility to mechanical damage
	Hygroscopicity
	Caking
At reconstitution	Wettability
	Sinkability
	Rehydration time
	Rehydration ratio
After reconstitution	Wholesomeness (eg microbiological quality)
	Nutritive value
	Colour and appearance
	Flavour

Depending on the type and use of the product, some of the quality factors will be more or less important than others. When producing dry foods for military supply or space programmes, low bulk density may be a highly appreciable quality parameter. This appreciation led to the development of densifying processes for manufacturing compressed vegetables (Rahman et al., 1969) and densified drum-dried products (Eskew and Drasga, 1962; Lazar and Hart, 1968). Wettability and sinkability are parameters certainly to be observed for dried instant liquid foods (coffee, milk, fruit juices). Modern agglomeration techniques (Jensen, 1975) try to match these requirements.

Yet most weight has to be placed on the third group of quality factors, ie on the quality of the final reconstituted product, and within this group, on <u>wholesomeness</u> and <u>nutritive value</u>. This does by no means say that colour, texture and flavour should be neglected, since, as Karel (1973) points out, 'food is more than just a safe packet of nutrients and thus food processing is aimed at the often difficult and elusive goal of maximising eating pleasure'.

Frequently, it will not be possible to reach a <u>maximal</u> value for all quality attributes. Quality factors can be inversely correlated. For example, rapid air drying of vegetable pieces yields a porous, low density product for which rehydration time is generally low. In contrast, slow air drying results in a high-density and thus much less voluminous product, but with considerably higher reconstitution time (van Arsdel et al., 1973a). Fortification of instant mashed potato flakes with iron might be highly desirable in point of view of nutritional standards for a specific consumer group. However, experiments showed that an addition of iron considerably impairs the storage stability due to faster oxidative changes of lipids (Sapers et al., 1974).

Maximal overall quality may also have to be disregarded due to economical reasons. This explains why freeze-drying will never exclude other, more economical drying methods in spite of the superiority of quality of some freeze-dried products. Energy consideration is yet another point which becomes more and more important when evaluating the relation between a drying process and the quality of the dried food (Flink, 1977). Thus, it is the goal for any evaluation of a given food process not to maximise, but to <u>optimise</u> for different quality factors, energy consumption, etc (Paulus, 1975a, b).

GENERAL RELATIONSHIP BETWEEN QUALITY CHANGES AND DRYING AND STORAGE CONDITIONS

In Table 2, possible changes occurring in foods during dehydration and storage and pertinent literature covering these changes are listed.

TABLE 2

CHANGES OCCURRING IN FOOD PRODUCTS DURING DEHYDRATION AND LOW-MOISTURE STORAGE

Changes	References	
Shrinkage of cellular material	Görling	(1958)
Loss of rehydration ability	Duckworth and Smith	(1962)
Migration of solids	van Arsdel et al.	(1973a)
Case hardening	LaBelle	(1966)
Loss of volatile constituents	Bomben et al.	(1973)
Non-enzymatic reations	Reynolds	(1963, 1965)
	Eichner	(1975)
(browning, oxidation)	Karel	(1975)
	Labuza	(1975)
Enzymatic reations	Acker	(1969)
Microbiological changes	Kempelmacher et al.	(1969)
Water sorption and desorption	Duckworth	(1975)

<u>Shrinkage</u> of cellular material, loss of <u>rehydration ability</u>, <u>migration of solids</u> and <u>case hardening</u> are phenomena having a decisive influence on texture of fruits, vegetables and potato pieces. Losses of volatile constituents and possibilities of increasing their retention were subject to numerous studies throughout the last years, predominantly with freeze-drying and spray-drying. Theories were put forth on <u>aroma retention</u>, some of them on a physical basis using mass-transfer laws (Bomben et al., 1973); some on the assumption that chemical association between volatile flavour and food matrix will also lead to an increased flavour retention (Solms et al., 1973).

Microbiological aspects are of great importance in so far that during most drying methods part of the micro-organisms which are present in the wet food, may survive and upon rehydration will be active again. Foods that were first pasteurised can be recontaminated during drying at improper conditions. That dehydration alone does not always provide enough microbiological safety, was clearly shown in an out-break of bacterial food infection some years ago, implicating drum-dried baby food as the infection source. The cause could be traced to the use of a contaminated yeast product that was added to the formulation prior to dehydration (Anon, 1975). On the other hand, it must be mentioned that selection of suitable drying conditions may still considerably improve microbiological standards. In a recent study on manufacturing Durum-wheat spaghetti, *Staphylococci* and *E. coli* could be excluded from the dry product by simply adjusting drying temperature and drying curve (Manser, 1977).

Undesirable enzymatic changes may occur during both drying and storage, provided that no blanching or scalding process precedes dehydration. Again, drying usually does not completely inactivate enzymes. After drum-drying wheat flour slurries without thermal pretreatment, more than 50% of the initial lipoxygenase activity was still present in the dry flakes (Schweizer, 1975). Notable exceptions, in which enzyme action during drying is desirable, are onions, garlic and other seasoning materials. Apparently, products of enzyme catalysed reaction yield important flavour compounds (van Arsdel et al., 1973a).

Non-enzymatic reactions (browning, vitamin degradation, lipid oxidation) and water sorption phenomena are especially important factors in impairing nutritional quality of foods and will thus be discussed more extensively. Many of these deteriorative reactions are of the first order (monomolecular) or second order (bimolecular) type according to the equations summarised in Table 3. For both reaction types, the rate constant is a function of temperature as controlled by the apparent energy of activation. Temperature dependence of the rate constant can be derived from the Arrhenius equation (Table

TABLE 3

DESCRIPTION OF FIRST AND SECOND ORDER CHEMICAL REACTIONS

(1)	First order chemical reaction $$A \xrightarrow{k} P$$ $$\frac{d(A)}{dt} = k(A)$$ where A = compound which reacts of form compound P
	           (A) = concentration of compound A
	           t = time
	           k = first order reaction rate constant

(2)	Second order chemical reaction $$A + B \xrightarrow{k} P$$ $$\frac{d(P)}{dt} = k(A)(B)$$ where A, B = compounds which react of form compound P
	           (A), (B), (P) = concentrations of compounds A, B, C.
	           t = time
	           k = second order reaction rate constant

(3)	Arrhenius equation $$\frac{d(\ln k)}{dt} = \frac{E_A}{RT^2}$$ where k = reaction rate constant
	           T = absolute temperature
	           E_A = activation energy
	           R = gas constant

In Table 4, typical apparent energies of activation for various reactions are mentioned. From the table it can also be seen that the apparent activation energy of moisture removal is in the range of 20 to 50 kJ/mole, which is close to the latent heat of vaporisation of water (King, 1971). Using these values, it is possible to compare relative temperature dependence of drying rate (or drying velocity) with that of nutritional losses. For example, an increase in temperature from 60°C to 80°C will increase the theoretical rate of moisture removal by a factor of 2, while the rate of non-enzymatic browning (activation energy around 125 kJ/mole) increases already by a factor of 10 to 15.

TABLE 4

SOME EXAMPLES OF APPARENT ENERGIES OF ACTIVATIONS (Labuza, 1972)

	kJ/mole
Enzyme reactions	40 - 65
Hydrolysis	65
Lipid oxidation	40 - 105
Non-enzymatic browning	105 - 210
Protein denaturation	335 - 500

With such comparisons, it is theoretically possible to optimise drying rate against rate of any undesirable chemical change. Also, the extent of change could be predicted for a given <u>residence time</u> in the dryer, or, vice-versa, the drying time for not exceeding a certain limit of undesired reaction. This procedure is widely used to evaluate sterilisation processes and blanching operations (Nehring, 1973). However, for dehydration the procedure cannot be applied directly, due to the fact that the extent of a reaction is also controlled by the <u>moisture content</u> and thus by the <u>water activity</u> of a food, the relation between moisture content and water activity being given by the sorption isotherm. Unlike sterilisation of foods in closed systems, the moisture content in dehydration is continuously changing along the drying curve. There are various explanations for the influence of water activity on chemical reactions in a food system, among them (1) increasing mobility of reaction partners and catalysts with increasing water activity (2) increasing energy of activation with decreasing water activity (3) changing concentration of water soluble reaction partners.

Therefore, for applying any system similar to the evaluation of heat processing, the following parameters must be known:
(1) time-temperature-moisture history of the product during

drying; (2) rate constant of reactions at any given temperature and moisture content (water activity). Hendel et al. (1955) and Kluge and Heiss (1967) used this approach to predict the extent of non-enzymatic browning in drying potato pieces and glycine-glucose-cellulose food models, respectively. Similar analyses on browning reactions during drum-drying of apple puree were carried out graphically (Escher and Neukom, 1970). In Figure 4, the extent of non-enzymatic browning (expressed as optical density of a methanol-water extract from apple flakes) at increasing steam temperatures in the heated drum was plotted against the residence time of the product on the drum dryer. Also plotted were lines of equal moisture levels. Depending on the desired moisture level for storage stability and on the tolerated extent of browning, the optimal combinations of steam temperature and drying time (ie product flow rate) can be selected.

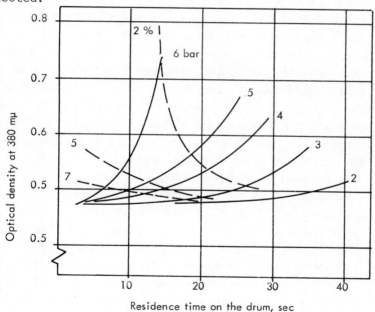

Fig. 4. Extent of non-enzymatic browning at different steam pressures as a function of residence time during drum drying of apple sauce (Escher and Neukom, 1970).
——— Optical density at 380 mμ.(Extent of non-enzymatic browning) at steam pressures of 2, 3, 4, 5 and 6 bar.
— — Lines of equal moisture content in the product (% on moist basis).

For evaluating storage stability after dehydration, **oxygen partial pressure**, and, in special cases, light intensity have to be added as controlling factors for undesired chemical reactions. Oxygen partial pressure and water activity are usually not constant throughout the storage period, as most packaging materials have a certain oxygen and water vapour permeability. Many investigations were made on predicting storage stability and loss of nutritive value during low-moisture storage by combining the reaction controlling factors mentioned above (Heiss and Eichner, 1971; Quast and Karel, 1972). Such prediction would be economically interesting, as storage tests are usually very time consuming.

Extensive studies were made on the relation between sorption behaviour, as described by the sorption isotherm, and storage stability, leading to the general concept in Figure 5. Microbiological action ceases at water activities of lower than approx. 0.7. Enzymatic reactions are usually observed only at water activities which give at least a monomolecular cover of the porous food system, although cases are known today where enzymatic lipid oxidations occur even at lower water activity (Acker and Wiese, 1975). Non-enzymatic browning reactions have a pronounced maximum at water activities around 0.7, due to the competitive influence of concentration of reaction partners and of energy of activation with changing water activity. Lipid oxidations are minimal at the monolayer point of the sorption isotherm. This minimum is again caused by a balance between catalyst hydration, mobility of catalysts, hydration of intermediate reaction products and free radical quenching (Karel, 1975)

Relationships between water activity and storage stability have become especially important in the development of the so-called intermediate moisture foods, ie of foods with a moisture content exhibiting water activities around 0.6 to 0.8 (Davies et al., 1976; Gee et al., 1977).

From the above explained derivations one might be tempted to assume that generally losses during drying and storage are well predictable. Unfortunately, all these rules have limited use in practice. By no means all reactions leading to losses in nutritive value are known to an extent that justify the assumption of first or second order behaviour. Too few data are available on rate constants of reactions at any given temperature moisture relation, not to talk of all other effects that might influence the reaction (acidity, presence of heavy metals, etc). To illustrate this point, the investigations by Speck (1976) are mentioned on carotenoid stability during storage of air-dried carrots which were pretreated in a 10% sodium chloride solution. Carotenoid stability was maximal not at moisture contents around the monolayer point of the sorption isotherm, as would have been expected, but at moisture contents considerably lower than this point. Similar results deviating from the general concept relating sorption behaviour and oxidative changes of carotenoids, were obtained by Farine et al. (1962), with conventionally air-dried carrots. One reason for these deviations might be the fact that most experiments relating sorption behaviour and stability of carotenoids were carried out with freeze-dried products which differ from air-dried products in structure and porosity (Speck, 1976).

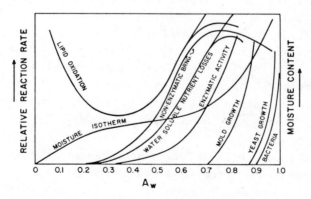

Fig. 5. General relationship between reaction rates of various changes in foods and water activity (Bluestein and Labuza, 1975).

The addition of other quality aspects (flavour retention, rehydration ability, etc) to the optimisation procedure will make theoretical predictions even more difficult. If theoretical relations were presented in this seminar in spite of the insufficient data basis, it is just because this field would offer possibilities for successful and meaningful co-operative research effort within the COST countries.

NUTRITIONAL LOSSES IN DRYING AND STORING PLANT FOODS AND DAIRY PRODUCTS

Loss of biological value of proteins

The loss of biological value as a consequence of the degradation of protein during dehydration is gaining importance due to the increasing interest in application of texturised vegetable proteins as food ingredients. A substantial part of texturised proteins is fabricated by extrusion cooking followed by various methods of drying. Some results on nutritional evaluations of texturised products have been published (Mustakas et al., 1964; Beetner et al., 1974). Yet more studies on combined extrusion-drying processes will be needed as use of extruded products increases.

The changes in biological value of proteins are caused by various mechanisms: (1) direct loss of protein biological value by heating and drying: (2) tying up amino acid side chains through various reactions, making them biologically unavailable. Quite a contrary effect, namely an increase in biological value may be caused by destruction of antinutritional substances such as the trypsin inhibition factor. Much work on loss of biological value has been done on milk proteins and will therefore be discussed in connection with dairy products.

Industrially, milk and milk products are mostly spray dried or drum-dried. The temperature-residence time relation is estimated to be 0.5 - 1 sec at $80^{\circ}C$ for spray-drying, and 2 - 3 sec at 100 - $130^{\circ}C$ for drum-drying, respectively. Depending on the mode of drying, the extent of milk protein denaturation varies considerably, as can be shown by immuno electrophoresis techniques (Figure 6). The number of precipitation arcs corresponding to the different whey proteins decreases steadily with increasing heat treatments due to progressive loss of native protein structure. Complete denaturation leads to the disappearance of the precipitation arc (Mauron and Blanc, 1965).

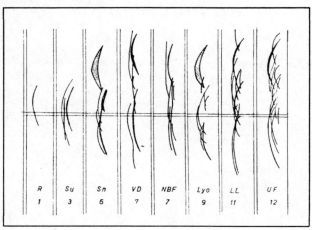

Fig. 6. Immunoelectrophoresis of milk powders. The lactoserum prepared by ultracentrifugation and concentrated by ultrafiltration to a protein content of 5%, was used throughout. Anti-serum: prepared from bovine lactoserum.

```
R    = scorched roller dried milk
Su   = standard spray-dried milk
Sn   = low temperature spray-dried milk
VD   = vacuum-dried milk (60°C)
NBF  = low temperature spray-dried 'humanised' milk
Lyo  = milk, freeze-dried on an industrial scale
LL   = milk, freeze-dried in the laboratory
UF   = fresh, fluid milk
1,3,6, ........number of precipitation arcs
```

(Mauron and Blanc, 1965).

Dehydration of milk has no influence on amino acid composition except for total lysine and available lysine content. Mauron et al. (1955) found a 3.6% and 30 - 70% decrease in spray-dried and drum-dried milk powders, respectively. Erbersdobler and Zucker (1966) showed that the availability of lysine in drum-dried skim milk powder varied beteen 75 and 95%. According to van der Bruel et al. (1971) losses in total and available lysine are not only depending on the drying method, but also on the conditions of preheating (Table 5). Using a 'whey protein nitrogen' index (WPN), which is an estimation of the total amount of soluble serum proteins, van der Bruel et al. (1971) classified spray-dried skim milk powder in 3 groups: low-heat (WPN > 6.0 mg/g), medium-heat (WPN between 1.51. and 5.99 mg/g) and high-heat (WPN < 1.5 mg/g).

TABLE 5

CONTENT OF LYSINE AND AVAILABLE LYSINE OF SKIM MILK POWDERS OF VARIOUS TECHNOLOGICAL STANDARDS (van der Bruel et al., 1971)

Product	Total lysine content mg/g N	Available lysine content mg/g N	Whey protein nitrogen index mg/g N
Skim milk	500	480	
Spray-dried skim milk			
preheated at 72°C / 15 sec	504	500	6.4
95°C / 30 sec	516	469	3.8
75°C / 30 min	493	422	1.65
95°C / 30 min	517	491	0.7
Drum-dried skim milk			
normal appearance	495	414	
slightly scorched	388	308	

Most milk powder for human consumption is instantised in combination with the spray-drying process. Potasi et al. (1974) reported only a 0 - 4% reduction in total lysine content during production of instantised spray-dried skim milk. Therefore, by using modern drying techniques, lysine retention can be expected to be over 90%.

No differences were found in protein efficiency ratios (PER) of spray-dried and drum-dried milk powders, unless drum drying was carried out at a reduced drum speed (Mauron and Mottu, 1958). Erbersdobler and Zucker (1966) found PER values of 3.54 and 2.63 - 3.44 for spray-dried and drum-dried milk powder, respectively. Van der Bruel et al. (1971) obtained PER values at 2.95 - 3.11, 2.62 and 1.67 for spray-dried, drum-dried (normal quality) and drum-dried (slightly scorched) products respectively.

Unfavourable selection of drying conditions may lead to undesirable overheating of the products and to a pronounced reduction in available lysine content. In this case, a close relation between available lysine content and PER value could be established (Mauron and Mottu, 1958; van der Bruel, 1971). Also, furosine can be detected in overheated milk powders. Furosine is a derivative of fructose-lysine which is formed by non-enzymatic browning reactions involving lysine and lactose. Fructose-lysine is not resorbed in the unhydrolysed stage, and thus nutritionally of no value (Erbersdobler, 1970). Losses of available lysine are especailly detrimental when milk protein is added to improve biological value of proteins containing insufficient lysine concentrations, such as cereals.

Storage of dried milk at low temperature and low relative humidity affects biological value and protein utilisation only after several months and only to a small extent (Richter and Schiller, 1958; Bock and Wünsche, 1965). Storage for 13 months at room temperature lowered total lysine content from 7.96 to 7.02 g/16 g N, and methionine content from 2.56 to 2.16 g/16 g N. At the same time net protein utilisation (NPU) diminished from 74.5 to 70.3%, biological value from 81.1 to 78.5%, and true digestibility from 90.2 to 89.2% (Eggum et al., 1970). Storage at increased temperature and relative humidity will lead to much higher losses in lysine content, regardless whether spray-dried or drum-dried milk powders are analysed. Huss (1974) found in a 14 months storage test that the relative loss of lysine is strongly temperature dependent: 0, 30 and 58% loss at 4, 10 and 30°C, respectively. Tests at constant temperature, but varying relative humidities indicated that availability of lysine is also depending on whether lactose is present in its amorphous or crystalline form (Huss, 1970). High temperature and relative humidity at storage again cause the formation of furosine.

Vitamin losses

Among the water soluble vitamins, <u>ascorbic acid</u> is one of the most readily destroyed and its retention was often used as an index of the severity of drying and storage conditions. Degradation rate is not only influenced by water activity and temperature, but also by the presence of oxidising enzymes, of heavy metals such as copper and iron, by light and by dissolved oxygen. As these factors are not very well controlled, the losses of ascorbic acid vary widely from product to product. Reference is made again to the introductory example in Figure 2 from which it became clear that losses during preparatory steps may match or even exceed losses during dehydration. The same relations hold true between dehydration and final cooking during rehydration, as Mapson (1956) demonstrated with air-dried and freeze-dried peas (Table 6). In discussing stability of ascorbic acid, difficulties with analytical procedures have also to be mentioned. Jadlev et al. (1975) found an apparent increase of ascorbic acid content during the add-back drying of mashed potatoes which was attributed mostly to traces of sugar-amino acid reaction products which analytically behave like ascorbic acid.

TABLE 6

RETENTION OF ASCORBIC ACID AT DIFFERENT STAGES OF PRODUCING AND USING AIR-DRIED AND FREEZE-DRIED PEAS (Mapson, 1956).

	Percent of initial concentration	
	Air-dried	Freeze-dried
Blanching	75	75
Drying	45	70
Cooking	25	35

Not many data are available on actual losses of ascorbic acid during dehydration. Harris and von Loesecke (1960) reported losses between 16 and 25% during air drying of vegetable pieces, Escher and Neukom (1969) losses of 5% and 25% for drum-drying apple sauce and starch-ascorbic acid model suspension, respectively. Similar values were obtained by Bose and Majunder (1950) for drum-drying date juice. For low-temperature vacuum drying of tomato-juice concentrate, Kaufmann et al. (1955) observed no loss of ascorbic acid.

More data have been collected on ascorbic acid destruction during storage of dried and semi-dried foods. Unfortunately, data have to be handled again with a certain precaution due to unreliable analytical methods, or due to the lack of reporting the environment in which the destruction rates have been measured (pH, presence of enzymes and heavy metals, etc).

The effect of temperature and humidity during storage was demonstrated in studies carried out by Vojnovich and Pfeiffer (1970) on fortified wheat flour and fortified drum-dried baby-food, and by Karel and Nickerson (1969) on orange juice powder. The data which demonstrate a high dependence of destruction rate to water activity, are summarised in Table 7. It should be noted that wheat flour underwent no pretreatment inactivating the enzymes naturally present, whereas the other two products did most probably not contain any active enzymes. Therefore, the mechanism of ascorbic acid destruction in the three products is not identical.

TABLE 7

ASCORBIC ACID HALFLIFE (MONTHS) IN WHEAT FLOUR, DRUM-DRIED BABY-FOODS AND ORANGE JUICE POWDER, STORED AT DIFFERENT TEMPERATURE AND MOISTURE CONTENTS (Labuza, 1972)

Wheat flour [1]	14.6% H_2O	13.7% H_2O	12.9% H_2O
45°C	1.2	3.2	10.6
37°C	12.0	17.3	39.1
26°C	50.4	78.2	156.0
Drum-dried baby-foods [2]	10.7% H_2O	7.0% H_2O	5.0% H_2O
45°C	0.41	0.47	0.79
37°C	0.52	0.86	1.47
26°C	0.68	1.56	3.39
Orange juice powder [3]	18.0% H_2O	7.4% H_2O	0.7% H_2O
37°C	13.0	1.87	1.0

[1], [2] Vojnovich and Pfeiffer (1970)
[3] Karel and Nickerson (1969)

The values from Table 7 and some additional experimental results were plotted by Bluestein and Labuza (1975) as shown in Figure 7. Destruction rate increases by a magnitude of 3 over the whole range of water activity. Furthermore, the extent of temperature influence on the reaction rate is increasing with increasing water activity. This indicates that during dehydration high temperatures at high humidities have to be avoided, by one of the following means: rapidly passing the first stage of dehydration; keeping initial temperature low; starting the drying at lower initial moisture content; lowering product viscosity through addition of a suitable solid.

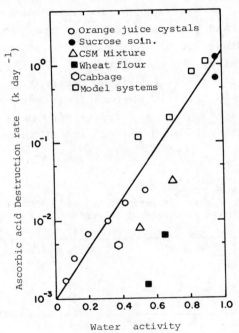

Fig. 7. Ascorbic acid destruction rate as a function of water activity (Bluestein and Labuza, 1975).

With the exception of <u>thiamine</u>, the destruction of <u>B vitamin</u> during dehydration was far less investigated than that of ascorbic acid. Data on storage stability in dry foods are almost lacking to-date. Average thiamine losses of 5 to 25% during dehydration of blanched vegetables were reported by Hein

and Hutchings (1971). Bluestein and Labuza (1975) concluded
that losses of thiamine and other B vitamins are generally not
exceeding 10% of the amount present in the blanched product.
Recently, Dennison et al. (1977) evaluated storage stability of
thiamine and <u>riboflavin</u> in low moisture food model systems and
confirmed this conclusion for water activities not exceeding
0.65 and temperatures not higher than 37°C (thiamine) and 30°C
(riboflavin). At higher temperatures and elevated water
activities, losses of both vitamins are substantial. Riboflavin
stability is also controlled by the amount of oxygen present in
the package. It is interesting to note that riboflavin stability
in real foods (breakfast cereals) was found to be better than
in model systems, indicating some interaction between the
vitamin and other food components (Dennison et al., 1977). This
observation shows again the difficulties in transferring results
from model systems to complex foods.

Changes in <u>carotenoid</u> content attained attention in
research due to the importance of these compounds both as
provitamins and colouring principles. Losses of these fat-soluble
compounds are caused by interaction with free radicals which are
formed by lipid oxidation. Prevention of lipid oxidation will
at least partially prevent degradation of carotenoids. Therefore, special precautions during storage have to be taken such
as addition of antioxidants, application of coating material
to exclude oxygen from the porous food material, packaging at
oxygen partial pressure.

In a recent study on air-drying vegetables, Speck et al.
(1977) could show that a pretreatment of carrots and potatoes in
10% sodium chloride solution and subsequent air-drying at 60°C
yields products with careteniod stability superior to that of
sulfited air-dried and freeze-dried products without sodium
chloride treatment. The mechanisms of increased carotenoid
retention is not quite understood. Yet it seems to be possible
to improve quality of dehydrated vegetables by fairly simple
and thus inexpensive methods.

Data on <u>vitamin A and D</u> losses are only available for milk. Apparently, losses during spray-drying and drum-drying are minimal (Hartman and Dryden, 1965). No studies are known on the effect of dehydration to <u>vitamin E</u>.

CONCLUSION

As dehydration will probably keep its important place among the various methods of food preservation, research and development efforts in this area should be given due consideration also in the future. However, research funds are becoming increasingly limited, and therefore careful selection of meaningful projects may be more necessary than in the past. In concluding this presentation some areas are mentioned in which a co-ordinated research effort within the COST countries seems to be justified.

There is definite need to shift research on food quality from dehydration as a unit operation to dehydration as a unit process, comprising all steps from selection and storage of raw material to reconstitution of the dry product. Although difficult to achieve, work on optimisation and on theoretical relations between process and quality should be continued. The enlargement of a valid data base will help to solve some of the optimisation problems. Such a data base has to comprise physical properties of foodstuffs which will be collected and revised in another COST project on food technology. Data on chemical changes during dehydration present another important group. For some of the data in this group, critical reviews of analytical procedures will be substantial, as has been shown in connection with changes in vitamin C content during dehydration processes.

Dehydration as a unit operation does not impair nutritional quality to a great extent. Again, losses are apparently more detrimental in the overall dehydration process, including raw material handling, pretreatment and reconstitution by the consumer. Thus, it would be appropriate to study possibilities of improving inexpensive methods of eg pretreatment and storage, rather than to try to make marginal progress in redesigning drying operations.

Finally, with increasing concern on energy consumption within the food industry, food quality research will have to compare dehydration with other preservation processes in point of view of energy requirement, convenience, food quality and costs. Based on such comparisons the relative importance of dehydration and thus of research and development in the field of food dehydration might well change in the future.

REFERENCES

Acker, L.W. 1969. Food Technol. 23 (10), 1257-1270.

Acker, L.W. and Wiese, R. 1972. Z. Lenensm. Unters. u. Forsch. 150, 205-211.

Anonymous 1975. Morbidity and Mortality Weekly Rep. 24, 65 (CDC,Atlanta, Ga.).

Beetner, G., Tsao, T., Frey, A. and Harper, J. 1974. J. Food Sci. 39, 207-208.

Bluestein, P.M. and Labuza, T.P. 1975. In: 'Nutritional Evaluation of Food Processing'(Harris, R.S. and Karmas, E., eds.). pp 289-323, AVI Publ. Co., Westport, Conn.

Bock, H.D. and Wünsche, J. 1965. Nahrung 9, 131-135.

Bomben, J.L., Bruin, S., Thijssen, H.A.C. and Merson, R.L. 1976. Adv. Food Res. 20, 1-111.

Bose, A.N. and Majunder, A.C. 1950. Food Technol. 4 (2), 54.

Davies,R., Birth, G.G. and Parker, K.J. (eds.) 1976. Intermediate Moisture Foods, Applied Science Publ. Ltd. London.

Dennison, D., Kirk, J. Bach, J. Kokoczka, P. and Heldman, D. 1977. J. Food. Proc. Preserv. 1, 43-54.

Duckworth, R.B. (ed.) 1975. Water Relations of Foods, Academic Press, New York, N.Y.

Duckworth, R.B. and Smith, G.M. 1963. Recent Adv. Food Sci. 3, 230.

Erbersdobler, H. 1970. Milchwiss. 25, 280-284.

Erbersdobler, H. and Zucker, H. 1966. Milchwiss. 21, 564-568.

Eggum, B.O., Nielsen, H.E. and Rasmussen, L.F. 1970. Z. Tierphysiol. Tierernähr. Futtermittelk. 27, 18-23.

Eichner, K. 1975. In: 'Water Relations to Foods' (Duckworth, R.B., ed), pp. 417-434. Academic Press, New York, N.Y.

Escher, F. and Neukom, H. 1969. Trav. Chim. Aliment. Hyg. 61 (5/6), 339-348.

Escher, F. and Neukom, H. 1970. Lebensm. -Wiss. u. -Technol. 4 (5), 145-151.

Eskew, R.K. and Drasga, F.A. 1962. Food Technol. 16 (4), 99-101.

Farine, G., Wuhrmann, J.J. Patron, A. and Vuataz, L. 1962. In: 'Proc. 1st. Intern. Congr. Food Sci.& Technol.' Vol. 3, pp. 603-613.

Flink, J.M. 1977. Food Technol. 31 (3), 77-84.

Gee, M., Farkas, D. and Rahman, A.R. 1977. Food Technol. 31 (4), 58-64.

Görling, P. 1958. In: 'Fundamental Aspects of the Dehydration of Foodstuffs, pp. 42-53, Society of Chemical Industry, London.

Harris, R.S. and von Loesecke, H. 1960. Nutritional Evaluation of Food Processing, John Wiley and Sons, New York, N.Y.

Hartman, A.M. and Dryden, L.P. 1965. Vitamins in Milk and Milk Products ADSA, US Dept. Agri., Washington, DC.

Hein, R.E. and Hutchings, I.J. 1971. Cited in Bluestein, P.M. and Labuza, T.P. 1975. p. 321.

Heiss, R. and Eichner, K. 1971. Food Manuf. 46 (5), 53-56, 65, (6) 37-38, 41-42.

Hendel, C.E., Silveira, V. and Harrington, W.O. 1955. Food Technol. 9, 433-438.

Holdsworth, S.D. 1971. J. Food Technol. 6, 331-370.

Huss, W. 1970. Landw. Forsch, 23, 275-289.

Huss, W. 1974. Landw. Forsch, 27, 199-210.

Jadlav, S., Steele, L. and Hadziyev 1975. Lebensm. -Wiss. u. -Technol. 8 (5), 225-230.

Jensen, J.D. 1975. Food Technol. 29 (6), 60-71.

Kampelmacher, E.H., Ingram, M. and Mossel, D.A.A. (eds.) 1969. Proc. 6th Intern. Symp. Food Microbiol., Intern. Assoc. Microbiol. Soc., Grafische Industrie, Haarlem, Netherlands.

Karel, M. 1973. CRC Crit. Rev. Food Technol. 3 (3), 329-373.

Karel, M. 1975. In: 'Water Relations of Foods' (Duckworth, R.B. ed.), pp. 435-453. Academic Press. New York, N.Y.

Karel, M. and Nickerson, J.T.R. 1964. Food Technol. 18, 1214-1218.

Kaufmann, V.F., Wong, F., Taylor, D.H. and Talburt, W.F. 1955. Food Technol. 9, 120-123.

King, C.J. 1971. In: 'Proc. 3rd Internat. Congr. Food Sci & Technol., pp. 565-574.

Kluge, G. and Heiss, R. 1967. Verfahrenstechn. 1 (6), 251-260.

LaBelle, L.R. 1966. In: 'Proc. Symp. Frontiers in Food Res'. pp. 109-114, Cornell University, Ithaca, N.Y.

Labuza, T.W. 1972. CRC Crit. Rev. Food Technol. 3 (2), 217-240.

Labuza, T.W. 1975. In: 'Water Relations of Foods' (Duckworth, R.B. ed.). pp. 455-474. Academic Press, New York, N.Y.

Lazar, M.E. and Hart, M.R. 1968. Food Technol. 22 (10), 124.

Manser, J. 1977. Bühler Diagramm nr. 64, 9-14, Bühler Broths, Ltd., Uzwil. Switzerland.

Mapson, L.W. 1956. Brit. Med. Bull. 12, 73. cited in Bender, A.E. 1966. J. Food Technol. 1, 288.

Mauron, J., Mottu, F., Bujard, E. and Egli, R.H. 1955. Arch. Biochem. Biophys. 59, 433-451.

Mauron, J. and Mottu, F. 1958. Arch. Biochem. Biophys. 77, 312-327.

Mauron, J. and Blanc, B. 1965. Nutr. Dieta 7, 69-81.

Mustakas, G.C., Griffin, E.L., Jr., Allen, L.E. and Smith, O.B. 1964. J. Amer. Oil. Chem. Soc. 41 (9), 607-614.

Nehring, P. 1973. Deut. Lebensm. -Rdsch. 69 (1), 12-20.

Paulus, K. 1975a. Lebensm. -Technol. 8 (1), 3-15.

Paulus, K. 1975b. Ernährungswirtsch./ Lebensm. -Technik 9, 538-552.

Potasi, L.P., Holsinger, V.H., de Vilbiss, E.D. and Pallansch, M.J. 1974. J. Dairy Sci. 57, 258-260.

Quast, D.G. and Karel, M. 1972. J. Food Sci. 37, 679-683.

Rahman, A.R., Taylor, G.R., Schafer, G. and Westcott, D.E. 1970. US Army Natick Lab. Tech. Rept. 70-52 F 1

Reynolds, T.M. 1963. Adv. Food Res. 12, 1-52.

Reynolds, T.M. 1965. Adv. Food Res. 14, 168-283.

Richter, K. and Schiller, K. 1958. Z. Tierphysiol. Tierernähr. Futtermittelk. 13, 345-354.

Sapers, G.M., Panasiuk, O., Jones, S.B., Kalan, E.B., Talley, F.B. and Shaw, R.L. 1974. J. Food Sci. 39, 552-554.

Schweizer, H. 1975. unpubl. results. Dept. Food Sci. Swiss Fed. Inst. Technol. Zürich.

Solms, J., Osman-Ismail, F. and Beyeler, M. 1973. Can. Inst. Food Sci. Technol J. 6 (1), A10-A16.

Speck, P. 1976. Doctor of Techn. Sci. Dissertation nr. 5753. Swiss Federal Institute of Technology Zürich, Juris Publ. Co., Zürich.

Speak, P., Escher, F. and Solms, J. 1977. Lebensm.-Wiss. u. -Technol. in press.

Thijssen, H.A.C. and Rulkens, W.H. 1968. De Ingenieur, CH 45, Nov 22, cited in King, C.J. 1971. p.573.

van Arsdel, W.B., Copley, M.J. and Morgan, A.l. (eds.) 1973a. Food Dehydration, vol.1., Drying Methods and Phenomena, 2nd ed. AVI Publ. Co., Westport. Conn.

van Arsdel, W.B., Copley, M.J. and Morgan, A.L. Jr. (eds) 1973b. Food Dehydration, vol. 2. Products and Technology, 2nd ed., AVI Publ. Co. Westport. Conn.

van der Bruel, A.M.R., Jenneskens, P.J. and Mol, J.J. 1971. Neth. Milk Dairy J. 25, 19-30.

Vojnovich, C. and Pfeiffer, V.F. 1970. Cereal Sci. Today 15 (9), 317-322.

SESSION 5

CHILLING/FREEZING/THAWING

Chairman: M.J. Cranley

INFLUENCE OF CHILLING, FREEZING AND THAWING ON FISH QUALITY - RECENT ASPECTS

W. Vyncke

Ministry of Agriculture, Fisheries Research Station, Ostend, Belgium.

ABSTRACT

1) Biological condition influences the processing possibilities of most fish species.

2) Much interest is shown in the resources of underexploited fish species, especially small pelagic species.

3) Refrigerated sea water systems and containerisation of the catch can improve quality of many fish species.

4) The most recent advance in freezing is the liquid dichlorodifluoromethane or LFF-process which produces a very fast freezing with an almost negligible weight loss.

5) Vaccum thawing and dielectric thawing, especially with microwaves, show promising results.

6) Minced fish has a great potential but many problems are still hindering its full utilisation.

7) Chilling and freezing result in little nutritional loss if the fish is handled correctly.

8) Main recommendations for further research are:

More systematic information should be gathered on seasonal and ground to ground variation of the most important fish species.

Quality changes during wet storage of raw materials especially of underutilised species should receive further attention.

The effects of ultra-rapid cryogenic freezing techniques on the quality of fish need further investigations.

The influence of vacuum thawing and dielectric thawing on the quality and yield of fishery products should be further studied.

More R & D is needed on the causes of quality changes and improvement of the recovered minced flesh.

More systematic studies on the influence of freezing and thawing on the nutrients of fishery products should be carried out.

INTRODUCTION

In a period in which European fisheries are faced with over-fishing, maximum attention should be given to achieving an optimal valorisation of the limited catches. For the same reason, the fish industry is under strongly increased pressure to make greater use of underexploited species for direct human consumption. Quality of the products however should be such as to guarantee sufficient consumer acceptance. In this respect, it should be stressed that quality standards are set at a higher level by today's consumers.

This paper deals with quality aspects of chilling, freezing and thawing of unprocessed fish (raw material). Although methods such as salting and canning can be employed for the preservation of the catch both on land and at sea, refrigeration remains the most important method of preservation. Research carried out since about 1970 on fish species and techniques of interest to European fisheries are reviewed with emphasis on raw material requirements, effects of chilling, freezing and thawing on the quality and nutritive attributes of the fish and deterioration during storage and commercial distribution.

REVIEW OF RESEARCH AND DEVELOPMENT

1) <u>Product requirements common to chilling and freezing</u>

In the seventies, more research work was carried out on the influence of biological condition and on the suitability of several underexploited or non exploited fish species for chilling or freezing.

<u>Biological condition</u>. Fish to be used as food are judged on the nature of the texture, flavour, odour, surface appearance and colour of the fish flesh. Investigations on cod *(Gadus morhua)* have shown that all of these attributes can vary according to the season and the place where the fish were caught, thus influencing their behaviour during processing.

The most important factor by far is their state of nutrition, heavy feeding during recovery from spawning producing toughness and gaping, while habitual good nourishment (probably) produces a stronger cold store flavour. Migration or a stationary habit seems to produce differences in the darkness of the flesh, and an actively swimming fish also seems much livelier on the deck - perhaps through having better immediate access to those muscle constituents which provide energy. A squashy non-resilient flesh indicates starvation, but this does not mean a fish which is unsuitable for freezing unless it is very watery indeed or unless it has already recommenced the heavy feeding (Love, 1974, 1975a, b; Love et al., 1974a, 1976).

Periodical tests made in Canada on fillets from fresh, post rigor, trawler-caught cod have shown that there is a slight decrease in salt extractable proteins during the spring spawning period. This was accompanied by a similar slight decrease in the lipid content of the muscle (Castell and Bishop, 1973).

Foda (1974) studied seasonal variations in approximate composition of Atlantic salmon *(Salmo salar)* for which data are scarce. Protein, fat and moisture varied with fish size, small fish having less fat and protein and more moisture than large fish. The levels of these components underwent seasonal variations, which could be attributed to fish size, growth rate and water temperature. Parr-smolt transformation was associated with a large decrease in fat, a decrease in protein and an increase in moisture.

Hardy and Keay (1972) performed approximate analyses and determinations of mean lipid unsaturation on monthly samples of mackerel *(Scomber scombrus)* caught off the Cornish coast (Scotland). Maximum and minimum total lipid levels were recorded in December and June respectively. The highest mean lipid unsaturation levels were recorded in December and June respectively. The highest mean lipid unsaturation levels were recorded in November and the lowest in May. An average linear relationship between lipid and water content with protein level remaining substantially constant was observed.

Underexploited fish species. Hake *(Merluccius spp.)* appears to have a considerable potential for filling the expanding needs for low-cost fish blocks. Several countries have carried out extensive research on these fish.

Crawford et al. (1972) evaluated the acceptance and frozen shelf-life characteristics of Pacific hake *(Merluccius productus)*. Their results indicated that fish portions prepared from frozen blocks of Pacific hake fillets possessed a relatively high degree of acceptance and frozen shelf-life stability when stored under vacuum in moisture-vapour proof film, indicating a good potential for utilisation of this species for human food.

A study of the effects of various delay periods and handling conditions before freezing on Cape hake *(M. capensis and M. paradoxus)* in the whole or eviscerated state was made by Burt et al. (1974) in order to assess the importance of these to the quality and acceptability of the final product. Increasing delays before gutting lead to increasing discolourations in sea frozen fillets and in fillets cut from sea frozen whole fish, even where a further delay for bleeding occurred before freezing. The extent of gaping apparent in fillets from whole fish was related to delay before freezing and could also be reduced by holding at chill temperatures throughout the pre-freezing period.

In recent years much interest has been shown in the resource of blue whiting *(Micromesistius poutassou)* in the North Atlantic. The blue whiting is a small white fish very similar to other gadoid species in chemical composition. Eating quality is good and close to that of cod and related species. Production of minced block seems to be a more promising processing method than fillet block (Dagbjarsson, 1975). While blue whiting is also used for the production of fish meal, there seems to be greater possibilities in its exploitation for direct human consumption.

Bussmann (1977) investigated the differences of biological condition of blue whiting from two fishing grounds (NE Faroes and Gulf of Biscay). Fish from the NE Faroes had a better texture and the flesh was whiter.

There is a rapidly increasing interest in the use of capelin *(Mallotus villosus)* for human consumption; 99% of the catch of this small fish occurring in large quantities in northern regions of the Atlantic and the Pacific is actually going into fish meal and oil. Studies have shown that fresh capelin is an excellent food fish and that capelin caught during spawning migration keeps well in frozen storage. The larger male capelin is especially suitable for that purpose (MacCallum et al., 1969; Jangaard, 1974; Hansen et al., 1974; Shaw and Botta, 1977). The fat and water contents of capelin fluctuate seasonally; the fat content can be as low as 1 - 2% following spawning but goes above 20% in the fall (Jangaard, 1974).

Other small fatty fish at present mostly converted to fish meal form an increasing part of the world catch of marine fish (especially Clupea, Sardinops, Sardinella, Engraulis, Brevoortia, Rastrelliger spp). Studies have been undertaken to evaluate their suitability for direct human consumption (Heen, 1974; Hanset et al., 1974). These small fish, however, are very delicate and susceptible to oxidative, enzymatic and bacterial spoilage. It is rarely practical to eviscerate them and the visceral enzymes may cause great losses of oil and protein in subsequent storage. Investigations have shown that rapid chilling to $0°C$ is essential for preservation (Hansen et al., 1974).

Several deep water species which are really not used commercially have been studied as possible substitutes for white fish species with declining catches (Anon., 1974;

Burgess, 1974). From these, the roundnose grenadier *(Coryphaenoides rupestris)*, which is already consumed in the USSR appears to be the main species of interest. Analyses indicated that the keeping time of at least 18 days was better than that of most species of whitefish. Roundnose grenadier seems to have a high potential for use as human food (Botta and Shaw, 1976). The closely related species roughhead grenadier *(Macrourus berglax)* appeared to be of lesser quality (Botta and Shaw, 1975).

Finally, increasing research on potential uses for human food of Antarctic krill *(Euphausia superba)* should be mentioned. This planktonic crustacean with high potential annual catches is characterised by a high content of trace elements, vitamins and essential amino acids (Lagunov et al., 1974; Flechtenmacher et al., 1976)

2) Chilling of fish

<u>Use of refrigerated sea water.</u> Refrigerated sea water (RSW) as a medium for cooling, storing and transporting fish has many advantages, such as a more rapid cooling of the fish, significant labour saving and elimination of damage and shrink losses due to excess pressures that often occur when fish are iced. This medium however has also disadvantages such as uptake of water by lean fish species and an increase in total salt. Controlling the growth of spoilage bacteria in fish stored in RSW also presents a problem (Barnett et al., 1971). The advantages of RSW however justified further research to develop modifications to eliminate the major deterrents.

Experiments carried out in the USA (Barnett et al., 1971; Nelson and Barnett, 1971) showed that storing fish in RSW treated with carbon dioxide inhibited the growth of bacteria and increased the storage life of rockfish *(Sebastodes flavidus)*, chumsalmon *(Oncorhynchus keta)*, halibut *(Hippoglossus hippoglossus)* and shrimp *(Pandalus jordani and borealis)* by at least one week.

Longard and Regier (1974) carried out similar experiments

on redfish *(Sebastes marinus)*. The colour of the skin was at least as good when the fish were held up to 12 days in RSW at $1^{o}C$ as when they were held in ice. Adding carbon dioxide to the RSW generally enhanced skin colour retention. The fish stored in RSW with CO_2 were kept in better condition, as measured by trimethylamine production in the fillet tissue, than in ice or RSW. The ionic diffusion, sodium uptake, and potassium loss between the fish and the RSW did not appear excessive but did not reach equilibrium in 12 days storage.

Further studies in Poland (Pielichowski and Nodzynski, 1974) showed that herring and horsemackerel *(Trachurus trachurus)* had a higher quality in RSW than in ice. For maximum quality and shelf-life of frozen products fish should be held in RSW for a maximum of one day.

Refrigerated sea water proved to be an improved method of holding mackerel *(Scomber scombrus)* (Lemon and Regier, 1977). The uniform lower temperature and reduction in available oxygen retarded the development of oxidative rancidity. Textural deterioration was also retarded. The addition of carbon dioxide to the RSW did not regularly affect the level of spoilage as monitored by the measurement of trimethylamine.

The use of carbon dioxide appeared to have limitations, as determined by the end product. Tomlinson et al. (1974) showed that because of the salt uptake by the fish (lingcod, sockeye and pink salmon) susceptibility of the fish to rancidity in frozen storage and, in salmon, impairment of flesh colour, the addition of carbon dioxide to the RSW was not suitable for these fish species to be subsequently frozen. The presence of dissolved carbon dioxide in the fish muscle of mackerel made the fish unacceptable for canning (Lemon and Regier, 1977).

The suitability of RSW systems also appeared to depend on fish species and biological condition.

Male spawning capelin were preserved ungutted in RSW both

with and without carbon dioxide. Contrary to results found
with other small pelagic species, RSW in both forms lead to a
rate of degradation considerably faster than that for similar
capelin held in ice (Shaw and Botta, 1975).

Instead of RSW, the use of sea water chilled with ice
(CSW) was experimented with extensively.

Trials have been carried out in the UK into a method of
transporting fish for food production in containers using
chilled sea water for cooling. The containers are insulated
aluminium tanks of 2.1 m^3 capacity and may be filled on board
fishing vessels and transported to the processing factories
by road. Herring stored in this manner was of superior quality
to herring of the same type stored by the conventional method
of icing in boxes (Eddie and Hopper, 1974).

Myers (1974) described a similar container system for
holding fish aboard small Norwegian fishing boats and subsequent
transport by road and reported an extended shelf life of the
fish. Herring and salmon *(Salmo salar)* with high oil contents
(up to 21%) had a shelf life of 80 h; fish with an oil content
of 4% could be held for 125 h. Bacterial counts after 60 h
were approximately one-tenth those of samples stowed by
conventional boxing system. The use of CSW for holding of white
fish *(gadoids)* on the other hand gave disappointing results.

Danish work showed no difference in quality of several
gadoids between storage in CSW and ice except for blue whiting
where a slightly better quality was recorded in CSW (Huss and
Asenjo, 1976).

Several tests were carried out in the USA to determine
the effects of CSW on the quality of mixed species. The
results of these tests indicated that there appeared to be no
detrimental effect on individual species when stored unsorted.
Shelf life was equal to that of traditional icing, therefore
giving several advantages over the CSW method (ease of unloading,

less bruising, etc.) (Baker and Hulina, 1977). The results of tests on bulkholding of herring in CSW carried out by the same authors (Hulina and Baker, 1977) indicated several advantages over normal practices, including a greater shelf life of the fish.

Use of boxes and containers at sea. The use of boxes or containers at sea is liable to improve quality of the catch.

Houwing (1971) and Peters et al. (1974) studied the effect of draining methods on the quality of boxed fish. Results of organoleptic, chemical and microbiological tests on several fish species stored with ice showed no significant difference in quality between fish from the upper and from the lower boxes. These results are important as they indicate that stowage procedures in the hold can be simplified without endangering quality of the fish.

Instead of boxes, trials with containers of larger capacities have been carried out in several countries.

A system with containers of 0.5 m^3 was described by Crépey and Le Berre (1974). Besides technological advantages, a better quality was claimed for the fish stored in these containers.

Use of containers of different sizes for transport and storage of mackerel for use in the canning industry is actually being experimented with in Denmark with promising results. Emphasis was laid on the need for low temperatures (Jensen, 1977)

Alterations during chilled storage and distribution
Work of a more fundamental nature (eg model studies) is not discussed here.

Attention was given in several countries to the chemical or organoleptic behaviour of species where data are lacking, but which are commercially important or could become so in the near future.

Spoilage in queen scallops *(Chlamys opercularis)* held in ice was studied by Thornson et al. (1974). Organoleptic results showed that the limit of edibility was reached some 8 - 10 days after capture. No difference was found in texture during storage. Total viable counts rose sharply over the first five days in ice. Hypoxanthine showed the most promise as a spoilage indicator.

Shaw and Botta (1975) measured the changes in chemical parameters in spawning male capelin held in ice for up to 14 days and compared them with changes in taste and texture. The point of rejection of these fish at 8 days storage gave capelin a longer storage life in ice than other major pelagic species used for food, such as herring, mackerel and sardines. Analytical values indicated that there were no changes due to development of rancidity in the fat. Studies on witch *(Glyptocephalus cynoglossus)* showed that ATP breakdown occurred at a slower rate than in most fish. After 18 days storage, the major components of the nucleotide degradation were still intermediates as production of hypoxanthine was low. Analysis of sensory and chemical results indicated that trimethylamine was a good indicator of quality and that while hypoxanthine correlated well with quality, its quantity was such that it was unlikely to have any effect on sensory evaluation (Shaw et al., 1977).

Vyncke (1970, 1977) studied the development of ammonia in several Elasmobranchs and reported that the borderline of acceptability could be set at 55 - 60 mg NH_3-N for dogfishes *(Squalus acanthias and Scylliorhinus canicula)* and at 60 - 70 mg for thornback ray *(Raja clavata)*.

Research was also continued on the bacterial reduction of trimethylamine oxide (TMAO) in general, which is an important phenomenon in sea fishes. All of the psychrophilic groups of organisms isolated from haddock *(Melanogrammus aeglefinus)* fillets during storage at $3^{\circ}C$ included organisms capable of producing trimethylamine (TMA) from TMAO. The percentage of TMA-producing isolates in the flora remained nearly constant during storage to spoilage, although the composition altered markedly. All *Pseudomonas putrefaciens* isolates were TMA producers and this organism was the most numerous TMA producer during storage (Laycock and Regier, 1971).

The TMAO-reducing activity appeared to have little relation to the initial viable cell counts and to the temperature at which the bacteria were able to grow (Sasajima, 1973).

The results of examination at subzero temperatures of the growth or viability and TMAO reducing activity of some psychrotrophic bacteria isolated from fresh fish indicated that the minimum growth temperature was -4 to $-5^{\circ}C$ and that, generally, there was no increase in bacterial cells nor any TMAO reducing activity at $-4^{\circ}C$ during 24 days (Sasajima, 1974).

Degradation of nucleotide has also received further attention, as this is an important aspect of fish spoilage.

Ehira and Uchiyama (1973, 1974) investigated 98 fish species with regard to the breakdown of nucleotides. They could distinguish hypoxanthine forming and inosine forming type species, showing that the pathways for degradation of nucleotides vary according to the fish species. For example, in cod, inosine-5-monophosphate (IMP) dropped remarkably within two days of ice storage whereas that of red bream *(Pagrus major)* persisted to some extent even after 16 days.

The rates of breakdown of IMP and of its product, inosine varied considerably in the muscles of cod, haddock, pollock, Atlantic halibut, American plaice, and winter flounder when stored in ice. Haddock and Atlantic halibut were noteworthy in having a very slow rate of breakdown of IMP in relation to the other species. Variations of the rate of dephosophorylation of IMP among individual winter flounder appeared to be due to differences in muscle pH. In homogenates of the muscles investigated, the maximum rates of breakdown of IMP occurred in the pH range 8.6 - 9.4, and were much greater than in the pH range of muscle. The effect of pH on the breakdown of inosine was less pronounced (Dingle and Hines, 1971).

Stone (1971) reported results on *Pandalus spp.* which showed that the major pathway for the degradation of the adenine nucleotides in the ice-held shrimp resulted in the accumulation of IMP rather than adenosine during the first 24 h of storage. After that it decreased. This result suggested that shrimp should preferably be processed within the first day of capture.

Additives and packaging. Regarding research on factors which could enhance the chilling effect and improve the keeping quality of fish, further work on the use of additives and adequate packaging should be mentioned.

Studies on EDTA showed that this compound had a markedly favourable effect on the shelf life of rainbow trout *(Salmo irideus)* and Baltic herring (Kuusi and Läytömähi, 1972) and hake (Doesburg et al., 1970).

Wessels et al. (1972) found that the use of ice containing 0.5% benzoate or 0.3% benzoate and 0.05% potassium sorbate increased shelf life of hake *(Merluccius capensis)*.

Russian research workers studied 50 new preparations of potential use for storage of fish. Based on these studies a new stabiliser (Polycycline) containing N-vinylbenzimidazole hydrochloride or bromonomylate, N-vinyl hydroxyaniline and polyvinylpyrrolidone was developed, which was claimed to preserve fish for three weeks at 0-2°C (Skvortsova et al., 1975).

Prepackaging is important as regards sales promotion of fish. This technique however raises problems as the wet fish cannot be stored or displayed in direct contact with ice.

Research which started in the sixties, was continued in recent years.

Huss (1973) reported new results of experimental packaging of plaice and haddock in polyethylene and polyamide pouches. It was concluded that in no case packaging of fish caused a decrease in fish quality and keeping time. Vaccum-packaging of fish in polyethylene pouches did not improve growth conditions for anaerobic bacteria.

Hansen (1972) showed that the storage life of prepacked trout *(Salmo irideus)* and herring greatly depended on the access of atmospheric air to the fish surfaces during storage. Vacuum-packing in air-tight films excluded fat oxidation and gave periods of storage life of 2 - 3 weeks. Trout and herring packed in this way mainly deteriorated because of microbiologica spoilage.

In Belgium, work has been carried out on prepacked whole cod (Devriendt, 1976) plaice (Devriendt and Declerck, 1977) and fillets of these species, on herring (Declerck and Devriendt, 1976) and on peeled brown shrimps *(Crangon crangon)* (Devriendt and Vyncke, 1976). The prepacked products were stored for up to four days under simulated retail sale conditions at three different temperatures: by day (from 2 to 18 h) at 1°, 4° or 8° C and by night at 1°C. The results indicated that a retail shelf-life of four days can be guaranteed at temperatures not exceeding 4°C.

3) Freezing of fish

Freezing techniques. Although work was also continued on air blast, plate, brine or liquid nitrogen freezing, the most recent advance in freezing is the liquid dichlorodifluoremethane process also called Freon, R12 or LFF-process. This technique produces a very fast freezing with an almost negligible weight loss. Better quality was also claimed in several technical articles owing to the ultra-rapid freezing (Aström, 1972; Morgan, 1974; Dew, 1974).

Freezing tests discussing in more detail the quality of fishery products frozen by the LFF-process compared to those frozen by conventional methods were also reported. Buchholz and Pigott (1972) concentrated on the qualities of king salmon *(Oncorhynchus tschawytscha)* steaks frozen by the LFF-process, air blast freezing and liquid nitrogen freezing. Instances of statistically significant differences between methods occurred infrequently but did not establish a trend and the authors concluded that no major differences in quality were shown by the three freezing methods employed.

Freezing by the use of the LFF-process was evaluated in comparison to air blast tunnel freezing and plate freezing on shrimps *(Pandalus borealis)*, scallops *(Pecten islandicus)* and lemon sole *(Microstomus kitt)* (Aurell et al., 1976). The superfast freezing of the LFF-process was confirmed as well as the very low weight loss due to dehydration. The lower reduction of spoilage bacteria caused by the faster-freezing was also confirmed. A statistical analysis of the sensory scores however pointed out only one significant difference between methods for odour on lemon sole, which was to the advantage of the LFF-process.

Crépey et al. (1976) carried out histological examinations on whiting *(Merlangius merlangus,)* plaice and saithe *(Gadus virens)* frozen by liquid R12, liquid nitrogen or air at $-40°C$. In spite of the high freezing rates at low temperatures no microstructural damage to the fish flesh was observed.

More research is apparently needed to evaluate quality aspects of this new technique.

<u>Alterations during frozen storage</u> As for chilled fish, studies of a more fundamental nature and model systems are not discussed in this paper.

Alterations during frozen storage are characterised by denaturation of the proteins and rancidity of the lipids of the fish. Both received further attention in the seventies.

a) Denaturation of proteins. The development of tough and dry texture during frozen storage of fish has been attributed to changes within their myofibrillar proteins. As a measure of the extent of such denaturation, differences in the amount of protein extractable from the muscle with neutral salt solutions has often been used.

At $-15^{\circ}C$, Love et al. (1974b) found differences in protein denaturation in frozen cod caught on different fishing grounds. Lowest and highest denaturation were observed on cod from Bear Island and Aberdeen Bank respectively.

Childs (1973a) found that during frozen storage of true cod *(Gadus macrocephalus)* the extractability of whole myofibrils decreased more rapidly than the combined extractability of component myofibrillar protein.

Myofribillar proteins from post-rigor unfrozen and stored frozen cod muscle have been studied by electron microscopy (Jarenbäck and Liljemark, 1975). In myofibrillar extracts a decrease in the number of actomyosin filaments and an increase in the number and size of large aggregates was found. A decrease in the length and amount of attached myosin was noticed in actomyosin filaments. The changes observed were most extensive at $-10^{\circ}C$, less pronounced at $-20^{\circ}C$ and hardly noticeable at $-30^{\circ}C$.

According to the appearance of the myofibril residues thick myofilaments had been almost completely extracted from unfrozen muscle and were increasingly resistant towards extraction the higher the temperature during frozen storage.

Further studies have also been carried out in several countries in order to elucidate the denaturation effect of dimethylamine (DMA) and formaldehyde (FA), formed by non-microbial breakdown of trimethylamine oxide (TMAO), on the muscle protein of gadoid fishes.

Castell et al. (1971 a, b) reported that during frozen storage at -5°C DMA was produced in the muscle of five gadoid species. The amount was lowest in haddock and increasingly higher in cod, pollock *(Pollachius virens)* cusk *(Brosme brosme)* and hake *(Merluccius bilinearis)*. When the dark lateral muscle was removed from the fillets before freezing, formation of DMA during frozen storage was either inhibited or greatly reduced. Under the same storage conditions no DMA was produced in the muscle of halibut *(Hippoglossus hippoglosus)*, American plaice *(Hippogloissoides platessoides)*, redfish *(Sebastes marinues)* or wolffish *(Anarhichas lupus)*.

The direct addition of FA to give concentrations of 0.001 to 0.05% caused marked reductions in the extractable protein. An important observation was also that gadoid species producing the most DMA and FA accumulated the least amounts of free fatty acids. Consideration should thus be given to the effect of FA in conjunction with unsaturated fatty acids.

Exposure to formaldehyde was found to cause a reduction in extractable salt-soluble proteins in fresh and frozen true cod *(Gadus macrocephalus)*. Tropomyosin and heavy chains of myosin were most easily insolubilised. Formaldehyde also decreased the quantity of extractable whole myofibrils in fresh and stored samples (Childs, 1973b).

Similar results were reported by Castell et al. (1973).

Tokunaga (1974) investigated the influence of the storage temperature on the rate of DMA and FA-formation and the effect of these substances on denaturation of proteins of Alaska pollock *(Theragra charcogramma)*.

In minced ordinary and dark muscle stored at various temperatures from -5°C to -40°C, the largest amount of DMA was produced at -10°C. In dark muscle, degradation of TMAO

was very rapid and DMA-N reached about 30 mg/100g after 10 days
storage at -10°C, but very little or no TMA was observed in the
same sample.

With the increase in DMA content, a gradual decrease in
extractable protein in frozen fillet was observed. In spite
of the fact that a wide fluctuation in the amount of DMA
produced was observed among individual fishes, there was a
close relation between DMA content and extractability of
protein. When the DMA-N value exceeded 2.0 - 2.5 mg/100g of
the muscle, the decrease in protein extractability became
apparent.

b) Rancidity of lipids. Lipid oxidation and hydrolysis
is the main problem during storage of frozen fatty fish.
During recent years mackerel *(Scomber scombrus)* received special
attention as a possible alternative for the declining herring
stocks.

A variable but high proportion of the total lipids of
mackerel appeared to be neutral lipids with substantial contents
of unsaturated C_{20} and C_{22} fatty acids (Ackman and Eaton, 1971;
Hardy and Keay, 1972). These acids are among the constituents
of fatty fish, in free or combined forms, most susceptible to
autoxidation.

In terms of controlling the onset of rancidity in lipids
of frozen fish the most important problem in quality preservation
would seem to be control of the initial oxidation reaction
before the chain can begin to propagate. When the initial
oxidation has occurred through direct interaction with oxygen
or/and a free radical mechanism (Labuza, 1971), the quality of
fish deteriorates, often very rapidly, and so far no practical
method has appeared for preventing the development of
rancidity after the lipid oxidation has proceeded beyond the
induction period.

Using low storage temperatures or vacuum packaging for

frozen mackerel could slow the monomolecular reactions of oxidation but did not eliminate problems completely (Ke et al., 1977).

Oxidation in the subcutaneous fat was found to be eight times faster than in the white and dark muscles from mackerel held at $-15°C$ for two months. This rapid development of rancidity in the skin was effectively inhibited by lowering the frozen storage temperature to $-40°C$.

Hardy and Smith (1976) studied storage conditions of mackerel and found that the rate of hydroperoxide formation in frozen fish increased in the following manner: vacuum-packed fillets, foil-wrapped fillets, ungutted fish, gutted fish.

Russian studies on the influence of freezing on discolourations showed that carotenoids penetrate quickly into the subcutaneous layers, causing yellow-orange pigmentation within 1 - 2 h prior to freezing of mackerel and within 3 - 4 h in herring. Pigmentation of the frozen tissues depended on storage conditions. Carotenoid content increased initially and declined later. Complete decomposition of carotenoids took place only when considerable oxidation had occurred (Lyubavina et al., 1972).

Work on the effect of antioxidants on the development of rancidity was continued and several favourable results were reported.

The use of nitrous oxide for prevention of oxidation of lipids in frozen mullet *(Mugil cephalus)* and eel *(Anguilla anguilla)* was investigated by Polesello and Nani (1972). The fish were packaged under vacuum or in air, nitrogen or nitrous oxide. Results showed that the protective effect of N_2O was greater than that of nitrogen or vacuum packaging.

Fat oxidation in mackerel *(Scomber japonicus)* was reported by Tanaka (1973) to be prevented by treatment with a 0.5% solution of tocopherol or a 0.01% solution of BHA.

An antioxidant treatment of 0.025% ascorbic acid and citric acid (1 : 1) before freezing of hake *(Merluccius productus)* improved waterbinding capacity and retarded fat oxidation (Andreeva, 1975).

Development of rancidity in mullet was retarded by treatment with ascorbic acid and/or monotertiarybutylhydroquinone (TBHQ) in combination with vacuum-packaging (Deng et al., 1977).

Effect of antioxidants appears to be largely dependent on fish species (especially fat content) and storage conditions (temperature, storage time, packaging) making it frequently necessary to consider each case separately.

4) Thawing of frozen fish

Rapid thawing of fish under controlled conditions has become a necessity to the fish industry. Indeed, large quantities of fish are block frozen at sea or on shore (peak landings) without further treatment and require further processing after thawing.

Although thawing in water or in moist air is still most commonly used, vacuum thawing and dielectric thawing, especially with microwaves have received special attention in recent years.

Excellent quality was reported for vacuum thawing owing to the high rate of heat transfer and a better temperature control (Everington and Cooper, 1972; Jason, 1974). Houwing (1973) found that brown shrimps *(Crangon crangon)* lost less flavour components and a lower bacterial count than similar shrimps thawed in water.

Dielectric thawing with macro- or microwaves, already experimented with in the 60s received a renewed attention owing to improved equipment. Not taking into account the high costs of the apparatus and energy requirements, the main drawback of these techniques appeared to be local overheating of the fish. The use of cold air in microwave thawing seems to overcome this disadvantage to a great extent (Schiffmann, 1975; Sale, 1976). These techniques appear to be especially suited for partial thawing (tempering) of fishery products.

German experiments carried out at 13.6 MHz in a dielectric thawing apparatus cooled with circulating water to avoid overheating showed that fillets of herring, cod, hake, saithe and redfish had a better quality than the corresponding fillets thawed in water. Especially the better appearance of the fish was stressed (Kietzmann, 1973; Flechtenmacher and Christians, 1973).

The best thawing method depends upon factors such as fish species, size, needed capacity, final product requirements etc.

5) Minced fish

Although minced fish is a raw material which has been obtained by a processing method and does not fall completely in the scope of this paper, some consideration should be given to it as it is becoming a very important product.

The introduction of machines that separate the flesh from skin and bones of fish created a considerable amount of interest in the process itself and in the economic potential of products arising from the use of these machines. Research into various aspects of the process have been actively pursued in many countries. One of the main reasons for the interest stimulated by these deboning systems is the possible utilisation of many currently non-utilised or underutilised fish species.

Many papers have already been published on this subject. Suffice it to refer to the proceedings of three conferences (Oak Brook seminar, 1972; Boston seminar, 1974; Torry Research Station Conference, 1976) and to a review by Bond (1975).

Many problems are still hindering the realisation of the full utilisation of minced fish.

One problem in particular is the relatively poor keeping quality during frozen storage and a tendency of the mince to become tougher. This appears to be largely due to the characteristic physical disintegration of the flesh resulting from the deboning process.

Other problems concerning the deboned fish as a raw material are related to the colour of the products, the presence of pieces of skin and bones, enhanced rancidity and occurrence of parasites.

Minced fish has a tremendous potential but much more research is needed in that field.

6) Nutritional aspects

Chilling and freezing of fish result in little nutritional loss if the fish is handled correctly. Only a small proportion of nutrients is lost through leaching during storage in ice (Burgess, 1971).

Thawing losses on the other hand can be severe and some loss of protein amino acids could occur (Kolakowki et al., 1972). DMA and formaldehyde are formed during frozen storage of some species as well as degradation products of lipids due to rancidity. The nutritional consequences of these reactions however have not been studied.

RECOMMENDATIONS FOR FURTHER RESEARCH

1) Influence of biological condition on the utilisation of fish

More systematic information should be gathered on seasonal and ground to ground variation of the most important fish species.

Further technological research should be carried out to establish the relation between the condition of the raw material and its suitability for processing including new end products in order to promote the valorisation of fish with a low value under present conditions.

2) Chilling of fish

More research is needed on RSW and CSW systems, in particular as regards influence of fish species and biological condition.

Further experiments on the use of large containers at sea should be carried out.

Quality changes during wet storage of raw materials especially of under-utilised species should receive further attention in order to provide basic information on which to design practical systems of handling, storage and processing of food fish.

3) Freezing of fish

The effects of ultra-rapid cryogenic freezing techniques on the quality of fish need further investigation.

Better information on the effect of quality of raw materials and other factors on the greatly varying shelflife and quality of frozen products, is required.

4) Thawing of fish

The influence of vacuum thawing and dielectric thawing on the quality and yield of fishery products should be further studied.

5) Minced fish

More R & D is needed on the causes of quality changes, influences of manufacturing processes and additives on quality, utilisation of small fish and improvement of the recovered minced flesh.

6) Nutritional aspects

More systematic studies on the influence of freezing and thawing on the nutrients of fishery products should be carried out.

REFERENCES

Ackman, R. and Eaton, C. 1971. Can. Inst. Fd. Technol. J. 4, 169.
Andreeva, T. 1975. Rybnoe Khozyaistvo (4), 70-71.
Anon. 1974. Fish. News Intern. 13 (1), 47.
Aström, S. 1972. Rev. gén. Froid (11), 1109-1114.
Aurell, T., Dagbjartsson, B. and Salomonsdottir. E. 1976. J. Fd. Sci. 41, 1165-1167.
Baker, D. and Hulina, S. 1977. Mar. Fish. Rev. 39 (3), 1-3.
Barnett, H., Nelson, R., Hunter, P., Bauer, S. and Groningen, H. 1971. Fish. Bull. 69, 433-442.
Bond, R. 1975. Background Paper on Minced Fish, FAO Fish. Circ.(332) 24 pp.
Boston Seminar. 1974. June 11/13 (Ed. R. Martin), National Fisheries Institute, Washington, D.C.
Botta, J. and Shaw, D. 1975. J. Fd. Sci. 40, 1249-1252.
Botta, J. and Shaw, D. 1976. J. Fd. Sci. 41, 1285-1288.
Bucholz, S. and Pigott, G. 1972. J. Fd. Sci. 37, 416-419.
Burgess, G. 1971. The Alternative Uses of Fish, FAO Fish. Rep. (117), 15.
Burgess, G. 1974. Fish Industry Rev. 4 (1) 8.
Burt, J., Dreosti, G., Jones, N., Kelman, J., McDonald, I., Murray, J. and Simmonds, C. 1974. J. Fd. Technol. 9, 235-245.
Bussmann, B. 1977. Information für die Fischwirtschaft 24, 77-80.
Castell, C. 1971a. J. Amer. Oil. Chem. Soc. 48, 645-649.
Castell, C., Smith, B. and Neal, W. 1971b. J. Fish. Res. Bd Canada, 28, 1-5.
Castell, C. and Bishop, D. 1973. J. Fish. Res. Bd Canada, 30. 157-160.
Castell, C., Smith, B. and Dyer, W. 1973. J. Fish. Res. Bd. Canada 30, 1205, 1246.
Childs, E. 1973a . J. Fd Sci. 38, 718-719.
Childs, E. 1973b. J. Fd Sci. 38, 1009-1011.
Crawford, D., Law, D. and McGill, L. 1972. J. Fd Sci. 37, 801-802.
Crépey, J. and Le Berre, Y. 1974. Bull. Int. Inst. Refrig. Annex 1. 235-247.
Crépey, J., Bécel, P. and Hadjadj, A. 1976. Bull. Int. Inst. Refrig., Annex 1, 603-610.
Dagbjartsson, B. 1975. J. Fish. Res. Bd Canada 32, 747-751.
Declerck, D. and Devriendt, H. 1976. Rev. Agric. 29, 1191-1199.
Deng, J., Matthews, R. and Watson, C. 1977. J. Fd. Sci. 42, 344-347.
Devriendt, H. 1976. Rev. Agric. 29, 1525-1536.

Devriendt, H and Vyncke, W. 1976. Rev. Agric. 29, 699-707.

Devriendt, H. and Declerck. D. 1977. Rev. Agric. 30, 691-706.

Dew, R. 1974. Ind. Alim. Agric. 91, 1121-1127.

Dingle, J. and Hines, J. 1971. J. Fish. Res. Board Canada, 28, 1125-1131.

Doesburg, J., Simmonds, C., Lamprecht, E., Elliott, M. and Rodrigues, J. de A. 1970. Ann. Rep. Fishg. Ind. Res. Inst. Cape Town, 24, 10.

Eddie, G. and Hopper, A. 1974. In : Fishery Products (Ed. R. Kreuzer), Fishing News (Books) Ltd, London, p. 69.

Ehira, S. and Uchiyama, H. 1973. Bull. Tokai Reg. Fish. Res. Lab. 75, 63-74.

Ehira, S. and Uchiyama, H. 1974. Bull. Jap. Soc. Sci. Fish. 40, 479-487.

Everington, D. and Cooper, A. 1972. Food Trade Rev. 42 (7), 7-13.

Flechtenmacher, W. and Christians, O. 1973. Auftauen von Gefrierfisch im Hochfrequenzfeld mit BBC-Auftauanlage-paper presented at the 4th meeting of the West-European Fish Technologists Association, Hamburg, Sept. 73.

Flectenmacher, W., Schreiber, W., Christians, O. and Roschke, N. 1976. Informationen für die Fischwirtschaft 23, 188-196.

Foda, A. 1974. Seasonal Variation in Proximate composition of Hatchery-Reared Atlantic Salmon *(Salmo salar)*, Techn. Rep. Ser. MAR/T-74-2. Fisheries and Marine Service, Canada, 12pp.

Hansen, P. 1972. J. Fd Technol. 7, 21-29.

Hansen, P., Olsen, K. and Peterson, T. 1974. In: Fishery Products (Ed. R. Kreuzer), Fishing News (Books) Ltd. London. p. 64.

Hardy, R. and Keay, J. 1972. J. Fd. Technol. 7, 125-137.

Hardy, R. and Smith, J. 1976. J. Sci. Fd Agric. 27, 595-599.

Heen, E. 1974. In : Fishery Products (Ed. R. Kreuzer) Fishing News (Books) Ltd. London. p. 144.

Houwing, H. 1971. Visserij 24, 442-443.

Houwing, H. 1973. Voedingsmiddelentechnologie 4, 140-141.

Hulme, S. and Baker, D. 1977. Mar. Fish. Rev. 39 (3) 4-9.

Huss, H. 1972. J. Fd. Technol. 7, 13-19.

Huss, H. and Asenjo, I. 1976. Annual report of the Research Laboratory of the Danish Ministry of Fisheries, Copenhagen. 32.

Jangaard, P. 1974. The Capelin *(Mallotus villosus)*, Bull. Fish. Rs. Bd Canada No. 186, 70 pp.

Jarenbäck, L. and Liljemark, A. 1975. J. Fd. Technol. 10, 309-325.

Jason, A. 1974. IFST Proceedings, 7, 146-157.

Jensen, J. 1977. Transport and Storage of Mackerel in Containers for Use in the Canning Industry - paper presented at the 7th Meeting of the West-European Fish Technologists Association, Tromsø (Norway), June.

Ke, P., Ackman, R., Linke, B. and Nash, D. 1977. J. Fd. Technol. 12, 37-47.

Kietzmann, U. 1973. Allgemeine Fischwirtschaftszeitung (1), 45-54.

Kolakowski, E., Fik, M. and Karminska, S. 1972. Bull. Int. Inst. Refrig. Annex 2, 65-72.

Kuusi, T. and Läytömähi, M. 1972. Z. Lebensm. Unters. Forsch. 149, 196-204.

Labuza, T. 1971. CRC Critical Rev. Fd Techno. 2, 355.

Lagunov, L., Kryuchkova, M., Ordukhanyan, N. and Sysoeva, L. 1974. in: Fishery Products (ED. R. Kreuzer), Fishing News (Books) Ltd. London. p. 247.

Laycock, R. and Regier, L. 1971. J. Fish. Res. Board Canada, 28, 305-309.

Lemon, D. and Regier, L. 1977. J. Fish. Res. Bd. Canada, 34, 439-443.

Longard, A. and Regier, L. 1974. J. Fish. Res. Bd. Canada, 31, 456-460.

Love, R. 1974. J. Cons. Int. Explor. Mer. 35, 207-209.

Love, R., Robertson, I., Lavéty, J. and Smith, G. 1974a. Comp. Biochem. Physio. 47b, 149-161.

Love, R., Muslemuddin, M., Ong, L. and Smith, G. 1974b. J. Sci. Fd. Agr. 25, 1563-1569.

Love, R. 1975a. J. Fish. Res. Bd. Canada, 32, 2333-2342.

Love, R. 1975b. Fish. News Int. 14 (4), 16-18.

Love, R., Yamaguchi, K., Créac'h, Y. and Lavéty, J. 1976. Comp. Biochem. Physiol. 55B, 487-492.

Lyubavina, L., Pakhomova, K., Khobotilova, L. and Dubnitskaya, G. 1972. Rybnoe Khozyaistvo (4), 67-69.

MacCallum, W., Adams, D., Ackman, R., Ke, P., Dyer, W., Fraser, D. and Punjamapirom, S. 1969. J. Fish. Res. Bd. Canada, 26, 2027-2035.

Morgan, J. 1974. Fd. Engin. (12), 67-68.

Myers, H. 1974. Tidsskrift for Hermetikindustri 60, 235-236.

Nelson, R. and Barnett, H. 1971. Proc. XIIIth Int. Congr. Refrig. Washington, Vol, 3. 57-64.

Oak Brook Seminar (1972) Mechanical Recovery and Utilisation of Fish Flesh, 21/22 Sept. (Ed. R. Martin), National Fisheries Institute, Washington, D.C.

Peters, J., Bezanson, A. and Green, J. 1974. Mar. Fish. Rev. 36 (2) 33-35.

Pielichowski, J. and Nodzynski, J. 1974. Prace Morskiego Instytutu Rybackiego, 17, 161-185.

Polesello, A. and Nani, R. 1972. Ann. Ist. Sper. Valor. Technol. Prod. Agr. 3, 67-76.

Sale, A. 1976. J. Fd. Technol. 11, 319.

Sasajima, M. 1973. Bull. Jap. Soc. Sci. Fish. 39, 511-516.

Sasajima, M. 1974. Bull. Jap. Soc. Sci. Fish. 40, 625-630.

Schiffmann, R. 1975. Food Engin. 47 (11) 72-78.

Shaw, D. and Botta, J. 1975. J. Fish Res. Bd Canada, 32, 2039-2046. 2047-2053.

Shaw, D. and Botta, J. 1977. J. Fish. Res. Bd. Canada, 34, 209-214.

Shaw, D., Gare, R. and Kennedy, M. 1977. J. Fd. Sci. 42, 159-162.

Stone, F. 1971. J. Milk Fd. Technol. 34, 354-356.

Svortsova, G., Domnina, E., Mansurov, Y., Kurov, G. and Glazkova, N. 1975. Rybnoe Khozyaistvo (1), 69-70.

Tanaka, K. 1973. Refrigeration (Reito), 48, 499-504.

Thomson, A., Davis, H., Early, J. and Burt, J. 1974. J. Fd. Technol. 9, 381-390.

Tokunago, T. 1974. Bull. Jap. Soc. Sci. Fish. 40, 167-174.

Tomlinson, N., Geiger, S., Boyd, J., Southcott, B., Gibbard, G. and Roach, S. 1974. Bull. Int. Inst. Refrig. Annex 1, 169.

Torry Research Station - Conference 1976. The Production and Utilisation of Mechanically Recovered Fish Flesh (Minced Fish) 7/8 April (Ed. J. Keay). Torry Research Station, Aberdeen (Scotland).

Vyncke, W. 1970. Med. Fakulteit Landbouwwetenschappen Gent 35, 1033-1046.

Vyncke, W. 1977. J. Fd. Technol. (in press).

Wessels, J., Lamprecht, E., Rodrigues, J. de A. and Simmonds, C. 1972. J. Fd. Technol. 7, 301-307.

EFFECTS OF FREEZING, STORAGE AND DISTRIBUTION ON QUALITY AND NUTRITIVE ATTRIBUTES OF FOODS, IN PARTICULAR OF FRUIT AND VEGETABLES

J.A. Munoz-Delgado
Instituto del Frio, Ciudad Universitaria,
Madrid 3, Spain

INTRODUCTION

A review of some of the books, monographs and papers dealing with this topic during the last 25 years leads to the following statements:

a) It is not sufficiently clear to what extent food quality is affected by the change of state of water resulting in mechanical action of ice crystals, physico-chemical effects due to the separation of water as ice, and increasing rates of enzyme-catalysed reactions in cellular systems.

Further research is needed to determine in which foodstuffs and to what extent the benefits of ultra-rapid freezing are not dissipated during subsequent storage, distribution, thawing and cooking, also, to what extent other factors involved such as type of product, species or variety, initial pH, physiological state, etc. are more or less important than freezing rate per se.

b) From TTT (time-temperature tolerance) studies it appears that TTT relationships are subject to wide variability because of variations in PPP (product, processing, packaging) factors and in panel sensory evaluation. It also appears that for certain products some dominating processes which result in quality loss in foods do not follow the Arrhenius equation so that counteracting influences which disturb the regular picture occur ('reversed stability'). Further research on kinetics of reactions at different temperature ranges seems necessary.

c) In the particular case of frozen fruits and vegetables, further research is advisable on the suitability of species and varieties for freezing; on the influence of growing and harvesting conditions, pre-treatments, freezing rate and packaging techniques. Further research is also required on quality, on the necessity and influence of blanching on the nutritive quality, on the chemistry of off-flavours in frozen vegetables, and of discolouration and browning of anthocyanins in frozen fruits.

d) Special attention should be paid to objective and subjective evaluation of quality attributes because of its possible influence on the variability of the results with regard to stability of frozen foods.

Although is is generally accepted that freezing under optimum conditions is a very good method of food preservation since it retains most of the original nutrient value and organoleptic quality attributes, it is also accepted that some changes may occur due to the freezing process and subsequent storage and commercial distribution.

The effects of low temperatures on food in general, and especially on the quality of fruits and vegetables during the mentioned stages, as well as the particular aspects requiring further research, are successively considered.

1. INFLUENCE OF FREEZING

1.1 Foodstuffs in general

Freezing may induce some detrimental changes in food quality which are generally irreversible. Fennema and Powrie (1964) state that this damage is due to ice formation rather than to the decrease in temperature per se. Two main mechanisms have been suggested to elucidate the damage caused by ice formation: mechanical action of ice crystals and physico-chemical effects of separation of water as ice.

As regards the former mechanism, the increase of volume occurring during the change of phase of water may affect the cellular structure. Haynes (1968) assumes that ice starts to form outside the cell (intercellular crystallisation) and that a water transport process is set up through the cell wall and membrane due to the higher vapour pressure of the supercooled water inside the cell. Partmann (1964) suggests that cell walls and membranes may act as barriers to the penetration of ice crystals inside the cell and subsequent seeding of the intracellular supercooled dispersion.

Low freezing rates may result in extracellular crystallisation with cell dehydration, large ice crystals causing cell separation and eventual cell rupture. In most vegetable foods an irreversible cell separation along the middle lamella and cell rupture occurs, due to the cell wall being less elastic than the membranes of animal muscles.

High freezing rates result in extra and intra cellular crystallisation with structural disruption in all morphological cell parts. Since crystal size decreases with increasing freezing rate, and with decreasing temperature, other conditions being equal, it is generally accepted that the lower the temperature and the higher the freezing rate, the less the size of crystals, then the less damaged will be the cell structure, the less extensive the migration of cell constituents and the less the amount of drip on thawing.

Many reported data (Partmann 1972, 1975; Love, 1966; Kondrup and Bolt, 1960) show a great variation in damaging effects of freezing rates depending on animal or vegetable species, type of tissue, variety, biochemical and physiological state (pre-rigor or post-rigor state of muscle tissue), etc. Thus a confusion exists about the real influence of the freezing rate on quality of frozen foods so that the differences which may result at different rates are not great enough in many products to be obvious to many investigators. However, most available data show that a microcrystalline structure (high freezing rates) may be considered as less harmful than a coarse one (low freezing rates).

The problem is more complicated if it is taken into account that the microcrystalline structure may become less uniform due to recrystallisation caused by temperature fluctuations during storage and distribution. Hence it appears that the real importance of the freezing rate influence on histological changes, crystal size, location of crystals and subsequent quality loss of frozen foods related mainly to texture, appearance and drip, have to be envisaged and judged within an overall survey of the various factors influencing the 'keepability' at a high quality level of frozen foods, and not in an independent context. It would be fruitless to make great efforts to improve the conventional freezing techniques, or develop new ones to achieve higher freezing rates, if, for instance, subsequent storage and distribution of frozen foods are not carried out at optimal and steady temperatures, or suitable packaging systems and materials are not used.

As regards the latter mechanism explaining the damage caused in food quality by ice formation, it is well known that separation of water as ice increases the concentration of solutes and suspended materials, induces changes in such properties as pH, titratable acidity and ionic strength which have some effects on the stability of hydrophilic colloids in the cellular systems of food, and leads to protein denaturation, lessening of the bound water, loss of the semi-permeability of membranes, loss of turgor (in fruits and vegetables), loss of texture (in fish and meat), etc., most of these changes being inter-related.

A typical example of this type of change is protein denaturation - a problem which has been thoroughly reviewed in recent years, especially by Dyer and Dingle (1961), Connell (1964, 1968), Fennema and Powrie (1964), Love (1966),Sikorski et al. (1976) and Sikorski (1977). Protein denaturation is of major importance for foodstuffs of animal origin.

As indicated by Ciobanu (1976) denaturation changes in proteins may include loss of biological specificity, changes in molecular shape, increase reactivity of side chain groups, decrease in water holding capacity, decreased solubility in 5% sodium chloride water solution and cell fragility (resistance to mild homogenisation in a dilute formaldehyde solution, in the case of fish protein).

Another effect of freezing on cellular systems of foods is the occurrence of increased rates of enzyme-catalysed reactions. A review by Fennema (1975a) on this subject supports the view that concentration of solutes during freezing is not a satisfactory explanation for this phenomenon and recalls that enzymes as well as enzyme activators and inhibitors exist in a compartmentalised state in cellular systems. Freezing damage would result in de-compartmentalisation of these constituents and in a likelihood of profound changes in the rates of enzyme catalysed reactions. Membranes of the nucleus, mitochondria, ribosomes, lysosomes, and the endoplasmic reticulum, can be damaged to varying degrees by freezing and thawing. Lysosomes are especially susceptible to freezing damage and mitochondria are moderately susceptible. The rupture of lysosoma membrane would permit the released hydrolytic enzymes and the substrates to interact more freely than would occur in undamaged tissue.

In the effects of partial freezing on rates of enzyme-catalysed reactions, several factors are involved: nature of the enzyme, composition of the medium, nature of the freezing treatment - namely rate of freezing and thawing, ultimate low temperature achieved during freezing, storage time and temperature. New research work is needed in this field to elucidate the real influence and importance of the factors mentioned.

1.2. Fruits and vegetables in particular

Plant tissues from which vegetable food is derived are composed of living cells that are highly organised entities containing the cell wall and protoplasmic organelles.

As to the influence of freezing on fruits and vegetable cellular structure, the increase of volume caused by the change of state of water into ice depends mainly on their content of freezable water and the amount of gas in their intercellular spaces. For instance, on average, whole strawberries increase 3% and whole raspberries 4% when frozen at $-18^{\circ}C$. Though the expansion caused by freezing of water will be compensated partially by air spaces, cell tissues will be exposed to strong mechanical forces by the increase of volume during ice formation and intercellular ice forces the cells apart rupturing the middle lamellae and tearing the cell wall. It seems that the more rapid the rate of freezing, the less is the amount of cell distortion and separation.

However, the cell wall withstands fairly well the pressure and changes in texture during freezing and the drip on thawing seems to be ascribed not only to the rupture of the cell wall but also to changes in protoplasm because of the high pressure. In plant cells containing large vacuoles a deformation of cell walls may occur during thawing leading to a breakage of cell walls.

The influence of freezing and thawing on structural changes in relation to texture and taste of fruit and vegetables have been investigated by several authors. Partmann (1975) in a review on this subject quotes the research work of Bassi and Crivelli (1969); Gutschmidt (1969a); Monzini et al. (1969); Nguyen (1969); Mohr and Stein (1969) and Mohr (1971, 1974) and concludes that most of them agree that no significant improvement may be obtained for most vegetable products by using freezing rates higher than those employed in industrial processing. Nevertheless the different cell types behave quite differently. In highly vacuolated large parenchyma cells a higher degree of disorganisation of protoplasmic structures is obtained than in smaller less highly vacuolated ones.

A rapid freezing and moderate thaw treatment causes the least extensive disorganisation and fragmentation of the protoplasm. However, after all freeze-thaw treatments, endoplasmic reticulum, ribosomes, Golgi bodies and mitochondria are unrecognisable in highly vacuolated cells, and rupture and fragmentation of membranes, with consequent loss of the liquid-retaining properties, are the damages mainly responsible for the changes in plant tissues during freezing and thawing.

The above mentioned problems are especially important in fruit freezing; in most vegetables freezing follows blanching and the structural changes induced by blanching are usually so marked that the freezing rate is of minor importance. In most cases freezing rates faster than 0.5 cm h^{-1} for retail packages, and faster than 5 cm h^{-1} for IQF products, may be considered as satisfactory. However, in fruit and vegetables, as well as in other foodstuffs, an important question arises concerning the particular freezing rate above which there is no significant rise of quality retention.

As was indicated for other types of food, the answer is not easy because many variable factors are involved. If the example of strawberries is considered, Woldord (1967) found that freezing with liquid nitrogen or R12 resulted in better texture and lower drip loss than for air-blast IQF produce. Gutschmidt's (1969b) data have shown that freezing rates exceeding 1.5 cm h^{-1} did not result in significantly better quality retention but freezing rates lower than 1.5 cm h^{-1} resulted in worse quality. Aström and Löndahl (1969) found a significant reduction in drip loss at freezing liquid nitrogen and R12 as compared to slower freezing in air.

This confirms what is generally accepted today that cryogenic freezing, with high freezing rates, may reduce changes in texture and excessive drip on thawing, particularly with mushrooms, melon and tomato slices, asparagus, green beans and strawberries.

On the other hand, many data show that other factors such as variety, maturity, growing area, seasonal variations, etc. may be involved and lead to different results exceeding in importance the freezing rate effects. Furthermore, if it is considered that freezing is always followed by storage and distribution, the benefits of ultra rapid freezing may even disappear for some products.

With respect to the storage effects, Gutschmidt's (1969c) data on strawberries and green beans show that quality differences induced by various freezing rates are maintained for six to twelve months at constant temperatures between -20 and -30°C, but whether these quality differences are maintained at fluctuating temperatures during the cold chain is still a matter of debate and further research is needed.

It appears that when choosing the freezing method or freezing rate it is necessary to consider not only the quality retention during the freezing process but also the total quality loss between harvesting and eating. More information on the influence of factors involved in this matter is desirable and necessary.

2. INFLUENCE OF STORAGE AND DISTRIBUTION

2.1 Foodstuffs in general

In order to keep wholesomeness and quality of frozen foods at a high level by minimising biological, biochemical, chemical, physico-chemical and physical changes resulting in loss of the sensory and nutritional properties, it is necessary to maintain the temperature at a sufficiently low level during distribution. What is this level? The answer is not easy. It is well known that most foods show an increased storage life when the storage temperature is reduced. The reaction rate of chemical processes resulting in changes of one or more quality attributes of foods is temperature dependent. Variation in reaction rate may be expressed by the Q_{10} temperature quotient according to the relation:

$$Q_{10} = \frac{k_{t+10}}{k_t}$$

in which k_t is the reaction rate at temperature t, and k_{t+10} the reaction rate at $(t + 10)°C$. Thus the Q_{10} temperature quotient shows how many times the reaction rate increases when the temperature rises by 10°C. As a rough guide the rate of reaction often doubles for every 10°C rise in temperature. However, as has been pointed out by Fennema (1975b) reaction rates at subfreezing temperatures cannot be obtained simply by extrapolating data obtained at above freezing temperatures, and many reactions either increase in rate during the early stages of freezing or decline in rate less than would be expected.

Frozen foods stability during distribution closely related to reaction rates, appears as a very interesting problem, the solution to which is still very far from being achieved.

The world wide known time-temperature tolerance (TTT) project initiated in 1950, aimed to establish the way in which frozen foods tolerate the effects of time and temperature. The general concepts and principles of the time-temperature tolerance of frozen foods have been dealt with in detail by Van Arsdel et al. who included in their book not only the results obtained in the investigations performed by the United States Department of Agriculture's Western Regional Research Laboratory in Albany on frozen fruits and juices by Guadagni (1969) and on frozen vegetables by Olson and Dietrich (1969), but the reviews of studies carried out in other countries, namely on meat and meat products by Jul (1969), on seafood by Bramsnaes (1969), on frozen poultry meat and eggs by Dawson (1969) and on bakery products by Pence (1969).

Excellent reviews on the cold storage life of frozen fruits and vegetables, frozen meat products and frozen fish have also been made by Guadagni (1968), Jul (1968) and Dyer (1968) respectively. The examination of the above mentioned publications shows that in principle when the time for a certain degree of change in a quality attribute to develop is plotted against storage temperature in a semi-logarithmic diagram, straight lines are obtained, and that the time required at a given temperature for a change in quality to be detectable was reasonably reproducible for a given product lot, even when different methods of presenting the samples to the judges and different statistical procedures were used to treat the data. The time for a just noticeable difference in a quality (JND) to develop (70 - 80% correct answers of experienced judges in triangular sensory testing) is referred to as high quality life (HQL), while the practical storage life (PSL) or commercial acceptability of the product for marketing or for consumption, is determined by means of sensory acceptability scales, and it is the time required to cause a consumer complaint or rejection. HQL is in the order of two to five times shorter than PSL.

From a study of the above mentioned publications it is also clear that TTT relationships were not mathematical functions but empirical data subjected to large variability because of variations in product, processing and packaging, the so-called PPP factors. Löndahl and Danielson (1972), considering this problem of variability, compared Danish and Swedish TTT investigations and found very large differences in the results obtained. Going into deeper comparison, they found that besides the three PPP factors there was another 'P' - the fourth P of utmost importance - the Panel sensory evaluation with regard to human beings which involved method of analysis and definition of storage life.

Bengtsson et al. (1972) in an attempt to systemise TTT data as a basis for the development of time temperature indicators, studied and tabulated with regard to PPP conditions, storage conditions and technique for sensory evaluation. The data for individual foods and food products appeared in 75 individual reports on TTT investigations, compilations and reviews. The data assembled were critically reviewed.

The results of Bengtsson's work (1972) confirm the assumption that the variability in storage life within the different food groups is probably mainly caused by variations in raw material quality, processing conditions and packaging, and to some extent also by differences in sensory evaluations and packaging, and to a further extent by differences in sensory evaluation technique and panel discrimination ability. This shows that legal regulations for quality assurance based on some fixed temperature limits will be very difficult to relate to actual product quality or quality changes because of the large variability in TTT data, as already pointed out by Olson and Morgan Jr. (1970). Therefore, further information is necessary on the factors other than time and temperature influencing the quality stability of foods so that in programming and planning new investigations, those factors being the same, the real effect of time and temperature on quality loss may be determined.

From all the TTT investigations just mentioned it seems that the lower the storage temperature the higher the storage life of frozen foods, however, this rule cannot be applied to every product. Recent publications such as those of Lindeløv and Poulsen (1975) and Lindeløv (1976, 1977) have shown that in some products dominating processes resulting in quality loss of foods follow the Arrhenius equation, resulting in straight lines in a semi-logarithmic plot for storage versus temperature. In other foodstuffs counteracting influences disturb the regular picture. Fennema (1975b) gives a systematic description of the various possibilities when the influence of the concentration of solutes during freezing competes with the influence of the temperature, whilst Lindeløv (1977) suggests dividing the products into three main groups: with normal stability, a temperature decrease results in a longer storage life; with neutral stability, the temperature has no influence on the storage life; with reversed stability, a temperature decrease results in a shorter storage life. This division, supported by further research, might lead to a revision of the storage temperatures for some products of animal origin, so that perhaps higher temperatures could be recommended to maintain quality during extended periods of storage.

Computer-aided mathematical models have been presented by Singh (1976) to allow prediction of quality during frozen storage of foods. These show very interesting possibilities but stress that an important prerequisite in attempts to simulate quality of foods during frozen storage is to have reliable information on kinetics of reactions resulting in quality degradation. Thus, further research on this very important topic is needed.

However, the effects of time temperature variations resulting in quality loss have two special characteristics - additivity and commutativity.

Thus, when the TTT curve of a given product is known, the loss in storage life may be calculated for any link of the cold chain. However, numerical or graphical calculations are possible only when the TTT curve refers to the specific PPP factors of the product concerned.

In order to predict with any precision the storage life of a product, one should first know the PPP conditions and to have access to a vast amount of TTT investigations in order to find those of them referring to similar PPP conditions. This is not an easy task and therefore taking into account all these difficulties, approximate guides for the practical storage life of typical frozen food available throughout the world have been considered necessary. One of these guides was drawn up by the International Institute of Refrigeration (1972) for three storage temperatures: -18, -25 and $-30^{\circ}C$.

On the other hand, the Joint EEC Codex Alimentarius Group of Experts on Standardisation of Quick Frozen Foods, aims to draft standards of quality for end products for the ultimate consumer. Van Hiele (1977), President of this Group of Experts, in a recent publication, states that there is general acceptance that the temperature of a frozen food at the end of the freezing process and during storage must be $-18^{\circ}C$ or lower. In practice, storage temperatures are often much lower for many products for technological and economic reasons and because of these lower temperatures there will be a temperature reserve in the handling transport and distribution stages.

It is also generally accepted that the temperature for quick frozen foods transported over long distances should not be higher than -18°C. Considerable work has already been done in this field by the Economic Commission for Europe which has an agreement on the International Carriage of Perishable Foodstuffs by Road Vehicles.

A suitable temperature for local distribution purposes has been more difficult to arrive at. There has been some support for a maximum temperature of -12°C on a basis that according to the results of the TTT programme and many other investigations there would be no appreciable loss of quality during the comparatively short time taken for local distribution. However, opposition to this figure has stemmed from the fact that there would be instances where products were transferred directly to retail cabinets which would only serve to maintain the temperature to -18°C which desirably had to be achieved as soon as practicable.

As to the temperature in display cabinets, results of many studies in various countries, for instance those of Bramsnaes and Bøgh Sørensen (1976) and Bøgh Sorensen and Bramsnaes (1977) indicate the impracticability of all packages being maintained at -18°C and even at -15°C, therefore a product temperature of -12°C is allowed by the Codex Group of Experts in the least cold pack. Nevertheless, new research work is needed to determine the real quality loss of frozen foods during distribution because of temperature increases during the operations of handling, loading and unloading, and to determine the measures to be taken to avoid them. Recent work by Spiess et al. (1977) and Løndahl (1977) has considered all these points.

2.2 Fruits and vegetables in particular
2.2.1 Influence of time and temperature

Guadagni (1969) has given clear and reliable information concerning the average stability of frozen fruits and fruit products, as is shown in Table 1. From TTT investigations it seems that for most frozen fruits no measurable changes occur for periods of five years or more when the products are held at $-29°C$ or lower.

TABLE 1

AVERAGE STABILITY OF FROZEN FRUITS AND FRUIT PRODUCTS HELD AT VARIOUS TEMPERAT

Product	Type of Pack	Stability - days		
		0°F (-18°C)	10°F (-12°C)	20°F (-7°C)
Apples	Pie filling	360	250	60
Boysenberries	Pie filling	375	210	45
	Bulk, no sugar	405	125	45
	Retail, syrup	650	160	35
Blueberries	Pie filling	175	77	18
Cherries	Pie filling	490	260	60
Peaches	Pie filling	490	280	56
	Retail, syrup	360	45	6
Blackberries	Bulk, no sugar	630	280	50
Raspberries	Bulk, no sugar	720	315	70
	Retail, syrup	720	110	18
Strawberries	Bulk, dry sugar	630	90	18
	Retail	360	60	10

Source: Guadagni, D.G. (1969) In: Quality and stability in frozen foods (Van Arsdel, W.B., Copley, M.J. and Olson, R.L. eds.) p.96. Wiley - Interscience, New York and London

If the average stability, expressed in days, is plotted on a logarithmic scale against temperature, an exponential relationship between stability and storage temperature is obtained, namely the logarithm of stability in days varies linearly with temperature within the range of temperatures -18, -12 and -7°C.

In all of the fruits studied in the TTT Project, except peaches, the q_{10} for colour changes was equal to or less than that for flavour, indicating that flavour is the quality attribute which is most affected by temperature changes. In peaches, however, colour was the most important factor that was degraded. The q_{10} values for colour change in frozen peaches were higher than for any other fruit studied, as indicated in Table 2.

TABLE 2

AVERAGE TEMPERATURE QUOTIENTS (q_{10}) FOR RATES OF QUALITY CHANGE IN FROZEN FRUITS

Product	Type of Pack	Temperature range, °F	Temperature quotient, q_{10}, for Flavour	Colour
Apples	Pie filling	10 - 20	4.0	2.4
Boysenberries	Bulk, IQF	10 - 30	3.1	2.8
	Retail, syrup	10 - 30	5.1	3.6
	Pie filling	10 - 20	4.3	3.1
Blueberries	Pie filling	10 - 20	4.4	2.2
Cherries	Pie filling	10 - 20	4.7	4.4
Peaches	Pie filling	10 - 20	5.0	8.8
	Retail, syrup	10 - 25		7.8
Blackberries	Bulk, no sugar	0 - 30	3.1	2.8
Raspberries	Retail, syrup	10 - 30	3.6	6.6
	Bulk, no sugar	0 - 30	3.8	3.2
Strawberries	Retail, dry sugar	0 - 30	6.0	6.0
	Bulk, dry sugar	0 - 30	6.0	6.0

Source: As Table 1.

In fruits, the quality change represented by completion of the stable period at temperatures in the -18°C to -1°C (0 - 30°F) range is approximately constant. This means that, for example, a 10-day exposure at -7°C (20°F) or a 60-day exposure at -12°C (10°F) is neither better nor worse than a year's exposure at -18°C (0°F) for frozen strawberries. Therefore, it is just as important to consider time as it is to consider temperature in guarding the original quality of frozen fruits.

As to the influence of fluctuating temperatures on quality, Guadagni (1969) points out that the effective temperature of a given fluctuating cycle depends on the amplitude of the cycle and the temperature coefficient or q_{10} for the particular quality attribute being measured. Thus, when exposed to fluctuating cycles, fruits with high q_{10} values change at a rate equivalent to a steady temperature well above the mean temperature of the cycle. On this basis, appreciable temperature fluctuations are undesirable, because as was mentioned, quality changes are additive and commutative.

As in the case of fruits, publications of the TTT series have given very valuable information concerning the stability of frozen vegetables, especially green beans, peas, spinach and cauliflower. Olson and Dietrich (1969), in Van Arsdel's publication, summarise the results of the TTT Project and conclude that frozen vegetables are basically stable enough to maintain high quality from one harvest season to the next when held at $-18^{\circ}C$ ($0^{\circ}F$) and that in five years at $-31^{\circ}C$ ($-21^{\circ}F$) the vegetables mentioned show no measurable or detectable loss of colour or flavour and no measurable changes in chemical constituents or attributes.

Plenty of work has been done on chlorophyll conversion to pheophytine, ascorbic acid loss and development of 'off flavours' in frozen vegetables as a measure of stability. From the TTT Project, Olson and Dietrich (1969) have assembled the results for frozen green beans, peas, cauliflower and spinach (Table 3). In storage at a given temperature, a colour change is detected before a flavour change, except in the case of spinach. Green beans with an initially high retention of chlorophyll after blanching have a relatively low rate of deterioration during subsequent storage in the frozen state. This relationship has not been found in peas.

TABLE 3

DAYS IN STORAGE AT VARIOUS TEMPERATURES REQUIRED TO BRING ABOUT A PERCEPTIBLE CHANGE IN QUALITY

$^{\circ}F$	$^{\circ}C$	Beans		Peas		Cauliflower		Spinach	
		Colour	Flavour	Colour	Flavour	Colour	Flavour	Colour	Flavour
0	-18	101	296	202	305	58	291	350	150
10	-12	28	94	48	90	18	61	70	60
15	-9.5	15	53	23	49	10	28	35	30
20	-7	8	30	11	27	6	13	20	20
25	-4	4	17	5	14	3	6	7	8

Source: Olson, L. and Dietrich, W.C. (1969) In: Quality and stability in frozen foods (Van Arsdel, W.B., Copley, M.J. and Olson, R.L., eds.) p.124. Wiley Interscience, New York and London.

Chlorophyll conversion in green beans and peas, and development of a tan colour in cauliflower, are generally linear with time. The rate of deterioration increases exponentially with storage temperature in the temperature range between -18 and -4°C (0 and 25°F). Chlorophyll deterioration for chopped spinach is about twice that for leaf spinach; it is thought that greater manipulation in processing creates conditions whereby chlorophyll deteriorates at a faster rate in subsequent storage.

The TTT study, as has been pointed out by Olson and Dietrich (1969) indicates that the just noticeable difference in the quality of many frozen vegetables occurs in less than a year at -18°C (0°F) and that the exponential temperature effect on deterioration rate means that a few degrees difference in temperature for long storage can cause a great difference in quality retention. On the other hand, since time is also an important factor, it follows that short exposures to temperatures above -18°C are not more damaging than long exposures at lower temperatures. Every bit of deterioration is, as in the case of fruits, additive and a prolonged rise in temperature during distribution stages can add up to a product showing a significant deterioration at the time of purchase.

From the TTT studies it may also be concluded that there is much variation in stability between lots of a particular frozen vegetable. This variability may be related to variety, harvest year, growing area, processing conditions or packaging. Further studies on this matter have been carried out in various countries to investigate interactions of all these factors. As an example, Steinbuch et al. (1977) have studied recently the stability of two frozen green bean varieties grown in Germany and in The Netherlands and have concluded that it does not seem possible to generalise regarding the storage behaviour of green beans because of the pronounced effect of variety, maturity at the point of harvest, growing area, pretreatments, etc. on the stability, so that information regarding these factors should always be given together with information about the time temperature related quality changes.

New research works aiming to elucidate the influence of these various factors which are involved in stability changes of frozen vegetables are necessary.

With regard to nutritional aspects, carbohydrates and the low amounts of fats and proteins of frozen fruits and vegetables do not change during normal storage at $-18^{\circ}C$ or lower temperatur and the same applies to mineral contents.

Carotene retention in vegetables is 80 - 100% during one year storage at $-18^{\circ}C$. The B vitamins will be retained to a high degree if the product is stored under conditions which retain ascorbic acid as the retention variable from vegetable to vegetable. During storage at $-18^{\circ}C$ ($0^{\circ}F$) for one year, accordin to the data of the IIR (1972), vitamin B_1 retention is as follows asparagus 90; broccoli 75; green beans 70; cauliflower 50; peas 90; and spinach 50%. Vitamin C loss in green beans and peas is linear with time and, as in the case of chlorophyll conversion, the rate of deterioration increases exponentially with storage temperature in the range between -18 and $-4^{\circ}C$ (0 and $25^{\circ}F$).

Martin (1976) quotes the Report of the Wisconsin Alumni Research Foundation (1956) and Fennema (1975c) complete review on nutritional evaluation of frozen foods. Referring especially to fruits and vegetables, he states that canning of vegetables and fruits often results in 2 - 3 times greater losses of niacin and vitamins B-1, B-2 and C than those occurring in vegetables during freezing, storing and cooking, or in fruits during freezing and frozen storage. Therefore it appears that the freezing process (pre-freezing treatments, freezing, frozen storage and distribution), if properly conducted, is generally regarded as the best method of long-term fruit and vegetable preservation, when judged on the basis of retention of sensory attributes and nutrients.

2.2.2. Influence of PPP factors

Factors related to product, processing and packaging, the PPP factors, appear as a wide field of research, because of their marked influence on quality and stability of frozen fruit and vegetables during storage and distribution.

As to the product factor, the suitability for freezing varies greatly with type and variety and one of the most fruitful avenues for minimising the textural-structural damage in frozen fruits and vegetables has been found in the selection and breeding of varieties. Many authors, such as Herrmann (1970); Philippon (1967);Cousin (1976); Risser et al. (1976); Latrasse (1976); Crivelli and Scozzoli (1975); Crivelli and Rosati (1975) and Crivelli et al. (1975), have studied this important matter.

Plant geneticists are trying to bring all the factors which are important to frozen product quality into a single variety but complete success is difficult to attain because it is also necessary to take into account economic and commercial factors such as yields, disease resistance, ease of harvesting and handling, size and appearance.

Further research is needed to extend the scope of vegetable products which can be successfully quick frozen: peaches, nectarines, raspberries, blackberries, mangoes, mushrooms and asparagus. Most of this detailed work should be envisaged to obtain more suitable varieties for freezing resulting in a higher quality end product, for instance, varieties lacking in polyphenolase enzyme or oxidisable phenolic constituents less susceptible to browning than those which are high in either or both of these factors.

Ulrich (1969) has pointed out the importance of time of harvesting, rainfall, humidity and application of fertilisers as factors influencing also the texture of the frozen product that are desirable fields of research.

As to maturity, most fruit and vegetables intended for

freezing are harvested at the same state of maturity as for fresh consumption, however, determination of the optimum time for harvesting is very important for some species and varieties. Sweetcorn, green peas, asparagus, raspberries and strawberries are among the most susceptible produce to both retarded harvesti and delayed processing.

As to the <u>processing</u> factor, preparatory or prefreezing treatments are usually necessary to obtain ready-to-use products and to provide the best conditions of preservation. Initial grading, cleaning, sorting, removal of defective produce and inspecting, are the main common operations for fruits and vegetables. Pitting, halving, dicing, slicing or mincing, are used only with certain produce.

Most preparatory treatments are mechanised and equipment and operation are constructed and performed so that minor bruising of produce occurs, thus preventing the development of abnormal colour and off-flavours.

Blanching is the most important prefreezing treatments for most vegetables; it inactivates catalase and peroxidase lightens texture, stabilises flavour and reduces total bacterial count.

Stoll (1970) has recently made an important study on the suitability of forty different vegetables for freezing and has determined optimum blanching conditions.

Blanching may be carried out in almost boiling water, in steam, or in a combination of both. Mechancial arrangements to obtain uniform heating of all pieces and control of piece size is important to shorten the heating time for adequate blanching. As to chlorophyll conversion, water provides highest retention and lengthy blanching lessens retention.

With regard to losses of carbohydrates, organic acids, mineral salts and vitamins, blanching is the worst offender resulting in a loss of nutritional quality of this type of food

In boiling water blanching losses of nutrients become greater as the time of contact is lengthened, whereas with steam blanching the loss becomes less important.

Conventional steam blanching at atmospheric pressure needs 30 - 50% longer duration than water blanching. However, as reported by Philippon (1975), new steaming methods, 'fluidised bed' and 'IQB' blanching, allow highly uniform blanching and shorter duration times of the process resulting in negligible loss of nutrients. Recently developed procedures have moved towards a short time, high temperature, water blanching, followed by rapid cooling, for better retention of colour and nutrients. However, the necessary blanching for most vegetables results in considerable loss in water soluble nutrients. Sulc (1975) indicates up to 35% for carbohydrates, 40% for mineral salts, and 40% for thermolabile vitamins such as, for example, vitamin C.

Volume of blanching water is a very important factor in determining the magnitude of vitamins B and C loss.

Reported results about nutrient retention are often conflicting but this is not surprising taking into account the many variables involved in the production of frozen fruits and vegetables and the variability of nutrients in the raw material.

More data are needed on nutrient losses for different vegetables and on blanching techniques so that control with respect to time-temperature effects on raw material during the blanching operation may be effective and ensure a high quality and nutritive frozen vegetable.

Some vegetables with a high natural flavour or low enzymic activity can be preserved by freezing without blanching. Recent works of Sulc (1975) and Kozlowski (1977) support previous reports showing that onions, leeks, cucumbers, green peppers, dill and parsley, can be preserved by freezing without blanching, resulting in better retention of nutrients. Further research is desirable to determine to what extent freezing without blanching may be extended to other vegetables and stability may be kept during subsequent frozen storage. This type of study is of the greatest interest because of possible energy saving and of nutritional improvement. With the exception of these products blanching appears necessary prior to freezing.

Sweetening is a prefreezing treatment of major importance for fruits. The addition of sugar and syrup has been discussed in terms of their ability to exclude oxygen from the fruit and thus help to preserve colour and appearance. Sugar syrup is considered as a better protecting agent than dry sugar. The use of sweeteners also aims at enhancing the natural fruit flavour, protecting the fruit against the action of enzymes.

Dry sugar may produce the effects mentioned if added one or two hours before freezing to allow sufficient solubilisation in the fruit juice, and in the proportion of 1 : 3 to 1 : 10 mass ratio depending on the type of produce and consumer preferences. Sugar syrup with a 30 to 60% sugar content must cover the fruit completely to prevent browning, acting as a barrier to oxygen transmission. Ascorbic acid is often added to the syrup in percentages ranging from 0.05 to 0.25% as an anti-browning agent. Oxygen oxidises the ascorbic acid and is used up before it can oxidise the phenolic constituents.

Guadagni (1969) reports that in products packed without sugar, the flavour differences are due to chemical or enzymatic changes within the fruit itself rather than to transfer of constituents by osmotic action. In strawberries and raspberries packed with sugar the physical changes arising from solids transfer appear to be faster than the chemical or enzymatic changes at temperatures above $-18^{\circ}C$ ($0^{\circ}F$) and the transfer of acids and other flavouring constituents from fruit to syrup or sugar phase is rather rapid as the temperature is increased also above $-18^{\circ}C$ ($0^{\circ}F$).

With regard to the <u>packaging</u> factor, Guadagni (1969) points out that q_{10} values for flavour changes in fruit pies

are somewhat higher than those for bulk pack fruits but lower
than those for syruped retail packs. Therefore it appears that
increases in temperature affect flavour changes most in the
retail packs, next in pies and least in bulk packs.

In cans, browning is much less than in metal and composite
cartons and improvement obtained by a hermetic container suggests
that partial vacuum packing further improves colour retention.
There is little doubt that packing in hermetic containers or in
sealed packages preventing oxygen access and water vapour loss,
not only preserves natural colour in peaches and cherries but
also gives better flavour and colour protection to strawberries
and retains better vitamin C levels during storage.

Recent and successful attempts (1976) have been made to
freeze peaches without the previous addition of sugar or sugar
syrup, by dipping peeled fruits in a cooling ascorbic acid
solution and subsequent cryogenic freezing and packaging in
'vacuum barrier boxes' with plastic interliners where oxidation
of fruit is prevented under vacuum sealing. This is a clear
example of the influence of packaging techniques and materials
on the maintenance of stability of frozen fruits and of the
possibilities and need of research in this field. Vacuum
packaging of blanched and non blanched frozen vegetables as a
means of improving stability during subsequent storage may
also appear as an interesting research field for some products.

The most common changes in the quality attributes of
frozen fruits and vegetables are poor texture and drip in most
fruits and some vegetables; diminished and off-flavours in
vegetables, and alterations of colour in fruits and vegetables.

Mushiness, flaccidity, loss of crispness and excessive
drip on thawing depend primarily on the suitability of the
produce for freezing and the variety concerned, but liquid R 12,
nitrogen or carbon dioxide and fluidised freezing may reduce or
retard such changes as well as cause weight loss during the
freezing process.

Cooked frozen beans are soft compared with the crisp texture of cooked fresh green beans; liquid nitrogen or fluidised bed freezing, for instance, can prevent this defect as well as the mushy aspect of berries on thawing. Delayed harvesting and processing may result in excessively softened fruits and toughened vegetables, as happens in the case of peas where not only toughening but the development of off-flavour may occur since the chemistry of these two deleterious changes is largely unknown. Further research in this field would be desirable.

Diminished flavour and sweetness in vegetables due to losses of water soluble compounds are caused, as was mentioned, by excessive blanching and subsequent cooling in water - the so-called 'leaching effect'. Leaching losses are affected by the particle size, the greater the surface area, the greater the loss of nutrients, particularly where cut vegetable surfaces are involved.

Off-flavours are primarily due to enzyme activity. The blanching of vegetables prior to freezing increases their storage stability, at least in part, because of the denaturation of enzyme proteins. Hydrolytic as well as oxidative enzymes may produce changes in composition and flavour in frozen vegetables and it has been demonstrated that some enzymes such as catalase and lipase may be even more active in the frozen state; in particular, lipase and lipoxidase have proved to be active at low temperatures, causing changes in the lipid constituents. Lee et al. (1955) demonstrated that rancidity development occurs in unblanched frozen vegetables in a relatively short time at $-18^{\circ}C$, with off-flavour formation. Besides acetaldehyde and ethanol, hexanol has been reported by Bengtsson and Bosund (1964) as a characteristic compound of autoxidised lipid material detected in unblanched frozen peas after storage. However, the flavour change which occurs in the frozen storage of blanched vegetables such as peas, snap peas, spinach and cauliflower, is not rancidity development because lipoxidase is rapidly inactiva by a short blanching time, as reported by Rhee and Watts (1966).

Peroxidase and catalase are highly resistant to heat inactivation therefore they have been used as an index of adequacy of heat treatment in the blanching of vegetables, catalase being much more labile than peroxidase, perhaps due to to locale; catalase occurs in the cytoplasm of the cells whereas peroxidase is associated with the cell walls.

The partial reversibility of the inactivation of peroxidase has been quoted by Leeson (1957) who showed, with many examples, the revival of peroxidase activity has been found to depend on three main factors: the test used for detecting the activity; the severity of the heat treatment and the storage conditions. According to Nebesky et al. (1950) it seems that the guaiacol test gives the most accurate results as compared to o-phenylendiamine and pyrogallol; it appears that regeneration is most likely to occur when the enzyme has not been wholly inactivated or has been heated just to the elimination point as judged immediately after heat treatment and when stored at room temperature, but not during freezer storage.

As a result of residual enzymatic activity, off-flavours will become more pronounced as storage time is prolonged and temperature is above $-18^{\circ}C$, thus quality of frozen vegetables may deteriorate even when the possibility of continuing enzymatic activity is minimal. However, neither peroxidase nor catalase have been specifically indicated as an initial cause of off-flavours but the enzymes probably involved in enzymatic off-flavour formation are closer to the stability of peroxidase. Off-flavours may also originate from delay in processing and from deterioration in mishandled or bruised produce before freezing.

Accordingly, the nature of the off-flavours and its mode of formation remain unknown and further research is needed in this interesting but difficult field.

Colour changes in frozen fruits and vegetables may be due to the loss of the true colour with the formation of an off-colour, to the loss of colour to the surrounding medium and to the development of an off-colour from originally non-coloured constituents.

The change of chlorophyll from bright green colour to duller and more yellow pheophytine is due to a simple hydrolysis where magnesium is removed and replaced by hydrogen. This can be minimised by use of a proper combination of time and temperature in blanching and by storage at temperatures not above $-18^{\circ}C$ ($0^{\circ}F$).

Brussels sprouts insufficiently blanched develop an undesirable pink colouration at the centres during frozen storage. It is undesirable to try to prevent this pink colouration by severe blanching since that would cause an excessive loss of the green chlorophyll colour. It is, therefore necessary to achieve a compromise between the two effects.

Enzymatic browning is due to polyphenoloxidase reacting directly with the oxygen in the air and takes place in most stone and many seed fruits (peaches, apricots, plums, prunes, cherries, apples, pears) and some vegetables (eggplant, potatoes, mushrooms) containing polyphenolic substrates which in the presence of the enzyme and oxygen are firstly changed to quinones and later to brown polymers. At the quinone stage the reaction can be reversed by reducing agents such as ascorbic acid. However, at the polymer stage, reducing agents are ineffective.

Polyphenoloxidase may be inactivated by heating or inhibited by chloride ions, acids lower than 2.5 pH or bases higher than 8 to 10 pH, however the acid or salt protection is only temporary since the amount needed to act permanently would make the produce unpalatable. In the case of apple slices, sulphur dioxide complexes form with the quinones and block the subsequent development of brown polymers.

The exclusion of oxygen in the package will prevent the darkening and prolong the high quality storage life of the product, therefore displacing the oxygen with sugar solutions or inert gases, consuming it by adding glucose-oxidase, use of oxygen impermeable films, and vacuum packaging, also prevent or retard browning as well as other oxidative colour changes. However, the problem is not completely solved and further research is needed, for example, on the influence of combining several inhibitors, time and temperature of storage.

Loss of colour to the surrounding medium occurs in frozen berries and red pitted cherries where anthocyanins pass into the sugar syrup. This change occurs very slowly at $-18^{\circ}C$ or lower temperatures but is greatly accelerated at higher temperatures and on thawing.

Discolouration and anthocyanin pigment changes have been studied recently by Jedrzejewska (1975) showing that the decrease of anthocyanin content in strawberries and raspberries during storage at $-18^{\circ}C$ and at higher temperatures is often accompanied by the development of unidentified brown substances, probably products of anthocyanin decomposition; strawberries being more sensitive to this change than raspberries. Variety has a great influence on this type of change. As polyphenolase is absent in strawberries, the colour change from bright red to a dull appearance cannot be caused by this enzyme - the mechanism of this change is unknown but it may also be oxidative.

Some vegetable products, particularly when stored under adverse temperature conditions, develop an off-colour from originally non-coloured constituents. This is the case with cauliflower where a water-acetone soluble dark colour of unknown origin is formed during long storage at temperatures above $-18^{\circ}C$ ($0^{\circ}F$).

As to the assessment of these changes in frozen fruits and vegetables, subjective and objective testing procedures are used.

It appears that unless the validity of the objective measuring techniques is established, stability comparisons

can be subject to large errors. The wide differences reported
by different investigators for the stability of the same commodity
can be ascribed partly to differences in the method of measurement
partly to the variability of the material. Therefore chemical
and physical changes occurring as frozen fruits and vegetables
deteriorate, can be measured and tested for their relationship
to changes in quality attributes and are meaningful only when
they correlate significantly with subjective testing procedures.
It still remains very difficult to correlate off-flavours with
specific chemical compounds so that changes in flavour cannot
so far be assessed objectively. Further research in this
difficult field would be very valuable and desirable.

REFERENCES

Anon. 1956. Quick Froz. Fds. 6, 127-130
Anon. 1976. Quick Froz. Fds. 3, 34-36, 104-106.
Aström, S. and Löndahl, G. 1969. Bull. Int. Inst. Refrig. Annex 1969-6, 121-127.
Bassi, M. and Crivelli, G. 1969. Rev. Gen. Froid, 60, 1239-1250.
Bengtsson, B. and Bosund, I. 1964. Food Technol. 18, 773-776.
Bengtsson, N., Liljemark, A., Olson, P. and Nilsson, B. 1972. Bull. Int. Inst. Refrig. Annex 1972-2, 303-311.
Bøgh Sørensen, L. and Bramsnaes, F. 1977. Meeting of the Commission C2 of the Internation.Institute of Refrigeration, paper no. 5.6., 6pp, Karlsruhe
Bramsnaes, F. 1969. In: Quality and stability in forzen foods (Van Arsdel, W.B., Copley, J. and Olson, R.L. eds.) pp 217-236, Wiley-Interscience, New York and London.
Bramsnaes, F. and Bøgh Sørensen, L. 1976. Compt. Rend. Colloque CENECA, Le froid en agriculture, rapport no.5226, 6pp, Paris.
Ciobanu, A., Lascu, G., Bercescu, V. and Nicolescu, L. 1976. Cooling technology in the food industry, pp 23-97, Abacus Press, Tunbridge Wells, Kent.
Connell, J.J. 1964. In: Proteins and their reactions (Schultz, H.W. and Anglemier, A.F. eds.) p.255, AVI Publishing Co, Inc., Westport, Conn.
Connell, J.J. 1968. In: Low temperature biology of foodstuffs (Hawthorn, J. and Rolfe, E.J. eds.) pp 333-358, Pergamon Press, Oxford.
Cousin, R. 1976. Compt. Rend. Colloque CENECA, Le froid en agriculture, rapport no. 5225, 6 pp, Paris.
Crivelli, G. and Stozzoli, R. 1975. XIIIth Int. Congr. Refrig. paper no. C2, 20, 4 pp, Moscow.
Crivelli, G. and Rosati, P. 1975. XIIIth Int. Congr. Refrig. paper no.C2, 18, 5 pp, Moscow.
Crivelli, G., Fideghelli, C. and Monastra, F. 1975. XIIIth Int. Congr. Refrig. paper no. C2, 19, 7 pp, Moscow.
Dawson, L.E. 1969. In: Quality and stability in frozen foods (Van Arsdel, W.B., Copley, J. and Olson, R.L. eds.) pp. 143-167, Wiley-Interscience, New York, and London.
Dyer, W.J. and Dingle, J.R. 1961. In: Fish as a food (Borgstrom, G. ed.) vol. 1, pp. 275-327. Academic Press, New York.

Dyer, W.J. 1968. In: Low temperature biology of foodstuffs (Hawthorn, J. and Rolfe, E.J. eds.) 1st edn. pp 429-447, Pergamin Press, Oxford and New York.

Fennema, O. and Powrie, W.D. 1964. In: Advances in food research (Chichester, C.O., Mrak, E.M. and Stewart, G.F. eds.) pp 219-347, Academic Press, London and New York.

Fennema, O. 1975a. In: Water relations of foods, (Duckworth, R.B. ed.) pp 397-413, Academic Press, London, New York and San Francisco.

Fennema, O. 1975b. In: Water relations of foods (Duckworth, R.B. ed.) pp 539-556. Academic Press, London, New York and San Francisco.

Fennema, O. 1975c. In: Nutritional evaluation of food processing (Harris, R.S. and Von Loesecke, H.W. eds.) AVI Publishing Co. Inc., Westport, Conn.

Guadagni, D.G. 1968. In: Low temperature biology of foodstuffs (Hawthorn, J. and Rolfe, E.J. eds.) 1st. edn. pp. 399-412, Pergamon Press, Oxford and New York.

Guadagni, D.G. 1969. In: Quality and stability in frozen foods (Van Arsdel, W.B., Copley, J. and Olson, R.L. eds.) pp 85-116, Wiley Interscience, New York and London.

Gutschmidt, J. 1968. In: Low temperature biology of foodstuffs (Hawthorn, H. and Rolfe, E.J. eds.) 1st. edn. pp 299-318, Pergamon Press, Oxford and New York.

Gutschmidt, J. 1969. Conserva, 17, 266.

Gutschmidt, J. 1969b. Kältech.-Klim., 21, (12), 355.

Gutschmidt, J. 1969c. Bull. Int. Inst. Refrig.Annex 1969-6, 147-150.

Haynes, J. In: Low temperature biology of foodstuffs (Hawthorn, J. and Rolfe, E.J. eds.) 1st. edn. p.79, Pergamon Press, Oxford and New York.

Herrmann, K. 1970. Ernährungsumschau, 17, 11, 458-463.

Van Hiele, T. 1977. Meeting of the Commission C2 of the International Institute of Refrigeration, paper no.5.1., 5 pp, Karlsruhe.

International Institute of Refrigeration 1972. Recommendations for the processing and handling of frozen foods, 2nd. edn. I.I.R. Paris.

Jedrzejewska, J. 1975. XIIIth Int. Congr. Refrig. paper no.C2, 16, 10 pp, Moscow.

Jul, M. 1969. In: Quality and stability in frozen foods (Van Arsdel, W.B. Copley, J. and Olson, R.L. eds.) pp 191-216, Wiley-Interscience, New York and London.

Jul, M. 1968. In: Low temperature biology of foodstuffs (Hawthorn, J. and
    Rolfe, E.J. eds.) 1st. edn. pp 413-428, Pergamon Press, Oxford and
    New York.
Kondrup, M. and Boldt, H. 1960. Bull. Int. Inst. Refrig. Annex 1960-3,
    309-330.
Kozlowski, A.V. 1977. Meeting of Commission C2 of the International
    Institute of Refrigeration, paper no. 3.2, 10 pp, Karlsruhe.
Latrasse, A. 1976. Compt. Rend. Colloque, CENECA, Le froid en agriculture,
    rapport no.5228, 6 pp, Paris.
Lee, F.A., Wagenknecht, A.C. and Hening, J.C. 1955. Food Res. 20, 289-297.
Leeson, J.A. 1957. Fruit and vegetable canning and quick freezing - Research
    Association Sci. Bull. 2.
Lindeløv, F. and Poulsen P. 1975. XIIIth Congr. Refrig. paper no. C2, 52,
    12 pp. Moscow.
Lindeløv, F. 1976. Bull. Int. Inst. Refrig., Annex 1976-1, 181-188.
Lindeløv, F. 1977. Meeting of the Commission C2 of the International
    Institute of Refrigeration, paper no.3.6, 13 pp, Karlsruhe.
Løndahl, G. and Danielson, C.E. 1972. Bull. Int. Inst. Refrig. Annex
    1972-2, 295-301.
Løndahl, G. 1977. Meeting of the Commission C2 of the International
    Institute of Refrigeration, paper no.5.3., 7 pp. Karlsruhe.
Love, M. 1966. In: Criobiology (Meryman, H.T. ed.) pp. 317-405,
    Academic Press, London and New York.
Martin, S. 1976. Quick froz. fds. 7, 14.
Mohr, W.P. 1971. J. Text. Stud. 2, 316.
Mohr, W.P. 1974. J. Text. Stud 5, 13.
Mohr, W.P. and Stein, M. 1969. Criobiology, 6, 15.
Monzini, A., Bassi, M. and Crivelli, G. 1969. Bull. Inst. Int. Froid,
    Annex 1969-6, 47-51.
Nebesky, E.A., Esselen, W.B., Kaplan, A.M. and Fellers, C.R. 1950. Food
    Res. 15, 114-124.
Nguyen, V.X. 1969. Bull. Inst. Int. Froid, Annex 1969-6, 41-46.
Olson, R.L. and Dietrich, W.C. 1969. In Qaulity and stability in frozen
    foods (Van Arsdel, W.B., Copley, J. and Olson, R.L. eds.) pp 117-141,
    Wiley-Interscience, New York and London.
Olson, R.L. and Morgan Jr., A.J. 1970. Proc. Third Int. Congr. Food Sci.
    and Technol. pp 580-584.

Partmann, W. 1964. Probleme der Ernährung durch Gefrierkost, p 32, Dietrich Steinkopf Verlag, Darmstadt.

Partmann, W. 1972. Bull. Int. Inst. Refrig. Annex 1972-2, 77-84.

Partmann, W. 1975. In: Water relations of foods (Duckworth, R.B. ed.) pp 505-537, Academic Press, London, New York and San Francisco.

Pence, J.W. 1969. In: Quality and stability in frozen foods (Van Arsdel, W.B., Copley, J. and Olson, R.L. eds.) pp. 169-199, Wiley-Interscience, New York and London.

Philippon, J. 1967. Rev. Prat. Froid. $\underline{20}$, 261, 2-9.

Philippon, J. 1975. Bull. Techn. Inf. $\underline{296}$, 49.

Rhee, K.S. and Watts, B.M. 1966. J. Food Sci. $\underline{31}$, 675-679.

Risser, G., Vaillen, J., Ferry, P. and Cabibel, M. 1976. Compt. Rend. Colloque CENECA, Le froid en agriculture, rapport no.5227, 10 pp, Paris.

Sikorski, Z.E., Olley, J. and Kostuch, S. 1976. Protein changes in frozen fish, Critical Reviews in Food Science and Nutrition, $\underline{8}$, 97.

Sikorski, Z.E. 1977. Meeting of the Commission C2 of the International Institute of Refrigeration, paper no. 02, 10 pp, Karlsruhe.

Singh, R.P. 1976. Bull. Int. Inst. Refrig., Annex 1976-1, 197-204.

Spiess, W.E.L., Wolf, W., Wien, K.J. and Jung, G. 1977. Meeting of Commission C2 of the International Institute of Refrigeration, paper no. 5.4, 5 pp, Karlsruhe.

Steinbuch, E., Spiess, W.E.L. and Grünewald, Th. 1977. Meeting of the Commission C2 of the International Institute of Refrigeration, paper no. 3.3, 9 pp, Karlsruhe.

Stoll, K. 1970. Schweiz. Landw. Forsch., IX, 3-4, 327-360.

Sulc, S. 1975. XIIIth Int. Congr. Refrig. paper no. C2, $\underline{15}$, 19 pp, Moscow.

Ulrich, R. 1969. Rev Gen. Froid, $\underline{60}$, 1, 23-28.

Wolford, E.R., 1967. Proc. XIIth Int. Congr. Refrig. vol. 3, 459-467.

SESSION 6

COOKING (DOMESTIC AND INSTITUTIONAL)

Chairman: M.J. Woods

EFFECTS OF DOMESTIC AND LARGE SCALE COOKING ON THE QUALITY AND NUTRITIVE VALUE OF VEGETABLES AND FRUITS

Rosmarie Zacharias
Institute of Domestic Economics
Stuttgart, W. Germany

Examination of consumption statistics and menu plans reveals that vegetables are consumed almost daily, both in the home and in catering institutions. The great variety of vegetables are available both fresh and preserved, and the many ways in which they can be prepared allow a great variety in menu composition. The daily consumption of fruit, especially fresh fruit, is also relatively high. Unlike vegetables, berries, stone fruits and apples are subject to less cooking both in the home and in large-scale institutions. In the majority of cases industrially processed quick-frozen or heat-treated products do not require any further treatment prior to consumption.

This paper is specifically concerned with the effects of cooking on the quality of fruits and vegetables. On the basis of the results obtained, future research requirements are indicated.

In view of the conditions under which meals are usually prepared in large-scale catering establishments coupled with the frequent criticism of the meal quality, possible changes in the nutritive and sensory value occurring during holding are discussed.

COOKING PROCEDURES FOR VEGETABLES, POTATOES AND FRUITS

The purpose of cooking is to convert by adequate heat treatment foods which have been already prepared for the kitchen into ready-to-eat foods. The typical cooking procedures for vegetables, potatoes and fruits are shown in

TABLE 1

COOKING PROCEDURES FOR VEGETABLES, POTATOES AND FRUITS

Cooking Procedures	Cooking conditions				
	Medium	Temperature °C	Time		
			Vegetables min	Potatoes min	Fruits min
Boiling	Water	Approx. 100	5 - 90	20 - 25	-
Stewing	Water + steam Fat	" "	5 - 75	-	2 - 20
Steaming	Steam	" "	5 - 45	20 - 30	-
Pressure-steaming	Steam	102 - 118	3 - 60	10 - 15	2 - 5
Deep frying	Fat	Approx. 100	-	Approx. 10	-
Frying in the pan	Fat	" "	-	10 - 15	-

Table 1. Of about 15 feasible procedures of heat treatment boiling, stewing, steaming and pressure-steaming are the most important; in the case of potatoes, however, deep-frying and pan-frying should be included. Recently also baking in foils and microwave cooking have assumed great importance. Table 1 further outlines the most essential characteristics of the individual cooking procedures (under the aspect of quality retention and of the specific reactivity of the nutrition-physiologically important nutrients) including:

1) Nature of the heat-transfer medium
2) Temperature of the heat-transferring medium
3) Duration of cooking.

The heat-transfer medium may be either water steam and/or fat. In addition boiling and stewing procedures differ in relation to the quantity of liquid used. By definition, the material to be boiled must be completely immersed in water, whereas substantially less water is used for stewing.

The temperature of the heat-transfer medium ranges from $100°C$ to $180°C$, depending on the procedure involved. With such cooking procedures as boiling, stewing, steaming and pressure-steaming, the temperature of the food material is roughly equivalent to that of the heat-transfer medium. In contrast, with deep-frying and pan-frying the temperature of the heat-transfer medium is reached only at the food surface. At the centre of the foods, the temperature seldom reaches $90°C$, due to the relatively poor heat conductivity and the continuous heat loss through water evaporation.

The duration of the heat treatment in the individual procedures ranges from 5 to 90 minutes depending mainly on the nature of the food, the degree to which the food was comminuted and the actual cooking temperature. Usually shorter times are required in the cooking or stewing of fruits than in the cooking of vegetables or potatoes. Due to an increase of the cooking temperature by 10°C in pressure-steaming, the cooking time to reach the same cooking degree is reduced by half.

QUALITATIVE EFFECT OF THE COOKING PROCEDURES ON NUTRITIVE AND SENSORY VALUES

The effect of heating on the cellular bonds and on the components of the vegetable food are numerous and may be partly recognised from changes in the external characteristics of the foods.

The data given in Table 2 on possible qualitative changes demonstrate that during cooking both negative and positive effects may occur (Zacharias, 1977). What may be described as negative effects on the nutritive and sensory value is intrinsic to the cooking procedure and cannot be avoided even if the food is treated most carefully. Due to thermal and oxidative influences as well as leaching, losses of protein, carbohydrates, minerals and vitamins are to be expected. In many instances also the actual appearance, especially the colour and structure of vegetables and fruits, may change in a characteristic manner. In fruits, for instance, colour intensity is reduced due to leaching of water-soluble anthocyanins. In green vegetable parts, the green pigments chlorophyll a and b are converted into the olive-green phaeophytine.

A positive effect of the thermal treatment on food quality is found with proteins and carbohydrates. It is well known that denaturation of proteins, gelatinisation of the starch grain and transformation of insoluble protopectines into soluble compounds enhances their digestibility and improves the utilisation of certain important components such as β-carotene, for instance. Cooking may also have positive effects on the sensory value by giving rise to the formation of certain aromatic and flavour substances. In addition, softening of vegetables and fruits results from the weakening

TABLE 2

INFLUENCE OF THE PREPARATION ON NUTRITION-PHYSIOLOGICAL AND SENSORY QUALITY

Quality Criteria	Preparing Negative	Preparing Positive	Cooking Negative	Cooking Positive	Re-heating Negative	Re-heating Positive	Holding Negative
Nutritive value							
Protein	X	-	X	X	-	-	-
Fat	-	-	X	-	-	-	-
Carbohydrates	X	-	X	X	-	-	-
Minerals	X	-	X	-	-	-	-
Vitamins	X	-	X	-	X	-	X
Sensory value							
Appearance	X	X	X	X	X	X	X
Taste	-	X	-	X	X	X	X
Consistence	X	X	-	X	X	X	X

X = Yes - = No, or not substantially, resp.

Note: Negative influence, for instance due to leaching, oxidation, effect of heat positive influence, in the sense of an increase of the cellular bonds as a consequence of the loss of integrity of the middle lamella.

As can be further seen from Table 2, holding has a negative effect on labile vitamins and on sensory characteristics (appearance, colour, taste and consistence). However, with proteins, fat and carbohydrates no losses are encountered.

Because of the extensive literature on nutritive changes caused by different cooking procedures and during holding, it is proposed to confine this paper to vegetables and potatoes. Fruits are predominantly consumed as fresh fruit and, moreover, only a small amount of data is available concerning the influence on them of thermal treatment in small kitchen units and large-scale catering institutions.

IMPORTANCE OF VEGETABLES AS NUTRIENT SUPPLIERS

Table 3 demonstrates the importance of frequent vegetable intake in the provision of the so-called "critical" nutrients* such as protein, calcium, iron, ascorbic acid, thiamin, riboflavine and β-carotene (retinol-equivalent) is concerned. If - as is demonstrated in this table - the content of these substances in the usual portion (200 g) of vegetable or potatoes, resp., is calculated on the basis of nutrient tables (Souci-Fachmann-Kraut, 1977) and related to the daily nutrient requirements (Deutsche Gesellschaft für Ernährung, 1975), the content of ascorbic acid is found to be mainly decisive for a high nutritive value of the food consumed. Nearly all vegetables listed contain more than 50% of the recommended daily supply of 75 mg per day. Also the four vegetables green cabbage, carrots, paprika and spinach are important sources of β-carotene. As far as the remaining components are concerned, these vegetables are either insignificant or noteworthy suppliers. There are a few exceptions to this (protein in green peas, calcium in green

---

*'Critical' nutrients are those which, under the usual consumption habits, may be present at minimum quantities only (Hötzel, 1969 and Spengler, 1970).

cabbage, iron in green peas and green cabbage, thiamin in green peas, riboflavine in green cabbage). Generally, however, the degree of nutrient supply is either below 10% or in the range from 10 - 25%.

The nutrition-physiological evaluation of foods should not however be based solely on the content of so-called "critical" nutrients. The data in Table 3 may also indicate focal points of future research. Besides the content of other minerals and important trace elements, it is also important - especially from the viewpoint of overweight - that the vegetables have a low calorific value. In this regard the energy content in 200 g ready-to-serve vegetables amounted to 180 kcal (maximum); the average value of about 10 vegetable meals analysed was about 120 kcal per 200 g (Bundesminister für Ernährung, 1974 and 1976).

EFFECT OF THE THERMAL TREATMENT ON THE NUTRIENT CONTENT OF VEGETABLES

Following these general remarks on the individual cooking procedures, their qualitative effect on the nutritive and sensory value and the nutritional importance of certain vegetables, the remaining sections of this paper shall be devoted to the quantitative changes which occur during cooking and holding. Thus the first question to be answered is:

"To what extent is the content of important components reduced?"

This is closely related to the second question:

"Is it possible to recommend, on the basis of the results available, cooking procedures or holding conditions which would largely guarantee the preservation of the nutritive value?"

TABLE 3

SUPPLY OF 'CRITICAL' NUTRIENTS THROUGH THE USUAL PORTIONS OF VEGETABLES AND POTATOES (200 g VEGETABLES, 200 g POTATOES, READY FOR KITCHEN PROCESSING)

Vegetable	Protein	Minerals Calcium	Iron	Ascorbic acid	Thiamine	Riboflavine	Retinol Equivalent
Cauliflower	X	–	X	XXX	X	X	–
Beans, green	X	X	X	XXX	–	X	X
Peas, green	XX	–	XX	XXX	XX	X	X
Cabbage, green	X	XXX	XX	XXX	X	XX	XXX
Potatoes	X	–	X	XX	X	–	–
Kohlrabi	–	X	X	XXX	–	–	X
Carrots	–	–	X	X	–	–	XXX
Paprika	–	–	X	XXX	–	X	XXX
Leek	X	X	X	XXX	X	–	
Brussels sprouts	X	–	X	XXX	X	X	X
Cabbage, red	–	–	–	XXX	–	–	–
Asparagus	–	–	X	XXX	X	X	–
Spinach	X	(XX)	XXX	XXX	X	X	XXX
Cabbage, white	–	X	–	XXX	–	–	–
Savoy cabbage	X	X		XXX	–	–	–

– = negligible (< 10% of the quantity recommended by DGE)
x = noteworthy (10-25% " " " " " ")
xx = (25-50% " " " " " ")
xxx= substantial (> 50% " " " " " ")

COOKING UNDER PRIVATE HOUSEHOLD CONDITIONS

In view of the data compiled in Table 3 it will be appreciated that in comparing cooking procedures the reaction of the ascorbic acid was mainly investigated both under the conditions of the private household and large-scale catering institutions. An additional consideration is the relatively high reactivity and water solubility of this substance. Ascorbic acid is frequently used therefore as an indicator of the changes which occurred under thermal treatments. With other thermolabile and/or water-soluble vitamins and minerals, the losses during cooking are either high or low. This evaluation of cooking procedures for vegetables and potatoes should not be solely confined to comparisons of ascorbic acid retention.

Comparison of the cooking procedures (boiling/stewing/steaming), using the same prepared vegetables, shows that ascorbic acid retention is highest with stewing (Table 4) where losses are about 34%. With the exception of bush-beans boiling and steaming at $100^{\circ}C$ both cause an almost equally high decrease of ascorbic acid (Zacharias, 1965). This result is contrary to some literature data which recommends steaming as the optimal cooking procedure (Hermann et al., 1973, Causeret and Mocquot, and Dienst, 1954). The results were however confirmed in further studies (Gordon and Noble, 1959, Lauersen, 1941 and Zobel and Wnuck, 1976).

These contradictory results may possibly be explained by the different degrees of comminution and the varying quantities of cooking water used. In finely cut beans, for instance, the ascorbic acid loss was about 20% greater. Similarly if cabbage is not cut into quarters, but cut into strips or is finely chopped, the losses increase by about 15% and 60% respectively (Wood-Crosby et al., 1953 and Wellington and Tressler, 1938).

TABLE 4

LOSS OF VITAMIN C DURING COOKING OF VEGETABLES AND POTATOES

Vegetable	Initial mg/100 g	Boil. Cooking %	Loss in total vitamin C Steaming %	Stewing %
Cauliflower	69.8	24	22	-
Brussels Sprouts	104.0	35	30	34
Spinach	47.0	-	50	35
Cabbage, white	45.8	65	-	40
Bush beans	19.5	43	30	36
Peas	25.5	40	41	26
Peeled potatoes	15.0	32	33	-
Unpeeled potatoes	15.0	11	13	-

The comparably high losses in boiling and steaming may be that in the relatively rapid heating of the food material the effect of the oxidation enzymes is limited, whereas leaching losses, on the other hand, are somewhat higher; the ascorbic acid loss was found to decrease by about 25% when the ascorbic acid content in the cooking water was taken into account (Zacharias, 1970). During steaming, however, it is assumed that ascorbic acid losses are caused more by oxidation and less by leaching. Precise evaluation of the three cooking procedures is difficult because only in a few cases are the test conditions precisely described and in the majority of cases statistical evaluation of the results is completely lacking.

Prolongation of the cooking time, for instance by 100%, resulted in an additional loss of 5 - 20%, depending on the type of vegetable (Noble, 1967 and Zacharias, 1970). All studies showed that ascorbic acid is reduced to a much greater extent through the actual cooking process than through the prolonged cooking.

Increasing the cooking temperature to 106°C (0.23 plus pressure or 1.23 bar, respectively) or to 116°C (0.8 plus pressure or 1.80 bar, respectively), such as is reached during cooking in the pressure cooker, had a different effect on the ascorbic acid content of various types of vegetables (Table 5)(Zacharias, 1974). Relative to cooking at 100°C, there was no substantial difference in vegetables which require only short cooking times (for instance, cauliflower, Brussels sprouts, kohlrabi), or losses were lower at 106°C, but higher at 116°C (bush beans, climbing beans). Better ascorbic acid retention with increasing cooking temperature was clearly apparent in stewed white cabbage and sauerkraut. These two vegetables require longer cooking times, and the reason for the better retention during pressure-steaming may be that the vegetables were exposed to an overall lower thermal stress- comparable to a high temperature - short time heating during sterilisation. In peeled potatoes, however, the losses were clearly higher under pressure-steaming than under steaming at 100°C.

TABLE 5

LOSS OF VITAMIN C DURING STEAMING/PRESSURE STEAMING OF VEGETABLES AND POTATOES

Vegetable	Vitamin C Initial value mg/100 g	Loss of vitamin C in %		
		100°C (1.0 bar)	106°C (1.23 bar)	116°C (1.8 bar)
Cauliflower	99	24	25	21
Brussels sprouts	118	32	29	28
Kohlrabi	93	36	31	37
Bush beans	26	30	21	38
Climbing beans	25	27	21	34
White cabbage	22	40	28	33
Sauerkraut	24	37	25	24
Potatoes, peeled	14	20	29	37

A decisive advantage in favour of pressure-steaming cannot therefore be demonstrated from the results available. Further studies will be required including a statistical evaluation of the results. It is certain, however, that pressure-steaming has the advantage of time-savings up to 80%, and in individual cases energy savings can be as high as 40%.

Studies of β-carotene changes during cooking demonstrated the considerable stability of this nutrient. The loss was 15% at the maximum, regardless of whether the fresh product was boiled, steamed or pressure-steamed (Harris and von Loesecke, 1973).

The losses of thiamine - as with ascorbic acid - were lowest after stewing and amounted to about 10%. Under steaming and boiling the maximum thiamine loss was 28% including the thiamine content of the cooking water, but 59% if the evaluation was based on the vegetable only (Zacharias, 1974 and Harris and von Loesecke, 1973). Increasing the cooking temperature to $116^\circ C$ did not lead to appreciable changes (Zacharias, 1974). It should be noted however that in contrast to the comprehensive results available on the behaviour of ascorbic acid, the present data on thiamine are based on a few investigations, most of them conducted some time ago.

The same applies to riboflavine for which losses similar to those of thiamin may be expected according to the literature data (Harris and von Loesecke, 1973). Reliable data on the reactions of other water-soluble vitamins such as niacin and folic acid are even more scarce. Thus further investigations are necessary in order to precisely evaluate the individual cooking procedures.

Very little data is available on the loss of minerals induced by different cooking procedures. Thus, comparable evaluation is not possible.

COOKING UNDER CONDITIONS OF LARGE-SCALE CATERING INS

Cooking of vegetables under the conditions of large-scale catering units is characterised mainly by use of larger quantities of foods and of different cooking equipment. Under certain circumstances, this may lead to longer heating times. The cooking procedures applied and hence the nature, quantity and temperature of the heat-transfer medium is assumed to correspond to the conditions of cooking under private household conditions. However, systematic investigations to confirm these assumptions have not been conducted as to the length of the processing times in relation to the type and quantity of vegetables and cooking liquid and the type and size of the cooking equipment.

The results available on the loss of ascorbic acid during the cooking of vegetables and potatoes under the conditions of large-scale catering institutions reveal differences in relating to the kind of vegetable and cooking procedure adopted. More literature contains contradictory observations (Zacharias, 1977).

Investigations with savoy cabbage, cauliflower and kohlrabi suggest a better retention of ascorbic acid during cooking in an automatic pressure-cooker at 115°C, as compared to cooking at 100°C or to pressure-steaming at 105, 110 or 120°C (Muskat, 1975). However, these results have not been statistically confirmed. The ascorbic acid loss during boiling and steaming of white cabbage, Brussels sprouts and cauliflower as a function of kind and size of the cooking equipment and cooking temperature is shown in Figure 1, where, in contrast to the results reported above, least losses were found under pressure-steaming at 105°C (Wood-Crosby et al., 1953 and Wood et al., 1946). It is noteworthy that no verified differences were found between cooking procedures for white cabbage and cauliflower. For instance, increasing the quantity of white cabbage from 2.2 kg in a cooker of 23 l volume up to 22.5 kg in a 95 l cooker, with double the quantity of water added in

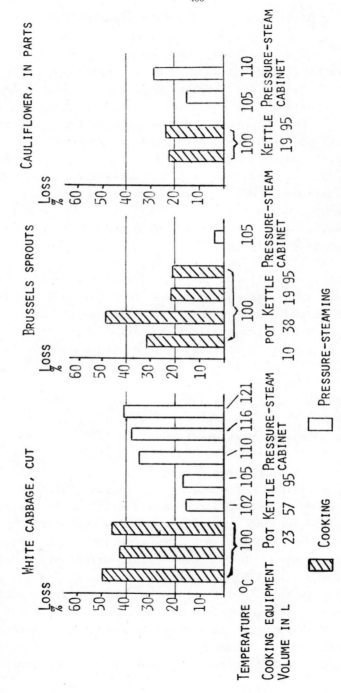

Fig. 1. Loss of ascorbic acid after cooking of vegetables under conditions of large-scale catering establishments.

TABLE 6

LOSS OF ASCORBIC ACID, POTASSIUM AND IRON DURING COOKING OF PEELED POTATOES

Cooking Procedure	Loss of Ascorbic Acid [2]				Loss	
	Aquila	Capella			Potassium [3]	Iron [3]
	Oct 1952 %	Oct 1953 %	Oct 1952 %	March 1953 %	%	%
Boiling [1] (Cooking time 25 min)	24	24	39	52	19.4	14.2
Steaming	27	40	35	47	13.1	11.4
Simmering (10 min boiling and 15 min simmering)	4	14	10	48	9.3	6.0

1) Quantity ratio potatoes : water = 4 : 1
2) Loss relating to ascorbic acid content of the raw potato
3) Variety and date of the investigation are not known

each case yielded the same results (Wood et al., 1946).

Finally, mention should be made of a cooking procedure which may assume importance from the viewpoint of "energy conservation". Vegetables and potatoes are simmered in a little water at a temperature below $100^{\circ}C$. The loss of ascorbic acid, potassium and iron during boiling, steaming and simmering of peeled potatoes is indicated in Table 6 (Zobel and Wnuck, 1976). With one exception (vitamin C loss in the variety Capella, March 1953) least losses were encountered with simmering. Further investigations using different vegetables confirmed the advantages of this cooking procedure (Somogyi, 1975). Whether the indicated energy savings of 40 - 50% can be achieved with existing equipment must be confirmed by corresponding measuring tests.

Few results are available on the behaviour of thiamine, riboflavine and β-carotene. Studies conducted some time ago indicate losses of thiamine during boiling, steaming and stewing of vegetables, up to 20, 15 and 28% respectively. However the average losses of 13, 6 and 15% respectively indicate a slightly better retention with steaming (Dienst, 1954). Due to the lack of precise information on the test procedures and analytical methods applied it is impossible to explain the differences in the results to those obtained under private household conditions.

Whether in fact there are inherent differences between the nutrient retention during cooking in private households relative to large-scale catering institutions will be considered taking the ascorbic acid loss during steaming of peeled potatoes at $100^{\circ}C$ as an example.

As can be seen from Figure 2, the available literature data confirms the advantage of cooking small quantities, as frequently emphasised in the pertinent literature (Weiss, 1974). The statistical difference of about 20% (21% during steaming under private household conditions, 43% during

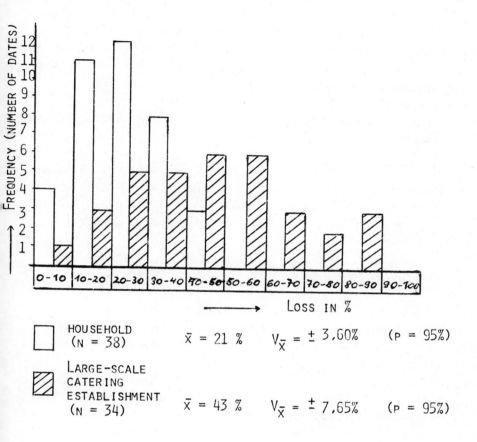

Fig. 2. Loss of ascorbic acid after steaming of peeled potatoes in households and large-scale catering establishments.

steaming under conditions of large-scale catering institutions) is about as high as the additional loss during cooking of other vegetables in large kitchen units, as indicated by many other authors (Zacharias, 1971). It is noteworthy that the confidence range of the mean value ($V_{\bar{x}}$) is twice as high during steaming under large-scale conditions, and results from wider fluctuations in individual values. The reason for this is uncertain. The wider range of fluctuation of the individual values may be due to a different quantity of potatoes and different degrees of comminution as well as differences in the total heating time. The analysis of the results demonstrates, however, that under large-scale preparation conditions it is possible to retain nearly as much ascorbic acid as with small quantities prepared in small kitchen units.

HOLDING OF VEGETABLE MEALS

There can be no doubt, however, that the unsatisfactory quality of meals served in institutional catering establishments as confirmed in numerous investigations and observations, is due to the holding of meals.

The mean vitamin loss in meals held at $70^\circ C$ (Table 7) demonstrates that after a holding time of 1 hour, ascorbic acid is the only vitamin decreased by more than 10% (Bundesminister für Ernährung, 1976). Thiamine and riboflavine are decreased by this amount after 3 hours and pyridoxine only after 5 hours. The range of fluctuations emphasises that differences are dependent on the meal composition. The ascorbic acid loss, for instance, was highest in peeled potatoes and least in spinach. After 3 hours' holding the ascorbic acid content decreased by 22% on average. If the absolute loss is related to the daily requirements, a decrease of up to 25% as a limit to the permissible holding time seems justifiable. Under these assumptions holding over a duration of 3 hours is tolerable, since only 30% of the meals analysed showed a loss of ascorbic

TABLE 7

VITAMIN LOSS IN MEALS DURING HOLDING AT 70°C

Vitamin	Number of Meals	Loss in % After					
		1 hour		3 hours		5 hours	
		x̄	range of fluctuation	x̄	range of fluctuation	x̄	range of fluctuation
β-carotene	1	8	-	9	-	11	-
Niacin	5	2	0 - 3	5	0 - 12	9	0 - 18
Pyridoxine	3	3	0 - 9	9	0 - 15	14	0 - 36
Riboflavin	5	4	0 - 22	11	0 - 34	17	0 - 42
Thiamine	11	7	0 - 18	14	0 - 32	21	0 - 64
Ascorbic acid	9	16	0 - 51	22	5 - 79	30	13 - 81

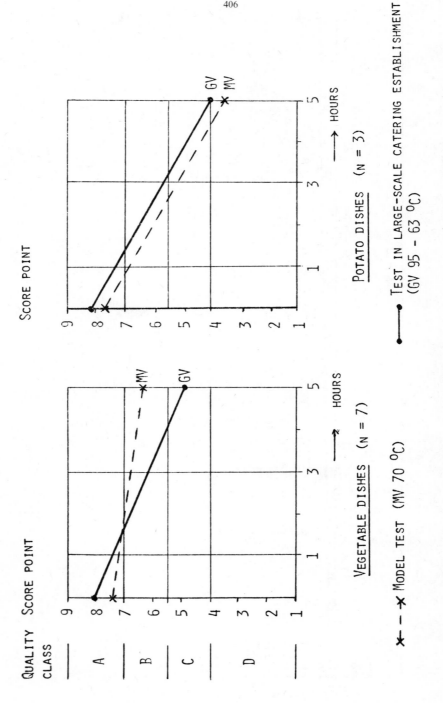

Fig. 3. Decrease of the score points as a function of holding time (regression analysis)

acid and 7% of the meals a loss of thiamine higher than 25%.
It seems advisable, however, to compensate an inadequate
vitamin supply that may arise with certain menu plans by
offering in addition dishes or beverages which are rich in
ascorbic acid or thiamine.

The investigations on sensory quality changes
(Bundesminister für Ernährung, 1976) showed that the most
sensitive characteristic "taste" decreases sharply - especially
with potato dishes - with increasing holding times (Figure 3).
The regression analysis of the score points for "taste"
demonstrated decreasing taste for vegetable dishes with 0.22
score points on average (fluctuation range from 0.07 to 0.49)
and for potato dishes with 0.80 score points on average
(fluctuation range from 0.65 to 1.05) per holding hour
(Table 8). Also these results show that a holding time of 3
hours at 70°C is tolerable. This limit, however, should not
discourage further consideration of possibilities aimed at
minimising holding periods.

CONCLUSIONS ON FUTURE RESEARCH REQUIREMENTS

If the present results on retention of nutritive and
sensory quality are considered from the viewpoint of future
research, it becomes immediately evident, particularly in the
field of cooking, that studies of thermolabile and water-
soluble ingredients, and of the sensory quality represent an
important requirement. An essential prerequisite however,
is the preparation of comparable fresh products, an exact
description of the investigation procedures and the
utilisation of identical analytic methods.

Statistical evaluation of the results is imperative for
the objective evaluation of the individual cooking procedures.
This is essential to enable recommendations to be made for the
application of optimal cooking procedures and to provide the
necessary guidance in decision-making to those concerned with
these problems in practice. The analytical data on the

nutritive values of cooked food products may also serve as a basis on which dietary intake of the nutrients taken can be established, particularly the thermolabile and water-soluble vitamins and water-soluble minerals.

TABLE 8

HOLDING OF VEGETABLE AND POTATO DISHES AT 70°C - REGRESSION ANALYSIS OF SCORE POINTS (SENSORICAL TESTS) AND ALLOCATION TO QUALITY CLASSES

Dish	Regression Analysis[1]			Quality class after Holding time in hours		
	A_o	B	R	1	3	5
Cauliflower	7.6	-0.19	-0.950	A	B	C
Beans, green	6.3	-0.21	-0.938	B	B	C
Peas, carrots	7.0	-0.11	-0.957	B	B	C
Brussels Sprouts	7.0	-0.24	-0.988	A	B	C
Cabbage, red	8.2	-0.19	-0.824	A	A	B
Sauerkraut	8.0	-0.07	-0.823	A	A	B
Spinach	8.0	-0.49	-0.956	A	B	C
Boiled potatoes	7.9	-1.05	-0.991	B	C	D
Mashed potatoes	7.2	-0.65	-0.985	A	B	D
Potato dumpling	7.5	-0.69	-0.993	B	B	C

1) A_o = Constant (calculated initial value = distance on ordinate for $x = o$)

   B = Regression coefficient

   R = Correlation coefficient

REFERENCES

Bundesminister für Ernährung, Landwirtschaft und Forsten (Hrsg) 1974 Schulverpflegung mit industriell hergestellten Gefriermenüs. Bonn.

Bundesminister für Ernährung, Landwirtschaft und Forsten und Bundesforschungsanstalt für Ernährung (Hrsg) 1976. Schulverpflegung mit warmgehaltenen Speisen aus Zentralküchen. Stuttgart 70 (Hohenheim).

Causeret, J., Mocquot, G. Thermo et radiosensibilité des vitamines. Ann Nutr. Vol. 5, Rapports Généraux, C 267 - C 354.

Deutsche Gesellschaft für Ernährung, 1975 Empfehlungen fur die Nährstoffzufuhr (3 Auflage), Frankfurt/Main

Dienst, C. 1954 Grossküchenbetrieb. Verlag f. angewandte Wissenschaften, Wiesbaden

Gordon, I. and Noble, I. 1959 Effect of cooking method on vegetables. J. Amer. Diet. Ass. $\underline{35}$, p 578

Harris, R.S. and von Loesecke, H. 1973 Nutritional Evaluation of Food Processing. 3. Published by The Avi Publishing Company, Inc., Westport Connecticut, USA, pp 418-442; 462-491

Hermann, K., Thumann, J., Suter, G. and Nebe, G. 1973 Einfluss der Gartechniken auf des Ascorbinsäuregehalt von Kohlrabi, Rosenkohl, Blumenkohl, Bohnen und Kartoffeln. Ernährungs-Umschau 20, p 438

Hötzel, D. 1969 Qualitätsbeurteilung aus ernährungsphysiologischer Sicht. Ernährungs-Umschau 16, p 132

Lauersen, F. 1941 Münch. Med Wschr. 88, p 1128

Muskat, E. 1975 Nährwertverluste bei der Zubereitung in der Grossküche. Mitteil. Sek. Lebensm. chem., gerichtl. Chem., Ges. Dt. Chemiker 29, p 11

Noble, I. 1967 Effect of length of cooking. Ascorbic acid and colour of vegetables. J. Amer. Diet Ass. 50, p 304

Somogyi, J.C. 1975 Einfluss der Zubereitungsweise auf den Vitamin C-Gehalt von Kartofflen und Gemüse. Ernährungs-Umschau 22, p 42

Souci-Fachmann-Kraut, 1962/1977 Die Zusammensetzung der Lebensmittel (Nährwerttabellen). Sachbearbeiter H. Bosch. Stuttgart 1962 and continuing work to 1977

Spengler, M. 1970 Der Nährwert von Obst und Gemüse und seine Veranderung durch industrielle Be- und Verarbeitung. Dissertation, Giessen

Weise, R. 1974 Beitrag zur Ermittlung des Vitamin C-Gehaltes in Kartoffleln in Abhängigkeit der Prozess-stufen in der Grossküche. Hauswirtschaft und Wissenschaft, 22. p. 97

Wellington, M. and Tressler, D.K. 1938 Food Res. 3, p 311

Wood-Crosby, M., Fickle, B.E., Andreassen, E.G., Fenton, F., Harris, K.W. and Burgoin, A.M. 1953 Vitamin retention and palatability of certain fresh and frozen vegetables in large-scale food service. Cornell Univ. Agricult. Exper. Stat. Bull. 891

Wood, M.A., Collings, A.R., Stodola, V., Burgoin, A.M. and Fenton, F. 1946 Effect of large-scale food preparation on vitamin retention: Cabbage. J. Amer. Diet. Ass. 22, p 677

Zacharias, R. 1965 Ascorbinsäureverluste bei der Zubereitung und Verarbeitung von Lebensmitteln. Wiss. Veröff. Dtsch. Ges. Ernähr. 14, p 187

Zacharias, R. 1970 Einfluss der Gardauer auf den Genusswert und Gesamt-Vitamin-C-Gehalt verschiedener Gemüsearten. Hauswirtschaft und Wissenschaft 18, p 16

Zacharias, R. 1971 Vitaminerhaltung bei der Verarbeitung von Lebensmitteln im Haushalt und in der Grossküche. Wiss. Veroff. Dtsch. Ges. Ernährung 19, p 118

Zacharias, R. 1974 Veränderungen bei der haushaltsmassigen Zubereitung Handbuch Ernährungslehre u. Diatetik. Bd. III Angew. Ernährungslehre Thieme, Stuttgart, p 350

Zacharias, R. 1977 Lebensmittelzubereitung unter dem Aspekt der Grossküche. Ernährungs-Umschau 24, p 304

Zacharias, R. and Hubner, U. (Eds) Lebensmittelverarbeitung im Haushalt, 2, Auflage Ulmer, Stuttgart, 1975.

Zobel, M. and Wnuck, F. 1976 Neuzeitliche Gemeinschaftsverpflegung. 11. Aufl. Bd. 1, VEB Fachbuchverlag Leipzig

THE EFFECT OF HEAT ON PROTEIN RICH FOODS

A.E. Bender

Department of Nutrition, Queen Elizabeth College, Campden Hill, London, England.

All the processes under discussion at this meeting (pasteurisation, blanching, canning, etc) involve some degree of cooking so it is not possible to separate factory processing from domestic cooking. Many of the nutritional losses incurred in the manufacture of processed foods are in place of rather than in addition to those incurred in domestic preparation from raw foods. Canned meat or beans, for example, require only short heating by the consumer, frozen and dried foods that have been blanched require a shorter cooking time than raw foods. Consequently the effects of heat on protein foods will be discussed without separating the effects of factory processing from those of domestic and institutional cooking.

PROTEIN CHANGES

It is not possible to describe precisely the effects of heat on protein foods because there is a great difference between what can happen and what, in practice, does happen. For example, there is considerable evidence that, except under severe conditions, proteins suffer little change, while at the same time there is a wealth of literature describing the damage that can be caused under even mild conditions (Bender, 1978).

Five types of processing change can be distinguished.

1) Mild heat treatment changes the tertiary structure – denaturation – which has considerable effects on functional properties but none on nutritive value.

2) Mild heat in the presence of reducing substances results in a linkage between the epsilon-amino groups of lysine and reducing substances which cannot be hydrolysed by the digestive enzymes (non-available lysine).

This is the browning or Maillard reaction (Hodge, 1967; Carpenter and Booth, 1973). The lysine is still present and is released from combination by acid hyrolysis, so the usual amino acid estimation (which is preceded by acid hydrolysis) yields false information. Similar losses can occur during storage at room temperature (Ben-Gera and Zimmerman, 1972).

3) More severe heat reduces the availability of other amino acids as well as lysine and can occur in the absence of reducing substances. Cystine is relatively sensitive and can be converted into compounds such as cysteic acid, methyl mercaptan, dimethyl sulphide and dimethyl disulphide at temperatures around $115^{o}C$. At the same time under such conditions digestibility is often reduced.

Reactions can take place within proteins themselves between the free amino groups of lysine and arginine and free acid groups of aspartic and glutamic acids, or with amide groups such as asparagine and glutamine.

Amino acids can react with sulphur groups, particularly cystine and to a lesser extent methionine and the imidazole ring of histidine. Phosphoester links can be formed between two hydroxyamino residues; a lactone ring may be formed between a terminal carboxyl and hydroxamino acid, and reactions can take place with the products of fat oxidation. (Finot, Viani and Mauron, 1968).

There are additional factors that may be related to nutritional changes. For example, amino acids are liberated more slowly from heat damaged proteins and it had been suggested but not verified that this could lead to a reduced rate of protein synthesis if all the essential amino acids are not made available to the sites of protein synthesis at the same time (Melnick, Oser and Weiss, 1946).

Other factors, not directly related to heat processing but doubtless affected by heat, also damage protein quality. Thus formaldehyde and acetal reduce available lysine in fish meal used as animal protein supplement (Lea, Parr and Carpenter, 1960); gossypol reduces the availability of lysine in cottonseed meal; fat solvents can have an effect on nutritive value (Hallgren, 1970).

4) At temperatures of 180 - 300°C such as are involved in roasting of meat and fish, and baking of biscuits, amino acids can be destroyed or racemised or form poly-amino acids (Hayase, Kato and Fujimaki, 1975). D-isomers of amino acids produce changes in flavour and are nutritionally unusable.

Destruction of amino acids is, of course, revealed by chemical analysis.

5) Alkali treatment and oxidation result in loss of nutritional value.

Effects on food

In general loss of available lysine appears to be the main cause of damage with cereal based foods, and loss of sulphur amino acids with meat and fish.

The loss of lysine without apparent damage to other amino acids was shown by the classical work of Block et al. (1946). A cake mix of flour, egg, yeast and lactalbumin had PER 3.5, reduced to 2.4 when baked at 200°C for 15 - 20 min, and to 0.8 when toasted at 130°C for 40 - 60 min. The addition of lysine restored the value to 3.5 so demonstrating that lysine alone had been damaged. (It is always possible, though there is no evidence on this subject, that damage may have been done to other amino acids present in sufficient surplus not to show up in biological evaluation).

Many samples of damaged fish meal and meat meal increase

in BV when supplemented with methionine (but not when supplemented with lysine). (It must be borne in mind that biological assay reveals damage only to the limiting amino acid and provides no information about those essential amino acids present in surplus (Bender, 1975)). For example, the NPU of canned, evaporated milk is the same as that of fresh milk (Bender, unpublished) yet, the colour indicates that some of the lysine has been damaged. Since the lysine was in relative surplus losses are not revealed in normal bioassays. Such damage would become of dietary significance in a mixture of cereal and milk, as commonly used for infant feeding, when lysine could be the limiting amino acid. (There appears to be little, if any, information on this topic in the literature).

More specific evidence of damage to cystine was provided by Bjarnason and Carpenter (1970) who heated proteins to $115^{\circ}C$ for 27 h and destroyed 50% of the cystine; at $145^{\circ}C$ almost all the cystine was destroyed but none of the methionine.

Other amino acids can also be affected. For example, Pieniazek et al. (1975) reported that when casein was heated with glucose for 24 h at $90^{\circ}C$ the availability of methionine, glutamic acid, threonine, serine, glycine, histidine and arginine were reduced to 25 - 30% of their initial values, and lysine and alanine to 85%. In the presence of 80% moisture all amino acids except tyrosine and phenylalanine were damaged and the addition of glucose damaged those as well.

Evans and Butts (1948, 1949) using soya meal and sucrose found losses of histidine, phenylalanine, threonine, leucine, aspartic acid and glutamic acid, as well as cystine and methionine, using enzymic hydrolysis in vitro as an index.

Physiological effects of heated proteins

Apart from reduced nutritive value there are a number of physiological effects of Maillard compounds produced by heat treatment of protein foods (Bender, 1978).

1) Reduced digestibility

2) Absorption of peptides that are later excreted in the urine, so accounting for observed falls in nutritive value not explained by reduced digestibility (Ford and Shorrock, 1971).

3) Partial inhibition of digestive enzymes by undigested peptides (Tanaka et al., 1976; Lee et al., 1976) together with inhibition of amino acid transport from the digestive tract. In practice this has shown up as diarrhoea caused by feeding severely browned apricots.

4) Possible toxic effects of Maillard compounds although the facts are not clear. (Adrian, 1974, Sternberg et al., 1975, Nutr. Rev. 1976).

Desirable browning

While it is usual to discuss the nutritional losses involved in heat processing of protein foods there are advantages. For example, almost all legumes contain toxins which are destroyed by heat treatment; bound niacin is liberated in cereals by alkali treatment (Clegg, 1963); roasting of coffee forms niacin.

In addition, many desirable flavours are formed from Maillard compounds such as those produced in baking, frying and toasting of foods such as biscuits, breakfast cereals, roasted peanuts, cocoa, meat extract and malt extract (Linko and Johnson, 1963, Adrian, 1973). The formation of the flavour compounds does result in the loss of some part of available lysine but this is generally nutritionally insignificant. The Maillard reaction can be exploited for particular purposes by the addition of reducing substances to the food. For example, the addition of milk powder to bread permits more even toasting at a lower temperature because of the presence of lactose.

Another example of the exploitation of amino acid-reducing reactions is in the production of chicken and meat-like flavours produced by heating amino acids with various sugars and aldehydes (Adrian, 1973).

On the other hand supplementation of cereal foods with milk powder intended as protein enrichment has resulted in damage. Clegg (1960) prepared a protein biscuit enriched with milk powder and found that the available lysine was half the expected level due to the reaction with the lactose. Enrichment with casein instead of milk powder avoids such loss. A parallel example is the production of Indian Multipurpose food (a mixture of groundnut meal and Bengal gram) which was subjected to a temperature of $120^{\circ}C$ to improve flavour. Sugars present in the groundnuts led to a fall in nutritive value (Rao, 1974).

CURRENT PROBLEMS

Much of the information available of protein changes through processing is derived from model, and therefore simple, systems in the laboratory. The foods themselves are obviously more complex and recent evidence suggests that mixtures of foods behave differently from single foodstuffs.

A. Meat and Meat Products

It has been well established over many years that when meat is canned, cooked or even roasted there is no change in nutritional quality, but there are discrepancies which may be due to methods of assessment. Beuk, Chornock and Rice (1948) showed that even after auto-claving for 24 h at $112^{\circ}C$ the only amino acid destroyed was cystine, but (1949) they found that PER was reduced from 3.2 to 2.6 after 2 h. Mayfield and Hedrick (1949) reported no change after roasting meat in open pans at $163^{\circ}C$ (internal temperature $80^{\circ}C$), browning in the oven for 30 min and canning, or salt curing (although there were some discrepancies in their pairs of assays). In a range of meat products subjected to cooking, canning and irradiation – bacon, pork, beef, chicken and shrimp – Thomas and Calloway (1961) found no loss of any essential amino acids.

However, there are two reports showing that damage can occur when the meat is heated in the presence of other foods. Hellendoorn, De Groot et al. (1969) canned six types of meat dishes and showed losses (Table 1) both from processing and also from subsequent storage over 3 and 5 year periods.

TABLE 1

EFFECT OF CANNING AND SUBSEQUENT STORAGE ON PROTEIN QUALITY OF CANNED MEAT MEALS (HELLENDOORN, DE GROOT ET AL., 1971).

Sample	Fall in protein quality		
	After canning	3 y. storage	5 y. storage
goulash and potatoes	0	17%	20%
spinach, pork, potatoes	12%	16%	20%
carrots, onions, beef, potatoes	18%	25%	31%
peas, minced meat, potatoes	0	15%	24%
dun peas with bacon	30%	33%	40%
white beans, bacon, potatoes tomato sauce	+30%*	14%	45%

* increase presumably due to effect of heat on the legumes

The second report is that of Bender and Hussaini (1976) who showed no loss of protein quality when meat was autoclaved alone but a fall when reducing substances (3% glucose and 6% starch, used to stimulate 'gravy') were added. Since the quality was restored to its initial value on adding methionine, the damage is to the sulphur amino acids (Table 2).

TABLE 2

EFFECT OF REDUCING SUBSTANCES (3% GLUCOSE + 6% WHEAT FLOUR) ON MEAT SUBJECTED TO HEAT (1 h at 115°C)

	NPU
Autoclaved alone	78
+ reducing substances	70
+ methionine	80

(Bender and Husaini, 1976).

A parallel finding is that of Skurry and Osborne (1976). They showed that 'all-meat' sausage containing 7% cereal as a binding agent had PER 2.5 when raw, falling to 2.0 on cooking. When 60% of the meat was replaced by soya or milk solids the PERs were 1.4 and 1.7 respectively when raw, with no change on cooking.

There is little information available on pickled and smoked products, and even less on the effects of home cooking such as stewing, roasting and frying on mixtures of meat with other foodstuffs for varying periods and temperatures (Table 3).

TABLE 3

EFFECT OF HOME COOKING ON NUTRIENT CONTENT OF BEAN PREPARATIONS IN GUATEMALA

Product	PER	Available lysine (g/16 g N)
Raw beans	0	5.8
cooked	1.24	6.3
fried	0.9	5.2

(Bressaini et al., 1972).

A major problem posed for all foods is that we do not know anything of the nutritional quality of foods as eaten, only that experimentally prepared or after leaving the factory or market.

A second problem is that of equating the newer vegetable protein meat analogues (TVP) with 'meat'. It has been suggested that such replacements should have the same nutritional value as meat, which immediately raises the question as to what is meant by the term 'meat'. Most nutrition textbooks that make any mention of the protein quality of meat state categorically that all cuts of meat have the same nutritive value and some refer this statement back to Mitchell and Carman (1926). In fact those authors stated the opposite of this, and suggested that cuts of meat rich in connective tissue should have a lower BV than those poor in connective tissue. Since collagen is so very low in sulphur amino acids this appears to be obvious and, indeed, Dvorak and Vognarova (1969) showed the large differences in collagen content of meats of different cuts.

However, until the report of Bender and Zia (1976) (Table 4) no-one appeared to have measured both connective tissue and protein quality on the same sample of meat. Their evidence confirmed what Mitchell and Carman had forecast half a century earlier but it does emphasise the question, 'what is meat?'. Meat in the diet in Great Britain may be of at least a dozen different types, including foods such as sausages which may be made from low quality manufacturing meat rich in connective tissue.

TABLE 4
QUALITY OF MEAT PROTEIN AND COLLAGEN CONTENT

Sample	NPU	Collagen 9/100 g protein
Shin	69	23.6
+ methionine	89	
Shoulder	78	13.5
Fillet	82	2.5
+ methionine	98	

(Bender and Zia, 1976).

B. Vegetable analogues of meat

A parallel problem is that of establishing the nutritional value of the vegetable protein products that may be used to replace (or extend) meat dishes.

The few assays that have been reported in the literature show NPU of extruded soya products (ie the whole soya flour without any fractionation of the proteins such as occurs when preparing the isolate) to be 0.40 - 0.50, whereas properly processed soya flour has NPU 0.70. Indeed legislation has been proposed that the protein quality of soya products should be 'not less than 70% of that of casein'. If casein has NPU 0.70 this ruling apparently accepts that soya products will have NPU only 0.50 (ie 70% of 0.70). Leaving aside the question of any dietary significance of such values it certainly poses a problem as to why the products should be so low.

Some of our own results certainly confirm that the NPU of many samples of extruded soya are low by reason of damage to the sulphur amino acids since their NPU can be increased by supplementation with methionine (Table 5). Moreover, Longenecker et al. (1964) showed completely inadequate quality control being exercised in factories in the United States. They purchased 21 samples of soya and soya isolates on the open market and found that 20 had been underheated (PER increased on further heating) while one had been overheated.

TABLE 5

PROTEIN QUALITY (NPU) OF SAMPLES OF SOYA FLOUR (MEAN OF DUPLICATE ASSAYS)

	BV	D	NPU	Available	
				Methionine	cysteine
Extruded soya sample A	64	84	53	0.92	0.60
B	68	85	58	1.20	0.66
Overheated soya flour (2 h at 121°C in steam)	49	80	35	0.80	0.32

C. Legumes

Legumes provide an appreciable part of the protein intake even in industrialised countries and often the major part in developing countries. Heat can have three effects, destruction of toxins, improvement of flavour, and loss of nutritive value through overheating (Table 6).

TABLE 6

BENEFICIAL EFFECTS OF HEAT PROCESSING - EFFECT OF HEAT ON PROTEIN QUALITY OF LEGUMES (MICHIGAN PEA BEAN)

	NPU	BV	Digestibility
Raw	15	37	41
Boiled 5 min	48	71	68
10 min	53	80	66
60 min	49	77	64
Heated $120^\circ C$ for 120 min	38	76	50

(Hellendoorn et al., 1971)

Since legumes are usually cooked the various toxins present do not appear to present a problem but a recent report in the United Kingdom of 10 cases of food poisoning thought to be due to toxins in soaked but uncooked beans suggests the need for further investigation. This may be particularly necessary through the introduction by immigrants of types of legumes not hitherto eaten in Europe.

The improvement of flavour appears to be due to the formation of Maillard complex which indicates some loss of nutritional value. Domestic cooking of legumes can extend from one to as much as 24 h, depending on the type of dish (eg Hungarian cholent) so that loss of available lysine might be a factor of some significance and worthy of further investigation. A small preliminary trial of our own has shown improved flavour and a fall in available lysine with increased cooking time in one legume, and no improvement of flavour and no loss of lysine in another variety. Little is known of the types of carbohydrates present in the various legumes, moreover, many of them are only 70% digested, both as regards protein and total dry matter.

D. Quality Control

There is clearly a difficulty in establishing quality control in the factory with regard to protein quality if the only reliable method is animal assay. Not only is this long

and expensive but the current argument between workers in the
United States who advocate the slope-assay ratio method, some
in Canada who advocate NPR, and those in UK who advocate NPU
by carcass analysis, leaves the food technologist without any
guidance.

In fact, as has been argued elsewhere (Bender, 1975)
bioassay does not provide the information required since it
yields only single values which cannot be used to forecast
the effect of adding the protein in question to the diet, nor
does it even reveal which of the amino acids is limiting.
What is required instead is figures of the content of available
amino acids in the food. Then one could forecast its value in
a diet, or as a supplement, or use the determination in quality
control.

A method for available lysine has been used for several
years (Carpenter, 1969, Carpenter and Booth, 1973) but is
inadequate in the presence of large amounts of starch, such as
in cereals. A recent modification (Hurrell and Carpenter, 1969,
1974) appears to be more promising.

As regards available methionine and cystine a method was
put forward by Pieniazek et al. (1975_1, 1975_2) and some of
their results are shown in Table 7.

So far as the present author can ascertain this method has
not been made use of, nor subjected to further examination by
other workers, but it is currently under investigation in our
own laboratories.

The original authors showed that like lysine, available
methionine in casein was most severely reduced by heat in the
presence of moisture and glucose; when dry (0.4% moisture)
heating at $90°C$ for 24 h reduced available methionine by 25%,
at 8% moisture it was reduced by 32%. In the presence of
glucose these values were 32% and 55% respectively. Available
cystine was not affected by moisture but the loss of 60%

in the absence of glucose was increased to 100% in the presence
of glucose. Total methionine and cystine were reduced to a
smaller extent.

TABLE 7

AVAILABLE SULPHUR AMINO ACIDS DETERMINED AFTER ENZYMIC HYDROLYSIS

	Methionine	Cystine	Total cystine
Fresh milk	100%	100%	
sweetened, condensed	85%	70%	
spray dried	100%	90%	10% fall
roller dried	80%	90%	
Fresh mackerel	100%	100%	
steamed	100%	100%	
sterilised at 115°C	100%	35%	
126°C	80%	25%	

(Pieniazek, et al., 1975)

Another potential index of change in nutritional value
may be the appearance on a chromatogram of substances derived
from the Maillard reaction. For example, acid hydrolysis of
the lysine-fructose complex yields both furosine and pyridosine
(Finot, Viani and Mauron, 1968). Ornithinoalanine, lysinoalanine
and lanthionine may serve as indicators (Bohak, 1964, Cuq
Provansal et al., 1976).

Within a range of protein foods of the same type tests
such as dye-absorption have been developed (Lakin, 1973).

CONCLUSIONS AND SUMMARY

1. There is very little information available of the
quality of the protein of manufactured foods. Much of what is
available is incomplete, eg amino acid data often exclude
cystine, total rather than available amino acids and chemical
rather than biological data are provided, unreliable biological
data based on single values are often published with no evidence

of reliability.

2. There is abundant evidence of inadequate quality control of factory processes involving protein foods, eg despite well established processing conditions examples are known of widely differing blanching times of peas, the quality of soya products indicates lack of control, canned evaporated milk appears, from its colour, to be damaged.

3. Almost nothing is known of the changes of protein quality in processes such as smoking of fish and the preparation of meat products. There is a tendency to regard all meat products as having similar protein quality despite the use of 'manufacturing quality' meat in many products. We know little of the effects of heat on the protein quality of mixtures of foods - meat heated alone suffers little damage but this is not so in the presence of other foods.

4. While there is some information available from model systems in the laboratory, and it is possible to examine products at the point of sale and that prepared in institutional kitchens, virtually nothing is known of the nutrient composition of foods eaten. Clearly the loss of nutrients in home cooking must cover an enormous range but until we know more of this we cannot conclude whether or not nutrient intake is adequate.

5. There is a need for relatively brief methods of estimating protein quality or to serve as an index of change in protein quality during processing in place of the biological tests which are not practicable for factory control.

6. We need to know (from the nutritionists) whether or not these problems have any practical importance in the diet as a whole.

REFERENCES

Adrian, J. 1973. Ind. Alim. Agric. 90, 559.
Adrian, J. 1974. World Rev. Nutr. Dietet. 19, 71-122.
Bender, A.E. 1975. Proc. 9th Internat. Congr. Nutr. Vol. 3. 310-320. Karger, Basel.
Bender, A.E. 1978. 'Food Processing and Nutrition'. Academic Press.
Bender, A.E. and Husaini. 1976. J. Fd. Technol. 11, 499-504.
Bender, A.E. and Zia, M. 1976. J. Fd. Technol. 11, 495-498.
Beuk, J.F., Chornock, F.W. and Rice, E.E. 1948. J. Biol. Chem. 175, 291-297. 1949. J. Biol. Chem. 180, 1243-1251.
Ben-Gara, I. and Zimmerman, G. 1972. J. Fd. Sci. Technol. 9, 113-118.
Bjarnason, J. and Carpenter, K.J. 1970. Brit. J. Nutr. 24, 313-328.
Block, R.J., Cannon, P.R., Wissler, R.W. et al. 1946. Arch. Biochem. 10 295-301.
Bohak, Z. 1964. J. Biol. Chem. 239, 2878-2887.
Bressani, R. and Elias, L.G. 1972. 'Nutritional Improvements of Legumes'. Proc. Symp. P.A.G., Rome.
Carpenter, K.J. 1960. Biochem. J, 77, 604-610.
Carpenter, K.J. and Booth, V.H. 1973. Nutr. Abstr. Rev. 43, 423-451.
Clegg, K.M. 1960. Brit. J. Nutr. 14, 325-329.
Clegg, K.M. 1963. Brit. J. Nutr. 17, 325-329.
Cuq, J.L., Provansal, M.P., Besancon, P. and Cheftel, C. 1976. Proc. IVTH Internat. Congress Food Sci. Tech. Vol. 1.
Dvorak, Z. and Vognarova, I. 1969. J. Sci. Fd. Agric. 20, 146.
Evans, R.J. and Butts, H.A. 1948. J. Biol. Chem. 175, 15-20.
Evans, R.J. and Butts, H.A. 1949. J. Biol. Chem. 178, 543-548.
Finot, P.A., Viani, R. and Mauron, J. 1968. Experimentia. 24, 1097-1099.
Ford, J.E. and Shorrock, C. 1971. Brit. J. Nutr. 26, 311-322.
Hallgren, B. 1970. Effect of processing on the nutritional value of fish protein concentrates. In 'Evaluation of Novel Protein Products'. Ed. Bender, A.E., Kihlberg, R., Lofqvist, B. and Munck. L. Pergamon Press.
Hayase, F., Kato, H. and Fujimaki, M. 1975. J. Agric. Fd. Chem. 23, 491-494.
Hellendoorn, E.W., de Groot, A.P., Slump, P. et al. 1969. Voeding 30, 44-63 (Fd. Sci. Technol. Abstr. 1, 967; Abstr. 8G333, 1969).
Hodge, J.E. 1967. Symposium on Foods. 'The Chemistry and Physiology of Flavours'. Schultz, H.W. Ed. Avi Publ. Co. Westport, Conn. Chap. 12.

Hurrell, R. and Carpenter, K.J. 1969. Brit. J. Nut. 32, 589.

Hurrell, R. and Carpenter, K.J. 1974. Proc. Nutr. Soc. 35, 23A.

Lakin, A.L. 1973. Evaluation of protein quality by dye-binding procedures. In. 'Proteins in Human Nutrition'. Ed. Porter, J.W.G. and Rolls, B.A. London, Academic Press. p. 179.

Lea, C.H., Parr, L.J. and Carpenter, K.J. 1960. Brit. J. Nutr. 14, 91-113.

Lee, C.M., Chichester, C.O. and Tung-Ching Lee 1976. Proc. IVth Internat. Congress Food Sci. Technol. Vol. 1.

Linko, Y. and Johnson, J.A. 1963. J. Agric. Fed. Chem. 11, 150-152.

Longenecker, J.B., Martin, W.H. and Sarett, H.P. 1964. J. Agric. Fd. Chem. 12, 411-412.

Mayfield, H.L. and Hedrick, M.T. 1949. J. Nutr. 37, 487-494.

Melnick, D., Oser, B.L. and Weiss, S. 1946. Science, N.Y. 103, 326-329.

Mitchell, H.H. and Carman, C.C. 1926. J. Biol. Chem. 68, 183.

Nutr. Rev. 1976. 34, 120-122.

Pieniazek, D., Rakowska, M. and Kunachowicz, H. 1975. Brit. J. Nut. 34, 163-173.

Pieniazek, D., Rakowska, M., Szkilladziowa, W. and Graberek, Z. 1975. Brit. J. Nut. 34, 175-190.

Rao, G.R. 1974. Ind. J. Nutr. Dietet. 11, 268-275.

Skurray, G.R. and Osborne, C. 1976. J. Sci. Fd. Agric. 27, 175-180.

Sternberg, M., Kim, C.Y. and Schwende, F.J. 1975. Science, 190, 992-994.

Tanaka, M., Amaya, J., Lee, T.C. and Chichester, C.O. 1976. Proc. IVTH Internat. Congress Sci. Technol. Vol. 1.

Thomas, M.H. and Calloway, D.H. 1961. J. Amer. Dietet. Assoc. 39, 105-116.

SEMINAR DINNER ADDRESS

Joel Bernstein

NUTRITIONAL NEEDS AND RESEARCH PRIORITIES:
REPORT OF THE U.S. NATIONAL ACADEMY OF SCIENCES

Joel Bernstein
National Research Council, Washington DC, USA

I am pleased and honoured to be invited to speak at the food technology seminar. I must say that I accepted with some trepidation, since I feel myself to be somewhat a fish out of water among this distinguished group of food technologists.

However, the organisers of the seminar felt that a new report by the US National Academy of Sciences, World Food and Nutrition Study: The Potential Contributions of Research, provides relevant context for the specialised work that is addressed in this seminar and also provides some relevant research recommendations. So here I am to discuss some of the relationships between the findings and recommendations of that report and your own work.

A NEW CONTEXT FOR FOOD TECHNOLOGY

Industrial food technology is concerned primarily with food processing, and with processing largely for markets in the higher-income areas of the world. Food processing in these areas has tended to become more and more sophisticated and costly, and to represent an ever-increasing share of the retail cost of food. The concerns of food technology with quality improvements have centred mainly on response to buyer taste and handling preferences, although there has also been some concern with nutritional impacts.

This state of affairs is natural and perhaps inevitable, since food industries and food technologists have not had much incentive up to now to focus their efforts on nutritional improvement. There has been little consumer interest in such improvement, and therefore little commercial incentive. Attention by health authorities to the effects of diet has been weak. And surprisingly little is known about the comparative effects on human health and performance of alternative diets, as I shall note further below, so that food technologists have very uncertain guidance on the changes of diet content that they should seek in order to improve markedly the impact of diet on human welfare.

Thus, I think it is fair to say that food technology has not played a very significant role in efforts to reduce world malnutrition. There have, however, been significant accomplishments in food technology that could play an important role in the future.

The world may well be entering a new era of concern about diet-health relationships and the effects of alternative public and private actions on the nutritional status of populations. Growing official concerns may foster rising consumer awareness and expectations in this sphere. These concerns are likely to spread among both high-income countries and the low-income countries that have been relatively unaffected to date by industrial food technology. There will be a growing convergence of interests, between these two categories of countries, in understanding better the factor relationships that underlie interactions between diet and human well-being. Until now, attention to the relevant factors has been fragmented so as to inhibit rational attack on the complex mosaic of variables involved in malnutrition: doctors, dieticians, agriculturalists, industrial food technologists, the relevant biological scientists outside of medicine and the food industry, and development planners and policy makers have worked in largely isolated compartments - each focusing on a relatively narrow band of concerns within the broad range of questions pertinent to the nutritional status of the various populations.

We can expect this situation to change in response to growing awareness of the importance of the composition of diet for the quality of life and growing expectations that governmen

can and must make greater efforts to improve diet. Human
nutrition is beginning to move from the wings towards centre
stage as a major focus of public policy and research attention.
I believe that this will create growing pressures for the
forward planning of food industries and their technologists to
take fuller account of the broad context of each society's
efforts to avert malnutrition, of what their role might be
in such efforts, and of what is being done in other segments
of the food chain (from seed to bodily absorption of nutrients)
that affects and interacts with the role of food industries.

What are farmers doing and how might this change so as
to improve the prospects for producing more nutritious food
products? What difference will alternative food processing
technologies make - both established and prospective
technologies? What new opportunities can changing public
practices open? How can shifts throughout the wide range of
relevant governmental policies change the incentives and the
opportunities for consumers and for food industries to make
action choices that will improve nutritional status: how can
more of the external costs and benefits of these decisions to
society be accrued to the decision making units (families,
companies, and so forth) so as to provide better action
incentives? How can the feasible action choices be improved
by research? Such questions are part of the broader range of
concerns that will, I believe, engage industrial food technology
in the future.

This emerging context for food industries - the multi-
faceted struggle to avert massive hunger and malnutrition in
the world over the next several decades and beyond - was
analysed in the World Food and Nutrition Study that I mentioned
at the outset. Let me turn now to a summary of some of its
findings.

WORLD FOOD AND NUTRITION STUDY

This Study was transmitted to President Carter on June

20, 1977 and has already had wide influence on action by US public and private organisations. The beginnings of some similar influences have also been reported from other countries The Study was done in response to a request from President Ford, after the World Food Conference of November 1974, asking the National Academy of Sciences for recommendations on how US research and development (R & D) capabilities could best be mobilised to help all countries produce more food and combat malnutrition.

The report assesses the world food and nutrition problem, recommends research areas of the highest priority for expanded US support in order to reduce world hunger and malnutrition over the next several decades, and recommends the steps needed to mobilise and organise research resources to obtain the most effective results.

The principal conclusions of the problem assessment follow, in highly condensed form.

The extent of hunger and malnutrition depends on complex interactions among several factors: adequacy of food production (quantity and quality), extent of poverty, stability of food supplies and prices, suitability of government policies effectiveness of institutions serving food systems, and rate of population growth.

The problem of world hunger and malnutrition can be solved only in the context of national and international policies around the world that support an integrated expansion of output, employment, and food supply - which are highly interdependent. This is the key to reducing poverty, which is the most important direct cause of malnutrition.

US strategy should be geared to mobilising financial and political support worldwide for this expansion and for specific policy and technological changes that would provide greater stability of food supplies and prices, another factor in world food problems.

The role of research is to provide improved options to those who make decisions affecting the status of food supplies and nutrition. The decision makers include farmers, consumers, agribusiness managers, health service personnel, government officials, and politicians.

Most of the hungry and malnourished are in the low-income countries, but there are severe health problems from inadequate or improper diet in the United States and other high-income countries. There is a growing convergence of the interests of both groups of countries in the problems that underlie diet-health relationships and in the specific technological improvements needed to assure an adequate, low-cost food supply in the decades ahead.

Eliminating hunger and malnutrition as major world problems requires, among other things, progress on two fronts:

1) strengthening R & D in order to open up a much better range of options for dealing with the underlying problems;

2) Mobilising the political will and also the resources, skills, and energy to apply the better solutions from among those that are available.

Existing technical options are simply inadequate for the scope of anticipated food problems. The need is to eliminate the massive current malnutrition in the world (estimated by UN sources to affect 500 million to one billion people) and also feed the additional two billion people expected by the year 2000. The low-income countries must roughly double their food production and imports by the end of the century and better distribute available food if this need is to be met. This growth of food supply requirements is unprecedented and cannot be met without accelerated R & D.

Most farmers in the low-income countries face severe crop stresses from variation and extremes in weather, from many kinds of pests, and from inadequate nutrients and toxic substances in the soil. These farmers have relatively little access to capital, to markets for farm inputs and outputs, and to other services. And they cannot afford to take much risk in applying their limited capital and labour supply. Their traditional farming technologies have tended to minimise risk as far as available knowledge permitted, but at a cost of low crop yields. Despite some claims to the contrary, we do not yet know enough to offer developing country farmers new options that are adequate to meet the anticipated food supply requirements for the next several decades. Sophisticated R & D beyond current capabilities, is needed in each country to adjust and combine known technologies so that they are useful to farmers, to develop new higher yielding technologies, and to design the policies and organisations required to effectively apply improved technologies.

The problem of insufficient options also faces <u>consumers</u> in the low-income countries and the other decision makers throughout food systems who determine what the consumers' options are. This is also true in the United States and in other high-income countries. There is great ignorance about how variations in the kinds and quantities of foods eaten affect the critical dimensions of human performance under different circumstances: work performance, susceptibility to disease, physical and mental development, fertility and lactation, school performance, and overall vitality. When we learn more about the effects of changing the composition of diet, we then need to know more about how these effects can best be improved in the varying circumstances of life around the world as it is actually lived - that is, by what combinations of growing more or different crops (including animals), socioeconomic changes that improve food selection and distribution, new methods of processing and marketing food, policy or institutional changes that affect the foregoing changes, or direct food distribution programmes.

PRINCIPAL STRATEGIC THEMES OF THE WORLD FOOD AND NUTRITION STUDY

There are a number of strategic emphases, based on the foregoing problem assessment, that underlie the recommendations of research priorities and new organisational steps throughout the Study. Four of these strategic emphases are particularly important and are not emphasised in most of the existing reports on food research priorities.

First, the report focuses on the nutritional needs of the individual human being as the centre of the problem. To develop efficient strategies for improving the human condition, it is essential to know more about the nutritional variables that I mentioned earlier in discussing the need to improve consumer options. The report first asks, 'Who is hungry and why?' and then builds its research and organisational recommendations from the findings and from the other strategic conclusions stated below.

Second, the report calls for a pronounced shift in US and worldwide strategies for producing more food. There are two ways to produce more food with a given amount of land and work force:

1) apply more capital (fertiliser, other chemicals, machinery, structures, etc);

2) improve the biological factors controlling plant productivity.

The United States has relied primarily on the first way, but needs to shift to a greater reliance on the second because of four trends:

1) higher energy costs

2) diminishing returns from applying additional capital in agriculture

3) levelling of yields of major US crops in recent years

4) increasing environmental contamination.

Substantially greater reliance on the second or biological route to higher crop yields requires major research advances. The developing countries also need to advance primarily via the second route because they have access only to moderate amounts of capital. Also, greater reliance on the second route improves their serious income distribution problems which are a primary cause of malnutrition. This improvement results because poor farmers can apply improved biological technology that can increase yields with little or no accompanying capital input, whereas they cannot afford the capital-intensive route. Improved biological technology can be combined with labour-intensive methods to provide the poor workers relatively more of the income from farming compared to the share going to the owners of capital. These same reasons also favour research to make the first route (applying more capital) more efficient, that is, to achieve given levels of yield with less use of energy, petrochemicals, and money.

Third, the report seeks to increase the divisions of labour and the sharing of results in food and nutrition research, both within the United States and worldwide. This will accelerate the gains from research in the United States, in the low-income countries, and in other countries. The existence in the low-income countries of competent research units that collaborate effectively with US research organisations is indispensable to applying US research findings in the major food shortage areas of the world. Actions that will help increase developing country R & D capabilities must be stressed. International research collaboration also provides US researchers with valuable research findings and materials that will advance their own research.

Fourth, the report stresses that government policies and the institutions affecting food systems and diet play a key role in providing both the incentives and the opportunities for individuals to make decisions that will improve food production and nutrition. The report also stresses improved R & D on the effects of alternative policies and institutional arrangements on hunger and malnutrition, including behavioural research. We still do not know nearly enough about the probable effects of alternative policies and institutions under the different circumstances found in particular countries. The report underlines the need for both the United States and the low-income countries to strengthen their social science research capabilities so that they can pursue such questions more effectively.

IMPACTS OF MALNUTRITION

What difference does all this make?

Reports on human disasters around the world regularly include shock reports about the threat of starvation for tens or hundreds of millions with pictures of emaciated children with swollen bellies. These conditions are real and they shame humanity: they are indefensible in today's world. They are not the major nutritional problem, however. They can and should be eradicated by effective local action - as some low-income countries have demonstrated - reinforced where needed by emergency help from abroad. The major problem cannot be dealt with so readily. It involves more moderate malnutrition that affects huge populations, variously estimated at 500 million to one billion people as I mentioned earlier.

The combined damage to human lives of nutritional deprivation is immense and as yet incalculable. There are many potentially calculable costs to society in the form of lost productivity and income, increased health service costs and spread of disease, higher death rates, losses of innovation and intellectual skills because of increased mental retardation and apathy, and other such functional costs. But the intangible essence of this human waste is an even larger dimension. Because of the sapping of their human vitality, the ability of hundreds of millions of people to enjoy the intangible satisfactions of life is greatly diminished. The great variety of sensory impressions, intellectual stimuli, and feelings of accomplishment and well-being that provide the greatest satisfactions in life are enfeebled or lost.

POSSIBILITIES FOR INTERVENTION

Effective action to improve food consumption choices from a nutritional point of view is frustrated by the pervasive lack of knowledge about what actions do and what actions do not have substantial effects on the nutritional status of various populations. Many effects on nutrition occur very subtly. There are tantalising indications that changes in the socioeconomic context in which people live, by affecting the sense of well-being of the individual in his family and community context, influence the individual's nutritional status. It is not surprising that we understand such relationships poorly. But even for what seem to be the most obvious and direct relationships - between actual diets and the nutritional status of the populations eating those diets as measured by human performance - relatively little is known, as I mentioned earlier. There are some relatively well-expolored areas such as the effects of changing protein intake on children and relationships between selected nutrient deficiencies or food toxins and some specific diseases, but these are exceptions from the predominant situation of ignorance.

So we are confronted with a chicken-and-egg relationship between our ignorance of these matters and the problem of lack of concern. The principal actors to not know what difference their actions make regarding nutrition partly because they do not care, and they do not care partly because they do not know what good or harm their actions may be doing.

One place to break into this vicious circle is on the knowledge side of it. The best persuader to stimulate concern and action may be a combination of research results

that clearly demonstrate the damage to the functioning of large
population segments caused by existing nutritional deficiencies
on the one hand, and effective programmes that result in
eliminating the deficiencies on the other.

Interventions in the food system to improve nutritional
impacts may occur at any stage from the genetic engineering
of seeds or manipulation of farming practices to affect the
quality of food produced, through selection of the mix of
crops and animals actually grown, through handling and processing (including nutrient fortification) of food between field
and consumer, through educational, informational or cultural
influences on consumer choices of diet, through health,
sanitation, family planning or other measures that affect
bodily absorption and reaction to what is eaten, through
choices between alternative research, financing or other
services that support these activity stages. Interventions
may also involve direct distribution of food to groups facing
the greatest risks of malnutrition: common examples are
school lunches, food distribution at maternal-child health
centres, food stamps for the poor, or food distribution as
wages for unskilled labour on public works projects.

Actions outside of the food system may have large
nutritional effects, such as import and export policies, interest
rate and credit allocation policies, tax and subsidy policies,
other major price influencing policies, regulatory policies,
changes in employment laws affecting women, location of roads
and other transportation facilities, public health measures,
formulation of national development strategies, and so forth.

SOME NUTRITIONAL RESEARCH PRIORITIES

To get at the crippling knowledge blocks to more
effective action for reducing global malnutrition, the World
Food and Nutrition Study proposes a wide-ranging R & D
programme for increased US support. The overall set of
recommended research priorities is summarised in Table 1,

TABLE 1

RECOMMENDED RESEARCH PRIORITIES

Priority area	Nature of research effort	Major effects	Sources of support
1. Nutrition			
Nutrition-performance relations	Determine damage caused by various kinds and levels of malnutrition: effects of diet patterns on levels of human functioning		
Rule of dietary components	Determine specific foods that best meet nutritional needs under differing circumstances: effects of individual nutrient levels, as consumed, on nutritional status	Nutritional improvement in short run; in long run dietary changes may benefit health and life expectancy of large population segments in United States: in developing countries nutritional interventions likely to be more effective for human health than comparable investments in medical care.	NIH, NSF, USDA, AID
Policies affecting nutrition	Improve effects of full range of government policies: effects on nutrition of policies and practices usually formulated with no consideration of possible nutritional consequences.		
Nutrition intervention programmes	Improve effects of direct intervention programmes: evaluate effectiveness of alternative programmes in reaching nutritional goals.		
2. Food production			
Plant breeding and genetic manipulation	Strengthen tools of genetic manipulation: plant breeding and 'classical' genetics: cell biology: genetic stocks.	Worldwide potential is immense: production increases from most lines of research expected in 10 to 20 years: instrumental for progress in other priority areas.	USDA, NSF, AID

TABLE 1 (continued)

Priority area	Nature of research effort	Major effects	Sources of support
Biological nitrogen fixation	Increase biological nitrogen fixation associated with major crops: improve recognised symbiotic associations; attempt to establish N_2-fixing associations with grains and other nonlegumes; transfer fixation capability from bacteria to plants.	Large potential world wide: results within 10 to 15 years for legumes and 15 to 25 years for nonlegumes.	USDA, NSF, AID, EPA
Photosynthesis	Increase amount of photosynthesis in major crops; reducing photorespiration and dark respiration; transferring traits from photosynthetically efficient plants.	Higher yields of food plants particularly in the tropics: substantial increases in potential yields after 15 years or more.	USDA, NSF
Resistance to environmental stresses	Improve resistance of major crops to drought, temperature extremes, deficiencies of acid soils, salinity rapid screening techniques; tolerance of acid soils; shorter season crops; larger root systems; better use of soil fungi; salinity resistance; farming systems.	Larger and more stable crop yields in 10 to 15 years, more efficient use of inputs, possible to grow crops in new locations.	USDA, NSF, AID
Pest management	Reduce preharvest losses due to pests; integrated pest management; specific control mechanisms.	Can eliminate large and pervasive losses due to pests, in short run by adapting known technology, in long run by biological control techniques.	USDA, NSF, AID, EPA

TABLE 1 (continued)

Priority area	Nature of research effort	Major effects	Sources of support
Weather and climate	Improve techniques for predicting weather and climate and using information to assist adaptation by farmers; reduce weather damage to food production.	Substantial analytical improvement in short run; can have pay offs many times the cost.	NOAA, USDA, NASA, NSF
Management of tropical soils	Improve management of tropical soils to increase crop productivity: soil classification; land clearing methods; correcting soil deficiencies maintaining desired soil characteristics; suitable cropping systems and technologies.	Annual crop production in humid lands can be 150 to 200 percent of temperature zone production per hectare.	USDA, AID
Irrigation and water management	Improve management of water supplies: adjustments in farming systems and management of water movement for optimal supply to crops; adapting farming operations to water availability.	May double crop yields in some areas, and make capital investments more effective.	USDA, AID
Fertiliser sources	Improve cost/return ratios of chemical fertilisers; new methods of producing nitrogen and phosphorus fertiliser; new fertilisers tailored to tropical conditions.	Advances in short and long run; new ways of producing nitrogen fertiliser will reduce energy consumption; new and more efficient fertilisers could have major effects on tropical food production.	AID, TVA, USDA, ERDA

TABLE 1 (continued)

Priority area	Nature of research effort	Major effects	Sources of support
Ruminant livestock	Increase product yields from ruminant livestock, particularly in the tropics; forages; priority animal diseases; genetics and reproduction.	Forage and disease research will open up new areas and will improve productivity and eliminate heavy economic losses.	USDA, AID, NSF
Aquatic food sources	Increase contribution of aquatic resources to world food supply; waste reduction and upgrading product through processing; aquaculture research in both breeding and seed supply and polyculture management.	Could double fish protein consumed by humans without increasing world catch; could raise potential aquaculture yields fivefold.	AID, DOI, DOC, NSF, USDA
Farm production systems	Improve production systems, particularly for small farms in developing countries; methodology for identifying appropriate farming system; multiple cropping; soil and water management; equipment-labour relationships.	Realise potential for two to four times present production in humid tropics; more modest increases in semiarid tropics.	AID, USDA, EPA, ERDA
3. Food marketing Postharvest losses	Reduce postharvest losses; nature and magnitude of losses; pest control after harvest; food preservation for humid and arid tropics.	Substantial reduction in losses that now range 15 to 50 per cent of production; encouragement of beneficial changes in cropping patterns.	AID, NSF, USDA, DOD

TABLE 1 (continued)

Priority area	Nature of research effort	Major effects	Sources of support
Market expansion	Extend market scope for consumers and farmers in developing countries; enhancing purchasing power; transportation; marketing institutions; managing marketing flows of major commodities.	Stimulate production and cut food losses; effects in 5 to 10 years.	AID, USDA
4. Policies and organisations			
National food policies and organisations	Improve policies and organisations affecting food production, distribution, and nutrition in developing countries; comparative performance in food systems; comparative studies to identify transferable improvement factors (decentralisation, local participation; staff development); interactions of income distribution with food production and nutrition; methodology of sector analysis.	Early results in improving effectiveness of policies and organisations relating to food systems and orienting selection and implementation of other biological and physical research, give farmers incentives for production and provide prices that will give more effective distribution.	AID, NSF, USDA
Trade policy	Improve effects of trade policy on food production and nutrition; studies on effects of trade liberalisation; consequences of international management of trade; optimum trading patterns.	Early effects on orientating country food policies for balance between own production and reliance on trade; improve diets, incomes, and national economic performance.	USDA, AID, State DOC

TABLE 1 (continued)

Priority area	Nature of research effort	Major effects	Sources of support
Food reserves	Improve role of reserves in relation to other measures for stabilising food supplies; improviding developing country food reserve practices; identifying improved mixes of reserves and other measures to stabilise food supply.	Relieve hunger and malnutrition due to production instability.	USDA, AID
Information systems	Improve flows of information in support of decision making on food and nutrition; producer information needs to use better technology; crop monitoring systems; international data bases on land uses and malnutrition: information systems design.	Large gains, especially in developing countries, from fewer wrong decisions and fuller use of available unproved technologies.	USDA, NASA, DOD NOAA. AID

extracted from the Study and attached to my remarks. Let me mention some of the recommended research that bears most directly on nutritional problems.

The sections on research priorities for food production stress research to improve plant protein synthesis and development of more powerful techniques for inducing genetic changes. The latter would permit better tailoring of plants so as to incorporate more desirable nutrient characteristics in harvested crops and also improved handling, storage and processing characteristics.

The sections on activity between the farmgate and the consumer stress the widening of consumer's access to existing food supplies and reduction of food losses. The importance of finding better preservation technologies for the socioeconomic and physical conditions of the tropics is given particular stress. The technologies of dehydration, compaction, irradiation, air-sealing, and protective packaging, for example, may all play major new roles in the future. R & D is also stressed on other factors affecting food distribution: national and local service institutions, stockpiling, trade policy, income distribution, and other social factors.

The central thrust of the report's focus on nutrition appears in its first section on research priorities. This section is concerned with national, local and international strategies for improving the impact of overall food systems on the nutritional status of populations. An integrated set of priorities for accelerated R & D is proposed which addresses specific aspects of five basic questions:

> What kinds of nutritional problems affect what segments of the populations with what severity and prevalence in which countries?

> What is the significance of each of these problems for the people affected?

What are the consequences to society of failure to reduce or eliminate these problems, and how can these consequences be usefully expressed in quantitative and qualitative terms?

To what extent would each or some of the problems be solved through increased production or imports of what foods, if existing distributional mechanisms were not changed?

How and to what extent would other kinds of direct or indirect food or nonfood interventions solve each or some of the problems and at what costs?

The specific recommendations for increased US R & D support on these nutritional questions assume continuing support at least at existing levels for research that is already proceeding on food production and distribution technologies, and on underlying fundamental problems addressed by the natural and social sciences, plus the accelerated R & D recommended elsewhere in the report.

CRITERIA FOR RESEARCH PRIORITIES

I am told that failure to establish adequate criteria for selecting research priorities has inhibited progress in ECC/COST efforts to establish an agreed programme of research priorities, and that the criteria used in the World Food and Nutrition Study would be of interest to this group.

The selection of priority research areas was based primarily on a consolidated weighing of the answers to two questions.

1) What is the best available consensus on the extent of the knowledge advances that specific areas of research are likely to produce, and what is the scientific or technological significance of these advances?

2) If the research succeeds in gaining the anticipated results, what is the likely effect on reducing global

hunger and malnutrition over the next several decades?

Subquestions were asked under each of these criteria. This consolidated judgment was modified in some cases, but not much, by judgments on the feasibility of gaining supportive action needed to achieve the research results, and also by considerations of the complementarity of different research areas: for example, a line of research that did not merit the highest priority on its own might have been recommended because of its complementarity with other lines of research that had been judged to have outstanding impact potential. More stress was given to the 'what difference would it make' criterion and to seeking empirical clues on this than is usual in reports on research priorities.

The priority judgments were made separately for likely research results and impacts up to 15 years and beyond 15 years, in order to assess better the time dimensions involved in putting together a package of research proposals.

The study employed a fairly complicated, multi-phased process for the application of these criteria to reach judgments on priorities, involving three different groups of people at the various phases. This process is described in Appendix V of the Study.

Note that the set of criteria used is _not_ adequate for an overall assessment of research priorities for the United States or any other country. The criterion of impact on world hunger was dictated by the nature of President Ford's request for this particular study. It is one important criterion to use in assessing US research priorities, given US support for the goal of eliminating world poverty, hunger and malnutrition. But there are other important domestic criteria that also need to be applied in the overall assessment of US research priorities, criteria that are outside the bounds of the report to the President. In other words, the appropriateness of the priority selection criteria depends on the purpose to be served by the research in question.

HOW TO GET THE RESEARCH DONE

I shall not take time today to review the recommendations of the World Food and Nutritional Study on steps that government departments and other US organisations should take to mobilise research resources more effectively in support of the recommended research agenda, except for one set of recommendations involving international action. However, your interest in reading the Academy's Study might be stimulated by knowing that several European research leaders have said that these organisational recommendations are relevant to their own countries.

Of particular interest is the Study's stress on expanding international R & D collaboration via 'networks' whose general characteristics are:

A focus on problem-solving (R & D) in discrete and widely pervasive problem areas;

Voluntary collaboration open to institutions all over the world that are doing R & D in a given problem area, including institutions working in different segments of the R & D spectrum ranging from experimental trials of technology under actual operational conditions through new technology development to basic research (eg collaboration via joint research projects, pooling and exchanges of research materials and results, programme coordination, provision of research sites, advisory or training services, exchange visits of professional staff, or use of common information services);

Availability of research results throughout the network in a common 'pool' from which all participants may draw whatever they can use for local adaptation as needed to fit their own situations;

Provision for performance of 'nerve centre' functions

needed for the network's operation by one or more participating institutions (eg leadership in organising collaborative programmes or projects, logistical support, and information or materials management services).

Although the research collaboration in such networks follows the traditional pattern of voluntary cooperation among scientists or organisations sharing common interests, the collaboration in the newly-emerging agricultural research networks tends to be more extensive, systematic, and sustained than the rather casual and intermittent international research cooperation that has long existed. Provision for the 'nerve centre' functions makes this possible.

Expansion of this type of international R & D collaboration in _nutritional_ research is also recommended in the Academy Study. The potential advantages include:

Providing low-income countries with access to worldwide S & T capabilities far beyond anything that they could provide or organise for themselves on a country-by-country basis, even if they receive large-scale aid;

Taking advantage of possibilities for international specialisation and divisions of labour so as to increase the power and rate of progress of R & D on global problems; appropriate divisions of labour permit economies of scale by concentrating much of the more expensive and difficult types of research at facilities that can serve international needs, particularly the needs of low-income countries;

Fostering technology flows in _all directions_ through each system as an automatic by-product of its operations - a most efficient and psychologically satisfying system for technology transfer;

Building the problem-solving capabilities of participating institutions and scientists from the low-income countries, who learn largely by working with professional colleagues around the world - again a relatively efficient and psychologically satisfying system for building institutional capabilities;

Providing practical, relatively low-cost and efficient means of achieving the foregoing results; bureaucracy and overhead structure can be minimised, as well as the complexity and 'least-common-denominator' tendencies that hamper programme efforts in formal intergovernmental organisations.

The element of psychological satisfaction is quite important. It is easier to convey a sense of partnership for mutual gain in collaborative R & D networks than in other systems of technology transfer. The participants are engaged in a 'positive-sum game' in which all are enriched, rather than in a 'zero-sum contest' to gain a larger share of a fixed total gain. There is interdependence accompanied by maximum autonomy and flexibility of choice by individual participants. There is minimum temptation to engage in disruptive polemics or ideological conflicts.

PROSPECTS

The World Food and Nutrition Study is cautiously optimistic about the prospects for producing a sufficient quantity of food to meet supply requirements over the next several decades. The steps required seem well within the financial and technical capabilities of the countries concerned, and the report is hopeful that there will be adequate political will in most countries to do what is required in agriculture.

However, the question of adequacy also depends on food demand factors and on the quality of food that reaches consumers. The challenge to nutritional scientists and food technologists is to balance progress in agriculture with progress in opening better options for consumers - providing them with foods that are nourishing, attractive, safe and low cost. If we can do this and also reduce environmental hazards that increase caloric requirements, the quantity of food needed to maintain good nutrition will be lowered and public health will be improved. How well the questions of diet composition are handled is not the secondary issue of human welfare that it once seemed to be. Indeed, it may well be that the potential for improving human health via improvements in diet is greater than the potential via advances in curative medicine.

COMMODITY STUDY PANELS

NUTRITION COORDINATION GROUP

PARTICIPANTS

Chairman:	G. Varela *(Spain)*	Fish Panel
Co-rapporteurs:	J.P. Kevany *(Ireland)*	
	F.M. Cremin *(Ireland)*	
Invited Experts:	D.H. Buss *(UK)*	
	G.B. Brubacher *(Switzerland)*	
	A. Dahlquist *(Sweden)*	
	G. Tomassi *(Italy)*	
	J.W.G. Porter *(UK)*	
	F. Fidanza *(Italy)*	
	W. Seibel *(W.Germany)*	
	J. Bernstein *(USA)*	
Commodity Study Panel Representatives:	J.H. Moore *(UK)*)	
	B. Blanc *(Switzerland)*)	Dairy Panel
	M. Caric *(Yugoslavia)*)	
	B. Krol *(Netherlands)*)	
	M. Jul *(Denmark)*)	Meat Panel
	A.E. Bender *(UK)*)	
	A. Hansen *(Norway)*	Fish Panel
	W. Spiess *(W.Germany)*)	Fruit and
	J. Solms *(Switzerland)*)	Vegetables Panel
	P. Linko *(Finland)*)	
	G. Fabriani *(Italy)*)	Cereals Panel
	D.A.T. Southgate *(UK)*)	
	J.L'Estrange *(Ireland)*	Observer

NUTRITIONAL PRIORITIES, WITH PARTICULAR REFERENCE TO FROZEN FOODS

D.H. Buss

Nutrition Section, Ministry of Agriculture, Fisheries and Food
Horseferry Road, London SW1, UK.

INTRODUCTION

Nutritionists can see both advantages and disadvantages to the preservation and thermal processing of foods. The main advantage of preservation is that it allows a wide variety of foods to be eaten throughout the year, regardless of distance from the areas of food production. Provided that the nutrients are well retained, this will increase the quality of the diet, often at comparatively little cost. Cooking, of course, improves the flavour, palatability and digestibility of many foods, thus enabling them to be eaten so that they can fulfil their basic function of providing a steady supply of nutrients to each one of us throughout our lives.

The disadvantages arise because many of the 40 or more essential nutrients are unstable and may be destroyed by excessive heat or during storage. They may also be dissolved out into cooking water.

Before these opposing effects can be balanced, much more needs to be known about the requirements of different people for all the nutrients (including minor B-vitamins and trace minerals), as well as about the losses which occur in different foods under different conditions. But there are so many factors which could be studied that some priorities must be established. These will differ from country to country, and even for different people such as infants and the elderly within each country, but the following considerations are important. They will be illustrated with particular reference to frozen foods, for this subject has recently been reviewed in a report about to be published in Britain.

FOODS OF IMPORTANCE

Nutritionally, those foods which are important are not only those eaten in large amounts but also those rich in certain nutrients. Quick-frozen food would not at first seem to be worthy of such study because on average little is eaten (only 10 - 20 kg per person per year out of the total food supply of 500 kg or more in northern European countries, and even less in the south with the exception of major cities where home-freezers are more common and an adequate distribution system exists). But increasing amounts of meat and fish are now bulk-frozen, and certain individuals (including some children and people in residential institutions employing a cook-freeze system) already depend heavily upon foods which have been frozen and thawed at some stage.

The foods which are most commonly frozen include a variety of vegetables as well as fish, meat (especially poultry) and meat products. Their exact importance will vary from country to country - thus in Britain peas, french-fried potatoes and fish sticks are the most popular quick-frozen items, but quick-frozen spinach (which is popular in Sweden and West Germany) is almost unknown. It is therefore on foods such as these, and quick-frozen dishes and complete meals, that research should be concentrated at first.

SOURCES OF LOSS IN FROZEN FOODS

The freezing process itself will have little or no effect on nutritional value. But losses can occur:

1) If new vegetable cultivars of unexpectedly low nutritional value are developed for commercial freezing;

2) During prolonged storage before freezing, for example, if over-ripe fruit and vegetables are preserved by this method in the home;

3) During blanching of vegetables in the factory or home, especially if prolonged or in excessive amounts of

water. On the other hand, blanching destroys enzymes which would otherwise reduce the amount of vitamin C present, and thus in effect may have a beneficial effect on nutritional value;

4) During storage, especially if prolonged or at an insufficiently low temperature. It is thought that fluctuating temperatures, which will occur when large amounts of fresh food are added to freezers, may be worse than steady temperatures. Some nutrient losses might, however, be reduced by air-tight and light-tight packaging;

5) During thawing. This may be especially important when large blocks of meat or fish are thawed in catering establishments either by immersion in water or in a refrigerator over periods extending up to 3 days;

6) During cooking. For example, blanched vegetables are partially cooked, and the final cooking time should therefore be reduced. If it is not, unnecessary losses may occur.

NUTRIENTS WHICH MAY BE LOST

The nutrients which are most likely to be affected are those which are water-soluble, for they will be lost in the water or steam during the blanching of vegetables, and in the 'drip' which results when frozen meat and fish are thawed. These nutrients include <u>all</u> the B-vitamins including folic acid, vitamin C, several trace minerals and even some amino acids and proteins. Some nutrients, such as riboflavin and vitamins A and E remain sensitive to light or oxygen even at $-18°C$, and the chemical form of some nutrients has been suspected as changing - which is important if certain forms are more readily absorbed into the body than others. The main factors which can lower the nutritional value of any food are summarised in Table 1.

TABLE 1

SUMMARY OF FACTORS WHICH MAY REDUCE THE NUTRIENTS IN FOOD

Nutrient	Heat	Light	Air	Water	Alkali	Other
Protein				(√)		
Vitamin A	(√)		√	√	√	Metals
Thiamin	√		√	√	√	SO_2
Riboflavin		√	√	√	√	
Folic acid	√		√	√	√	
Vitamin C	√		√	√		Metals Enzymes
Minerals				(√)		

An example of these considerations is provided by frozen potato products. Potatoes are shown by our National Food Survey to provide, on average, more than 20% of the vitamin C and 10% of the thiamin (vitamin B_1) in Britain; it may therefore be important for some people if much vitamin C is lost during the blanching and subsequent storage of quick-frozen french-fried potatoes, or if the thiamin is destroyed by sulphur dioxide added to preserve their whiteness. Fish is an important source of iodine, and any loss in frozen fish could increase the prevalence of goitre. On the other hand, it would not matter if all the nicotinic acid were lost from frozen peas because peas are a very minor source of this nutrient and our diets on the whole provide a large excess of this nutrient over the intakes which are recommended for health.

CONCLUSION

Ideally, all foods should be processed and cooked for the shortest times, at the lowest temperature, and using the least water possible. They should also be stored for the shortest times necessary, at low temperatures and with the exclusion of light and air. This applies as much in the home as in the factory. Where this is not practicable, it is most important to preserve the nutrients which are in shortest supply for at-risk groups of the population, especially in the foods which are likely to be the main providers of those nutrients.

IMPORTANCE OF MEAT PRODUCTS FOR THE SUPPLY OF MICRONUTRIENTS IN HUMAN NUTRITION

G. Brubacher

Department of Vitamin and Nutritional Research of
F. Hoffmann-La Roche & Co. Ltd. Basle, Switzerland.

In investigations into the influence of the processing of food on its nutritional content, priority should be given to such nutrients for which the food in question is an important source; among these nutrients priority should be given to those where problems exist in the supply of population or segments of the population.

This point will be illustrated with the following three examples:

1) In industrialised countries, like Switzerland, it is quite usual that people consume a high proportion of their daily energy requirement in the form of empty calories. A normal diet for instance is composed of:

15	calorie percent	sucrose
5 - 10	"	white flour, rice and products from refined starch
20	"	refined edible fat and oils
5 - 10	"	alcoholic beverages

It means that 50 to 60% of calories are eaten as empty calories and that the micronutrients such as vitamins and minerals should be contained in the remaining 50%, or in about 1200 kcal. Under certain circumstances it is difficult to find enough food items which contain these micronutrients in the necessary concentration to fulfil this condition. Above all, this holds true for vitamin B_1. The vitamin B_1 supply is just on the borderline in most of the European countries. Should, however, the trend continue that more and more empty calories are consumed or that more and more vitamin B_1 is destroyed during processing, serious problems could arise in certain segments of the population. It can be concluded that with a proportion of 50 calorie percent of empty calories, only food with a nutrient density of at least 0.8 mg vitamin B_1 per 1000 kcal can be considered as a source of vitamin B_1. Meat and meat products are amongst the main food items which meet this condition; this is shown by the following Table.

TABLE 1

VITAMIN B_1 CONCENTRATION IN MEAT AND MEAT PRODUCTS (EXTRACTED FROM GENERAL FOOD COMPOSITION TABLES)

Food item	Nutrient density mg vit. B_1/1000 kcal.	
beef heart	4.1	
pig's heart	3.8	
pig's heart	3.3	rich
beef liver	3.3	
pork	2.5	
veal	0.9	
beef	0.3	poor

With the exception of beef all these food items would be suited to supplementing a diet consisting of 50 calorie percent of empty calories. From the Table the conclusion can be drawn that, with regard to the vitamin B_1-content, large differences may exist between the various meat products. Therefore, the exact knowledge of these values is necessary in planning an adequate diet. Some research should be undertaken to complete our knowledge of the vitamin B_1 content of meat produced under different conditions.

The main problem, however, in using meat and meat products as a vitamin B_1 supplement lies in the fact that up to 70% of the vitamin B_1 can be destroyed by thermal processing. In this field it is very important to have exact figures, because the knowledge of these figures could immediately be used for better planning of diets and as a lead for improving kitchen practice.

2) In most of the industrialised countries the supply of proteins has not changed quantitatively in the last 30 years, but the proportion of animal protein has risen whereas the proportion of protein deriving from potatoes and bread has drastically dropped. There exists a relationship between the amount of protein consumed and the vitamin B_6 requirement in that, the higher the protein consumption the higher is the vitamin B_6 requirement. In Table 2 the vitamin B_6/protein ratio is given for a series of protein sources. From this Table it can be derived that, with the drop in consumption of potatoes and bread made from whole meal flour, and with the rise in consumption of meat the vitamin B_6/protein ratio has diminished. If vitamin B_6 is destroyed by thermal processing of the meat, the ratio can be lowered to a critical point, where an adequate supply of vitamin B_6 will no longer be guaranteed to the population. Therefore it is very important to carry out all cooking processes extremely carefully. This is only possible by understanding all factors leading to a destruction of vitamin B_6.

3) If we consider the most urgent nutritional problems to be solved on earth we see that in addition to calorie-protein deficiency and vitamin A deficiency, we also have iron deficiency. About 20% of the world's total population suffer from this illness, regardless of whether the people belong to industrialised or to developing countries. We should not discuss the reason for this fact but bear in mind that in our diet meat is one of the most important sources of iron. The daily requirement of iron is probably only 1 mg, but the absorption of iron from most food items is very poor; therefore most experts in this field recommend a daily dietary allowance of at least 10 mg

TABLE 2

VITAMIN B_6/PROTEIN RATIO FOR VARIOUS PROTEIN SOURCES (EXTRACTED FROM THE DUTCH FOOD COMPOSITION TABLES 1971)

Food item	vitamin B_6/protein ratio mcg vitamin B_6/g protein	
	raw	cooked or ready to eat
potatoes	150	110
whole meal bread		22
beans	17	6
white bread		8
milk	12	< 12
eggs	13	12
veal	17	9
beef	11	4
pork	19	6

of iron. Only about 1.5% of iron, for instance, is absorbed from spinach with an iron content of about 3 mg/100 g whereas from veal 20% is resorbed with an iron content of 0.4 - 3 mg/100 g. Therefore, the daily requirement of iron can be better met with veal than with the same amount of iron in spinach. We have recently shown that with the preparation of meat in the kitchen no losses of iron occur. However no information is available on the bioavailability of iron during thermal processing, including home cooking practices. This is an open question which should be investigated in a general research programme on changes of nutritional value of food during thermal processing.

It has been demonstrated, in the case of three micronutrients, which problems have to be considered in establishing a programme to investigate changes in the nutritional quality of meat and meat products during thermal processing. Similar consideration has to be given to all other commodities.

It has been shown that such a research programme is very important in connection with the problems of adequate supply of certain micronutrients to the population as a whole or to certain segments of the population. This programme should concentrate on such micronutrients as are really relevant from the point of view of public health.

THE NUTRITIONAL VALUE OF PROTEIN IN BREAKFAST CEREALS

Arne Dahlqvist, Nils-Georg Asp and Gunilla Jonsson,
Department of Nutrition, University of Lund, Chemical Centre,
Box 740, 220 07 Lund, Sweden.

During recent years the influence of industrial processing on the nutritive value of different selected foods has been studied.

Of great interest are the results obtained in a study of breakfast cereals where some alarming figures were found both regarding the composition in general, and the nutritive value of the protein in particular. The breakfast cereals studied were such as could be obtained in Swedish shops and 14 different kinds were examined. In addition a home-made mixture, quite popular among students called 'Putte' was studied.

The total protein content of the cereals varied between 6 and 20%. Some of the lowest figures were obtained in frosted flakes, in which sucrose amounted to nearly half of the weight.

Chemical analysis of the amino acid composition showed that in all cases lysine was the limiting amino acid. This was, of course, to be expected in cereal products. The chemical score was, however, in many cases remarkably low. Several of the cereal brands had a chemical score around 15, which is much lower than in the raw materials used for preparation of the products (compare corn meal, 50; rice, 64; oat meals, 74). Most remarkable was one kind of breakfast cereal which according to the declaration on the package contained rice enriched with gluten, wheat germ and skim milk powder, but nevertheless had a score value as low as 31.

The biological value of the protein may be still lower than indicated by the lysine assay, since on chemical analysis after acid hyrolosis some biologically unavailable early products of the Maillard reaction between lysine and carbohydrates are hydrolysed into free lysine. We therefore also investigated the biological value of the protein and some of the breakfast cereals in rats, using the PER (protein efficiency ratio) method. In many cases the results indicated a still lower nutritive value of the protein than was found by lysine analysis. The highly enriched cereal product mentioned above, with a chemical score of 31, did not even maintain the weight of the small rats during the assay period. Apparently a considerable degree of Maillard reaction had occurred in this cereal.

When the cereals were mixed with milk, however, (equal part of milk protein and cereal protein) the nutritive value of the mixture improved. This is due to the fact that the milk contains a large excess of lysine, and this lysine compensates for the low lysine content and the loss of lysine in the cereals.

These findings are considered to have serious nutritional implications. A plate of breakfast cereals with milk is regarded as a good contribution to breakfast. In spite of the high protein content of the milk, the cereals will usually provide more than half of the total protein in an ordinary portion of cereals and milk. It is concluded that,

> Firstly: If the cereals are consumed without milk, which occasionally happens due to dietary advice on decreased milk consumption, the protein value will be very low.

> Secondly: The difference between the lysine content of the raw material and the available lysine in the product indicates that an important degree of Maillard reaction has occurred during the industrial preparation. This is especially true for cereals which are enriched with lysine-rich proteins, but which nevertheless do not contain enough available lysine to maintain weight in rats. There is justification for the belief that Maillard products can exert a toxic effect, strongly suggesting the need to avoid such treatment.

METHODS FOR THE DETECTION OF PROTEIN QUALITY IN TRADITIONAL AND NEW PROTEIN FOODS

G. Tomassi

Instituto Nazionale della Nutrizione, Rome, Italy.

The assessment of the nutritive value of proteins and protein foods is of great interest from the nutritional and industrial point of view, as the recommended dietary allowances and nutritional labelling can be influenced by its exact determination. Methods for measuring protein quality are of a chemical or biological nature, since protein quality depends both on amino acid composition and digestibility. Chemical methods are based on the determination of amino acid composition and on calculation of a 'chemical score' (Mitchell and Block, 1946), or of an 'essential amino acid index' (Mitchell, 1954) in relation to a reference protein or to a standard amino acid pattern. These methods evaluate the protein when fully digested and therefore tend to overestimate the true biological value of the proteins, but nevertheless are useful for a rapid prediction of the potential protein quality and are widely used.

The results of chemical analysis, however, are influenced by the procedures followed in the preparation of protein hydrolysates and by the experimental conditions such as dilution levels, the presence of oxygen and the time of hydrolysis. These conditions can produce variable degrees of amino acid destruction (Menden and Cremer, 1970) leading to different amino acid content values (Porter et al., 1968; Lunven et al., 1973).

The specific behaviour of each protein, with regard to the aforementioned analytical conditions, makes it necessary to establish the most suitable procedures for a large range of them, and particularly for new proteins or new protein foods. Intensive research in this direction represents one of the current topics of the Italian National Research Council programme under the heading 'New protein sources and new food-formulations'.

Another aspect which needs international investigation and clarification is the establishment of rapid, precise and reproducibile methods to estimate the availability of amino acids other than lysine, such as methionine, theronine, tryptophan, isoleucine. The interlaboratories collaborative study undertaken by the Protein Quality Groups of the Agricultural Research Council of the United Kingdom (Zuchermann, 1959) could serve as a useful basis for a European collaboration in this regard.

A fundamental problem, which is still the subject of much debate, is the choice of the reference amino acid pattern for the calculation of chemical scores. Controversy arises between the use of the amino acid pattern of egg protein, (which is rich in the sulfur amino acids) and the FAO pattern. However, chemical methods, to be acceptable need to be highly correlated with biological methods, which measure the response to protein intakes in laboratory animals. The latter are considered the only methods suitable for the estimation of the true nutritive value of the proteins.

The protein efficiency ratio method (PER), largely used to date and still the official method of AOAC (1975), has been recently criticised, because of non-linear variation in animal response with the amount of ingested protein. As a consequence, the protein quality appears to depend more on the ingested amounts rather than on the effective amino acid protein composition. Other methods like the Net Protein Ratio (NPR) (Bender and Doell, 1957), Net Protein Utilisation (NPU) (Bender and Miller, 1953) and Biological Value (BV) (Thomas, 1909) are considered to give more valid results but, like the PER value, they have the disadvantage of being based on a single dose response. This procedure assumes the existence of a linear response from the zero protein level to the protein dose used, which is often not true, especially for low-quality proteins and particularly for lysine deficient ones. Other methods based on the multiple dose/response relationship have been proposed, and are achieved in practice by using different levels of protein in the experimental diets (Allison, 1964; Hegsted and Chang, 1965). Using this approach protein quality is given by the slope of the dose/response curve in the linear tract, which describes the variations of either the body weight or the carcass nitrogen or water content as a function of the amounts of the ingested protein, divided by the dose/response curve of a reference protein (relative protein value, RPV). The RPV method, as well as the other mentioned biological methods, has the limitation of not distinguishing the influence played by various degrees of protein digestibility on the animal response. This limitation may be particularly important when assessing new protein sources whose digestibility gradient is often unknown, and may be lower than that of conventional proteins. In spite of the above mentioned limitations a recent collaborative study, prompted by Hegsted, in which 8 different laboratories in the US participated, shows that the RPV method (using albumin as reference protein) gives better results than the PER and NPR methods with respect to variability (Sammonds and Hegsted, 1976).

This has led the joint FAO-OMS-FNB committee to recommend that the RPV method should replace the currently accepted PER assay (in press). The general acceptance of the RPV method, however, needs additional support, by way of experiments incorporating various factors, the relative importance of which have yet to be defined. Of these the most important is agreement on the appropriate protein reference standard, together with a definition of its identity and stability characteristics, and the experimental design and conditions, such as animal strain and weight, duration of experiment, body weight gain or carcass nitrogen or water content measurements. Two-week RPV tests, carried out in the laboratories of our Institute give satisfactory results with respect to variability and response (Corcos-Benedetti et al., 1977).

The general inconvenience of biological methods is evidenced by the time lapse involved in obtaining the end response (4 weeks for PER and 2 weeks for RPV). Consequently rapid bioassays which utilise an 'in vitro' enzymatic digestion procedure have been proposed as alternative methods.

In theory, these methods which take into account the digestibility of the protein, appear to overcome the disadvantage of chemical and 'in vivo' biological methods. Several methods have been developed, differing with regard to the amount of the digestive enzymes used, the experimental conditions for the digestion of the protein and measurement of the liberated amino acids. Limitations, associated with these enzymatic 'in vitro' methods include the difficulty of reproducing physiological digestive conditions, of separating the amino acids as rapidly as they are liberated and of autodigestion of the enzymes.

Using this approach, the evaluation of protein quality has been made on the basis of various amino acid indexes, calculated in relation to a reference protein. The pepsin-pancreatin-digest index (PPD) (Sheffner et al., 1950), the pepsin digest residue index (PDR) (Akeson and Stahamann, 1964), the pepsin-pancreatin dialysed digest index (PPDD) (Mauron et al., 1955) have all been proposed. More recently the enzymatic ultrafiltrate digest amino acid index (EUD) utilising a new enzymatic digestion system and an ultrafiltration cell seems to correlate well with other biological methods (Floridi and Fidanza, 1975).

These rapid bioassays are very promising, but need more research to be validated by in vivo biological methods and particularly with the RPV method, using different proteins and protein foods. These aspects are under current investigation in Italy as part of CNR programme and could be expanded to encompass other European states.

The needs of the food industry for rapid, low-cost,

reliable and officially acceptable tests for assessing protein quality is generally recognised (The Midlands Conference, 1977). Consequently more research should be encouraged to produce results which fulfill this necessity.

REFERENCES

Akeson, W.R. and Stahamann, H.A. 1964. J. Nutr. 83, 257.
Allison, J.B. 1964. Mammalian protein metabolism, Vol. II, Munro N.H. and Allison, J.B. edtrs. Acad. Press. New York.
A.O.A.C. 1975. Official methods of analysis - 12th Ed. Assoc. of Official Agricultural Chemists, Washington, D.C.
Bender, A.E. and Miller, D.S. 1953. Biochem. J. 43, Vii.
Bender, A.E. and Doell, B.H. 1957. Brit. J. Nutrition 11, 140.
Corcos-Benedetti, P., Tagliamonte, B., Di Felice M., Gentili, V. and Spadoni, M.A. 1977. Atti del Congresso SINU Siena.
Floridia, Fidanza, A. 1975. Riv. Sci. Tecn. Alim. Nutr. Um. V. 13.
Hegsted, D.M. and Chang, Y. 1965. J. Nutr. 87, 19.
Joint, IUNS-FAO-FNB Committee (in press). Evaluation of protein quality - 2nd edition. Young ed. NAS-NRC.
Lunven, P. - Le Clement de Saint-Marcq C., Carnovale, E. and Fratoni, A. 1973. Brit. J. Nutr. 30, 189.
Mauron, J., Mottu, F., Bujard, E., Egli, R.H. 1955. Arch. Biochem. Biophys. 59, 433.
Menden, E. and Cremer, A.D. 1970. Newer methods of nutritional biochemistry. Vol. IV pag. 146 ed. by Albanese, Acad. Press.
The Midlands Conference (1977) New concepts for the rapid determination of protein quality - Nutr. Rep. Intern. 16, 157.
Mitchell, H.H., Block, R.J. 1946. J. Biol. Chem. 163, 599.
Mitchell, H.H. 1954. Symposium on methods for evaluation of nutritional adequacy and status. Spector H. Peterson M.S. Friedman, T.E. edtrs. pag. 13. National Res. Conc. Washington, D.C.
Porter, J.W.G., Westgarth, D.R. and Williams, A.P. 1968. Brit. J. Nutrition 22, 437.
Samonds, K.W. and Hegsted, D.M. 1976. Evaluation of protein food Provision of Publ. 1100 National Acad of Sci. Washington, D.C.
Sheffner, A.L., Eckfeldt, G.A. and Spector, H. 1956. J. Nutr. 60, 105.
Zuchermann, S. 1959. Nature (London), 183. 1303.

EFFECTS OF THERMAL PROCESSING ON THE NUTRITIVE VALUE OF MILK AND MILK PRODUCTS

J.W.G. Porter

National Institute for Research in Dairying, Shinfield, Reading RG2 9AT, England.

Milk is an almost complete single food and milk and milk products make an important contribution to our daily needs for energy, protein, calcium and certain vitamins. Thermal processing of milk is necessary to prevent microbial spoilage. Such processing has little or no effect on the fat, fat soluble vitamins, carbohydrates and minerals of milk but it may impair nutritive value by damaging the proteins and by destroying the more heat-labile water soluble vitamins. The effects of thermal processing depend on the method used and are generally smaller with processes using high temperatures for short times.

Studies of the losses of vitamins during traditional procedures have established that pasteurisation, UHT sterilisation and the preparation of dried milk causes little loss but that in-bottle sterilisation and the sterilisation of evaporated milk may cause considerable destruction of thiamine, vitamin B_6, vitamin B_{12}, folic acid and vitamin C. Concentration of milk by reverse osmosis causes little loss of vitamins but with ultrafiltration only the vitamins that are firmly protein-bound, such as vitamin B_{12} and folic acid, are completely retained. Further losses of these vitamins, and also of riboflavin and vitamin A, may occur during storage of liquid products if they contain dissolved oxygen or are filled into containers that are permeable to oxygen or transparent to light. Less is known about the effects of storage on the vitamin content of ice cream and other frozen desserts, especially when large amounts of air are incorporated in such products.

The proteins of milk are valuable nutrients, well endowed with essential amino acids, well digested and of high biological value. Casein has a slightly lower biological value than the whey proteins because it has a slight deficiency of the sulphur-containing amino acids methionine and cystine. Both casein and the whey proteins have a high content of lysine. This means that in a mixed diet milk proteins can be of particular value in enhancing the value of other proteins, such as those from cereals, which have a low content of lysine.

It is well established that the whey proteins are denatured to some degree during heat treatment of milk. Denaturation may inactivate immune globulins and may affect milk clotting but it does not affect the biological value or digestibility of the whey proteins for the rat or for the human infant. However, for reasons that are not yet clear, calves and baby pigs do not thrive when given diets in which half or more of the whey proteins have deen denatured.

Chemical interactions leading to measurable loss of nutritive value occur during severe heat treatments such as in-bottle sterilisation and some roller-drying processes. Loss of nutritive value may derive from protein-protein or protein carbohydrate interactions and results primarily from a lowering of the availabilities of the sulphur-containing amino acids and lysine, respectively, though protein-protein interactions may also cause the availability of all the amino acids to be reduced. Further work is needed to establish whether the products of these interactions interfere with the normal processes of protein digestion and amino acid uptake.

Protein-carbohydrate interactions are of the Maillard type, the free ε-amino group of lysine combining with the aldehyde group of lactose. Although the resulting lactosyl-lysine is released on enzymic digestion, it is largely biologically unavailable to the animal and is excreted in the urine. It is, however, split by acid hydrolysis, so that conventional analysis reveals only a small loss in the total lysine content.

Maillard reactions, leading to inactivation of lysine, occur during in-bottle and in-can sterilisation of liquid products, during prolonged storage of UHT milk, during roller-drying and in stored dried milks if the moisture content is allowed to rise to over 5%. Little is yet known about the extent of the Maillard reaction in recently developed products prepared after enzymic hydrolysis of lactose to glucose and galactose.

From the nutritional standpoint, if milk is to be used as a sole food the losses in the limiting sulphur amino acids due to protein-protein interactions are of greater concern than the loss of lysine through Maillard reaction, for milk has an abundance of lysine. On the other hand, in a mixed diet in which lysine is the limiting amino acid such inactivation of lysine is the more serious factor.

Besides the role of the major milk proteins as valuable nutrients, attention must also be given to those minor components such as the immune globulins (IgG), lactoperoxidase and the vitamin-binding proteins which may act as protective factors helping to regulate the establishment of a normal gut flora in the neonate. These minor components are heat sensitive and although some may survive pasteurisation they are all inactivated by more severe thermal processing.

A less desirable property of cows' milk proteins is their ability to cause allergic reactions in a small minority (0.5 - 1.0%) of infants during the first two years of life. Each of the major milk proteins has been shown to possess antigenic activity and to produce symptoms in allergic patients when tested by oral challenge. Further work is needed to clarify the presently confused picture of the effects of thermal processing on the antigenicity of milk proteins.

The indications from this brief survey are that further R & D could usefully include.

1. Monitoring of the vitamin content and protein quality of new products made from milk or the components of milk.

2. Studies of the nutritional significance of digestion products from proteins damaged during processing.

3. Elucidation of the problems of milk allergy.

THE ENZYMATIC ULTRAFILTRATE DIGEST (EUD) AMINO ACID INDEX FOR PROTEIN QUALITY EVALUATION

F. Fidanza

Istituto di Scienza dell'Alimentazione - Universitá degli Studi Perugia, Italy.

A rapid, accurate, widely-applicable method for food protein quality evaluation is urgently needed, as the recent Midlands Conference (1977) has stressed. This method has to be used by the food industry for purposes of monitoring raw materials and ingredients, processing conditions, final formulations and products, as well as shelf life stability. Such a method could be of great help also for plant breeders engaged in screening programmes.

Among the available methods, the protein efficiency ratio (PER) is time consuming and inaccurate. In fact it requires 4 weeks, is non linear in its estimation of protein quality, yields nonreproducible data among different laboratories (probably because also slight procedural modifications are introduced), tests a single protein and not a complete food at a fixed level in the diet. Notwithstanding that the PER is the official method for protein quality evaluation in some countries. The biological value (BV) and the net protein utilisation (NPU), besides being costly and time-consuming, show measurable values with proteins totally lacking in one or more essential amino acids. The chemical score does not give any evaluation of bio-availability of essential amino acids and this becomes quite an important problem in processed foods.

Recently we became interested in the enzymatic methods for direct protein quality evaluation (Floridi et al., 1972). The past literature has been reviewed by Mauron (1970). Instead of using very long incubations or unphysiological conditions or dialysis, we carried out the enzymatic hydrolysis in an ultrafiltration cell. The sample, diluted to contain

1 g of protein, is incubated at 37°C for 6 h, under stirring, with pepsin (16 mg of 3 x, grade B) at pH 1.8. After correction of pH (7.8) and bringing to volume (50 ml), an aliquot of 10 ml is transferred to an Amicon stirred ultrafiltration cell, thermoregulated at 37°C, with UM2 Diaflo Membrane. After addition of 0.2 ml of a 1% solution of trypsin grade A and 2 ml each of pancreatin and erepsin extracts (500 mg each are homogenised in 15 ml of 0.1 M boric acid - 0.9 M NaCl buffer at pH 7.8. After centrifugation the clear supernatant is brought with the same buffer to a final protein concentration of 4 mg/ml) the ultrafiltration is carried out for 6 h at a speed of 8 ml/h. The collected ultrafiltrate is brought to volume (50 ml) with H_2O; after hydrolysis of glutamine and asparagine with HCl, the available amino acids are determined chromatographically according to Spackman et al. (1958) with an amino acid analyser. After chemical hydrolysis, total amino acids of the same protein are used to calculate an integrated index, using egg protein as reference, according to the method of Sheffner (1967) for the Pepsin Digest Residue (PDR) amino acid index, in which the most important factors are the pattern of essential amino acids in the enzymatic and residue stages and the amount of amino acids as a group containing the particular pattern. The release of amino acids with ultrafiltration in vitro has been checked with that one in vivo in the rat and a good agreement has been observed. Then the EUD amino acid index has been compared with other biological or chemical methods using the same or different preparations of protein source. In all cases the agreement has been very good (Floridi and Fidanza 1975).

With this method at hand, which allows results in about two working days, we have analysed samples of raw, cooked and processed foods. In bread, as shown in Table 1, we have observed a decrease of 7% in protein quality in comparison with raw mixture. In rusks this decrease is of 20% and in biscuits between 9 and 14%. Canned baby foods (Table 2) with steer and chicken meat have shown a mean decrease of 10%

TABLE 1

EUD OF PROCESSED FOODS BASED ON CEREALS

Sample	Biscuits		Rusks	Bread
BU-raw	57	67	56	
BU-processed	49	61	45	52
GE	58			
PL	61			
DE	60			
ME	59			

TABLE 2

EUD OF CANNED BABY FOODS

Sample	Steer	Chicken
PL	70	64
RPL	71	69
BU	70	67
GE	72	65
DE	68	55
Raw food	77	72

TABLE 3

EUD OF RAW AND COOKED FOODS

	EUD
Chick-peas	58
Pasta	52-54
Pasta and chick-peas	65
Canned meat	61
Stew with peas (frozen)	68
Egg plant in parmesan (frozen)	68
Neapolitan pizza (frozen)	55

in comparison with the raw food.

As shown in Table 3 for a cooked combination of pasta and chick-peas we have observed an increase of protein quality of 16%. Also of interest are the results with some frozen ready meals which show the complementary efficiency in essential amino acid pattern of the component proteins in the meal.

The criteria for a perfect method for protein quality evaluation have been indicated in the aforementioned Midlands Conference. With the EUD amino acid index we can fulfil many of them. We still need comparison with the PER for areas where this is the official method. Also the inter-laboratory reproducibility has to be checked and it has to be validated against the results of human studies.

REFERENCES

Floridi, A. and Fidanza F. 1975. Riv. Sci. Tecn. Alim. Nutr. UM. 5, 13-18.

Floridi, A., Simonetti, M.S. and Fidanza, F. 1972. Sci. Tecn. Alim. 2, 289-294.

Mauron, J. 1970. In evaluation of novel protein products (Bender, A.E., Kihlberg, R., Lofqvist, B. and Munk, L. eds.) pp 211-234. Pergamon Press, Oxford.

Midlands Conference: New concepts for the rapid determination of Protein Quality (1977) Ntr. Rep. Int. 16, 157-226.

Sheffner, A.L. 1967. In:Newer methods of nutritional biochemistry (Albanese A.A. ed.) vol. 3 pp 123-195. Academic Press, New York.

Spackman, D.H., Stein, W.H. and Moore, S. 1958. Anal. Chem. 30. 1190-1206.

BAKING CONDITIONS AND NUTRITIVE QUALITY OF DIFFERENT BREAD VARIETIES

W. Seibel

Institute of Baking Technology, Federal Research Centre of Cereal and Potato Processing, Detmold, West Germany.

INTRODUCTION

In the Federal Republic of Germany there are nearly 200 bread varieties. This high figure arises from the use of wheat and/or rye flour and/or wheat and rye meal, alone or in different ratios for bread baking. Furthermore many compounds like milk products or vegetable products, such as oil seeds or soya semolina, may be included in the bread recipes. All special breads are described in a bread catalogue (Seibel et al., 1977). The different bread types can be produced as tin bread or free standing loaves. The third possibility is to bake the dough pieces side by side.

The size and weight of the different breads, the use of tins and the baking conditions determine the crumb/crust ratio of the different bread types, within certain limits.

CRUMB/CRUST RATIO

It has been very difficult to determine these ratios, due to the absence of a method to determine the crust percentage. In 1974 Bruemmer and Seibel (1975) developed a simple method for the purpose. The crust percentages of different breeds were determined with the following results:

Variety	Crust % (dm)
Baguette (French wheat bread)	39
German roll (50 g)	35
Wheat bread (500 g, tin)	29
Rye mixed bread, 1000 g baked side by side	18

The crust percentages of different German breads vary by more than 100%. With increasing temperature the thickness of the crumb increases too. The prebaking process with temperatures between 400 and 500°C, used in West Germany, causes thick crust.

CRUST AND NUTRITIVE QUALITY

Heat treatment of food stuffs influences the digestibility of the different nutritive compounds (Greaves and Morgan, 1934). Denatured protein is less digestible than natural protein. Menden (1971, 1975) has tested the nutritive effect of bread crust and bread crumb by feeding rats. For these trials with growing rats, different bread varieties were produced in our institute and the bread was separated in crust and crumb. After the 12 weeks feeding trial the weight per rat with the same diet was as follows:

    with crust: 50 g
    with crumb: 120 g

With the crust diet, there was no weight increase with growing rats. Nevertheless the animals remained healthy.

VIEW ON RELATIVE RESEARCH NEEDS

In most countries bread is still a staple food. About 50% of the worldwide food energy supply is derived from bread and other cereal products. In connection with over-nutrition in developed countries and under-nutrition in developing countries it is necessary to know the exact nutritive quality of daily bread

In developed countries it could be very interesting to treat a bread during baking in such a way that the digestibility and the available energy (measured in joule) is decreased. On the other hand baking conditions should be developed especially for developing countries - which improve the nutritive quality of the bread. Currently the effect of steam treatment on nutritive value during baking is still unknown.

It is therefore necessary to investigate the influence of the temperature during baking on the digestibility of protein and carbohydrates in bread.

REFERENCES

Bruemmer, J.M. and Seibel, W. 1975. Ernährungsumschau 4, 107-109.

Greaves, E.O. and Morgan, A.F. 1934. Cereal Chemistry, 8, 728-730.

Menden, E. 1971. Brot and Gebäck 4, 224-226.

Menden, E. and Elmadfa, J. 1975. Getreide Mehl und Brot 10, 253-257.

Seibel, W., Bruemmer, J.M., Menger, A. and Ludewig, H.G. 1977. Brot und Feine Backwaren, Arbeiten der DLG, Band 152, DLG-Verlag, Frankfurt.

DAIRY PANEL

PARTICIPANTS

Chairman:	J.H. Moore *(UK)*	
Rapporteur:	P.F. Fox *(Ireland)*	
Invited Experts:	B. Blanc *(Switzerland)*	
	J.A.F. Rook *(UK)*	
	M. Naudts *(Belgium)*	
	R. Negri *(Italy)*	
	G.C. Cheeseman *(UK)*	
	M. Caric *(Yugoslavia)*	
	H.G. Kessler *(Germany)*	
Other Participants:	W.K. Downey *(Ireland)*	Plenary Author Session 1
	J. Foley *(Ireland)*	Plenary Author Session 2
	H. Burton *(UK)*	Plenary Author Session 3
	J. Porter *(UK)*	Nutrition Coordination Group
	I.F. Vujicic *(Yugoslavia)*)	
	V. Veinoglou *(Greece)*)	Observers
	D. Martin *(Spain)*)	
Panel Coordinator	R. Brew *(Ireland)*	

MILK TECHNOLOGY WITH REFERENCE TO HUMAN NUTRITION

B. Blanc

Federal Dairy Research Institute
CH-3097 Liebefeld-Bern, Switzerland

ABSTRACT

The influence of typical thermal treatments (pasteurisation at $72^\circ C$ and $92^\circ C$, treatment at ultra high temperature, direct or indirect) applied to milk was studied. Samples were taken monthly throughout one year and were subjected to a multidisciplinary study of over 60 factors: biophysical, biochemical, chemical, microbiological, hygienic and nutritional. The results are accompanied by studies at present being carried out on the influence of storage on principal properties of milk. It is not yet possible to draw complete conclusions from the results of these studies. However, some of the main findings are presented here.

The first area of nutritional interest relates to proteins and other nitrogen fractions. Neither the total nitrogen nor the non protein fractions varied during the treatments. The effect on the β-lactoglobulin fraction is shown in Figure 1. As can be seen, pasteurisation at $72°C$ is the mildest and UHT indirect the most severe treatment. Large differences exist between milks for the supernatant various nitrogen fractions. They all have a similar trend to that of β-lactoglobulin, eg the non casein nitrogen (ie, the total whey proteins and the non protein nitrogen in milk), the non casein nitrogen measured in the supernatant after ultracentrifugation, total albumin nitrogen, total nitrogen in the supernatant after ultracentrifugation (whey proteins with remaining casein), free ammonia, the amount of denaturation of whey proteins determined by measuring the importance of the endothermic peaks during calorimetric analysis, the distribution of the casein fraction and lastly, the changes in the free sulfhydryl groups. These criteria allow one to establish the following sequence of increasing changes:

Fresh milk > pasteurised $72°C$ > pasteurised milk $92°C$ > = UHT milk direct > UHT milk indirect

During storage, the levels of non protein nitrogen remain more or less constant up to the 16th week. However, the non casein nitrogen which decreases to a large extent due to UHT treatment was found to start increasing again after the fourth week when milk was stored at $5°C$.

It is difficult to know what importance should be attached to these variations since it was found that even with the obvious denaturation, as shown in Figure 1, the amino acid analysis (total and free) did not change after thermal treatment. Even lysine, which has a tendency to form denatured complexes, was found to decrease only to a very slight extent (approx. 1% No denaturation product, for example, lysino-alanine, could be detected in the milks.

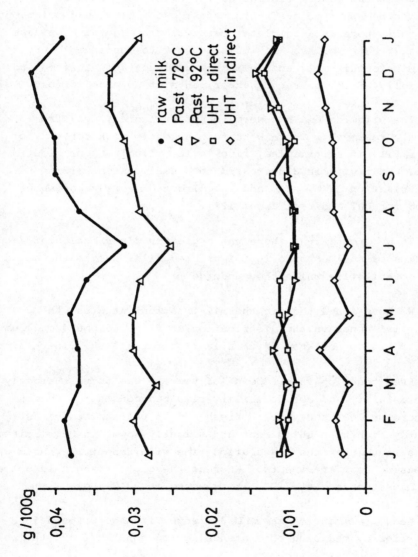

Fig. 1 Monthly variations of β-lactoglobulin nitrogen (raw milk - various heat treatments)

Although some enzymes are completely resistant to thermal UHT treatment, for example, ribonuclease (Alais, 1977), we noted that amylase was only partially so. Most of the enzymatic activities, even though they remain to a certain extent at $72°C$ are destroyed by the other thermal treatments. In Figure 2 we show the effect of thermal treatments on peroxidase activity throughout the year. Similar curves have been found for alkaline phosphatase and xanthine oxidase. We should point out here that there is a complete destruction of xanthine oxidase in heat treated milks. It would be vain to continue to try to incriminate this milk enzyme in the aetiology of atheroma and especially after the recent experiments of Ho and Clifford (1977).

The vitamins did not seem to suffer a great deal from thermal treatment. There were no changes in milk contents of either vitamin A, thiamine, riboflavin, vitamin B_6 or nicotinamide. Only vitamin C decreased to a small extent (for pasteurisation $72°C$, $92°C$ and UHT direct a decrease of about 5% and for UHT indirect about 12%).

It was found that there was a greater distribution of fat globules of smaller size in the treated milks due to the amount of homogenisation which they undergo.

We have found that in thermal treatment studies, there was no influence on the distribution of free and bound calcium. However, we do not know if this is true for all the minerals.

From the biophysical point of view it was found that only the osmolarity and conductibility remained constant. The pH was more or less the same in fresh and in pasteurised milks, slightly higher in UHT direct and slightly lower in UHT indirect. It was found that the coagulation time with rennet increased with increasing heat treatments. Whereas the clot strength is better in milk pasteurised at $72°C$, it is worse in $92°C$ and UHT milks.

Heat treatment of raw milk destroyed its bacteriostatic properties by reducing the latence period for *E.coli* growth.

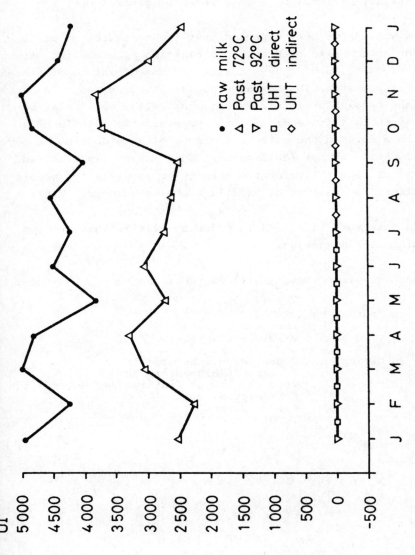

Fig. 2 Monthly variations of peroxidase activity (raw milk - various heat treatments)

The different indices calculated on the basis of the content of essential amino acids such as the chemical score, the modified essential amino acids index and the RAO index, were the same for raw and thermally treated milks. Also, it has been found so far that there is no significant effect in feeding rats the various types of milks in relation to growth rate or the fertility of females, even after nine generations.

It would have been expected that the denaturation of the proteins accompanied by changes in clotting time and the strength of the clot etc. would have facilitated the digestion of heated milks. The changes in the fat globules (homogenised milk has fat globules similar in size to maternal milk) would also be expected to facilitate digestion. However, the absence of a difference between the different groups of rats does not imply that digestion is not facilitated. This may be important in some cases, especially for the newborn or in diets for people who suffer from gastric or hepatic troubles (Renner, 1976).

Further work is needed in order to clarify this problem of facilitated digestion.

More complete reports will be published soon:

- in French, by   SIRMCE,
  rue du Sceptre 5,
  B-1040 Brussels.

- in German, by   Oesterreichische Gesellschaft
  für Ernährungsforschung,
  Physiologisches Institut der Universität
  Schwarzspanierstr. 17,
  A-1000 Vienna 9.

REFERENCES

Alais, C. 1977. Personal communication

Ho, C.Y. and Clifford, J., 1976. Digestion and absorption of bovine milk xanthine oxidase and its role as an aldehyde oxidase. J. Nutr. 106, pp 1600-1609.

Renner, E. 1976. Ernährungsphysiologische Veränderungen der Milch durch Be- und Verarbeitung, Deutsche Milchwirtschaft, 33, pp 1021-1031.

FACTORS AFFECTING THE HEAT STABILITY OF MILK AND OF CONCENTRATED MILK AT STERILISATION TEMPERATURES

J.A.F. Rook

The Hannah Research Institute, Ayr, Scotland, KA6 5HL.

Milk is secreted as a complex mixture of fat globules and micellar casein dispersed in a solution of lactose, whey proteins and mineral salts. Before the milk reaches the consumer either in liquid form or as a processed product, it is generally subjected to some form of heat treatment. This treatment reduces the level of bacterial contamination, in- activates milk (but not necessarily bacterial) lipoprotein lipase and, depending on its severity, induces a number of other physical and chemical changes, some of which affect the stability of the protein of milk during processing and storage. The commercial consequences of instability are either a wastage of product or a reduction in the efficiency of processing. The following is a brief review which covers work on this topic being undertaken at the Institute and an indication of the need for future research.

Heat stability of raw milk

When coagulation time (CT) of raw milks is measured over a range of pH, two different relationships may be observed. In some milks (type B) the CT increases progressively with pH over the range 6.4 to 7.3; these milks are characterised by a low ratio of β-lactoglobulin to casein. The more common relationship (type A), however, is that the CT increases to a maximum at pH 6.7, decreases to a minimum at pH 6.8 and then rises again as pH is increased further. The shape of the relationship is temperature-dependent and milks that show a type B curve at 120° in some instances show a type A curve at 140°. At least two different types of mechanism appear to be involved since under type B conditions and under type A conditions out- side the minimum, coagulation occurs according to simple second-order kinetics. Within the minimum under type A conditions, coagulation is a two-stage process; there is first a 'premature' coagulation of a portion of the protein and then a second coagulation stage for which the kinetics are similar to those for type B conditions.

The denaturation of the serum proteins, particularly the β-lactoglobulin, which occurs at 70 to 80° is thought to have an important effect on coagulation. On denaturation, the β-lactoglobulin reacts with the κ- fraction of the casein, and it is assumed that this alters the stability of the micellar suspension. If the denatured protein is distributed throughout the entire micellar framework, then the surface properties of the micelles may be affected comparatively little. If on the other hand the denatured protein binds in such a way that the surface charge density of the micelle is radically altered, then there may be a marked reduction in stability.

Other factors also have been shown to affect the stability of milk when it is heated. Urea increases stability, the effect for type A milks being least marked in the region of the minimum of the CT - pH profile. The mechanism of the effect is not understood. Other additives (eg aldehydes) which are known to alter the overall micellar charge through chemical modification produce a change in stability expected on theoretical grounds. In a recent survey, the heat stability of raw milks at their natural pH was found to be influenced by seasonal factors, and much of this seasonal variation was explained by natural variations in the concentration of urea.

Heat stability of concentrated milks

Fewer studies have been made of the heat stability of concentrated milks. The CT-pH profile for concentrated skim milks (22½% total solids) is roughly bell-shaped, with a maximum CT at a pH of about 6.6, and is similar in form to the first half of the type A profile for raw milks. This mechanism also appears to involve an interaction between the caseins and denatured β-lactoglobulin, since the addition of sulphydryl-blocking agents increases the CT. Addition of urea, however, has no effect and survey results suggest that about 80% of the variation in stability is accounted for by variation in the ratios of certain minerals and of β-lactoglobulin to casein.

The mechanism has certain similarities to the premature coagulation observed in type A milks in the region of the minimum of the CT-pH profile and the stability should be sensitive to the charge properties of the micelles. Increasing the net negative charge by the addition of aldehydes which react with the lysine residues of the caseins does increase stability, and a similar effect is produced by iodination of the protein which specifically reduces the pK of iodinated tyrosyl residues.

Future research

Clearly, it is possible to increase the stability of both raw and concentrated milks by the addition of suitable additives and, especially in the case of concentrated skim milk, stabilisation by addition of natural products such as erythrose or glyceraldehyde may prove to be an acceptable industrial process as it should not introduce undesirable contaminants. Many of the observations made to date, however, have arisen more from empirical studies than from a detailed understanding of the mechanisms involved. Further empirical studies are projected. More detailed information is to be obtained on the seasonal interrelationships between heat stability and physical and chemical characteristics for both raw and concentrated milks, and on the post-synthetic changes in milk caused by bacterial degradation and by processes used in the manufacture of milk products which are known to affect stability. There is an urgent need, however, for more fundamental studies into the composition, structure and properties of the casein micelles, of their interactions with the inorganic constituents of milk and of the kinetics of particle formation in heat-treated milks so that the mechanisms of protein destabilisation may be described and the possibility of more efficient means of controlling protein instability explored.

SOME AREAS FOR FUTURE RESEARCH INTO HEAT TREATMENT AND LONG LIFE PROPERTIES OF MILK, MILK POWDER AND YOGHURT

M. Naudts

Government Dairy Research Station of Melle, Belgium

This paper deals with some aspects of dairy research concerning the interaction between quality and heat treatment.

HEAT TREATMENT AND SENSORY QUALITY OF LONG-LIFE MILK

Both in-bottle sterilisation and UHT-treatment result in a more or less objectionable 'cooked' flavour, and in many instances in instability of milk proteins during storage, due to chemical changes.

Concomitant with the development of a 'cooked' flavour many sulfhydryl groups previously masked within the native protein structure, are exposed by the heat denaturation of milk proteins and thus become reactive.

Many investigations have suggested that a direct cause-and-effect relationship exists between the exposure of the sulfhydryl groups and the development of 'cooked' flavour. Furthermore some authors suggest the existence of an enzyme-sulfhydryloxidase - which acts specifically on the SH-groups. The application of this enzyme, possibly by immobilisation or a support neutralising this cooked flavour merits attention.

Some studies suggest that slow sulfyhdryl-disulfide interchange reactions between milk proteins, catalysed by the reactive sulfhydryl groups, may lead to protein instability and gelation, especially in UHT-heated milk. Concerning age-thickening and gelation there is still much uncertainty.

Alternatively gelation may be initiated by either

enzymatic action - proteolysis - or it may simply reflect physico-chemical changes in the casein complex.

The mechanism of gelation in UHT-heated milk and milk products is not very well known and should be further studied with particular attention to the possible involvement of heat resistant enzymes.

During heat treatment aggregates can be formed which subsequently separate out during storage, leading to bottom deposits as well as the accumulation of a fatty mass on the top of the liquid milk. Studies of such separation phenomena have concentrated mainly on homogenisation (formation of a fat/protein complex). The influence of other factors, such as heat stability, salt composition, casein aggregation and κ-casein/β-lactoglobulin ratio is not sufficiently understood. Thus more extensive studies of these facets is advocated.

Heating causes partial or complete denaturation of some milk proteins. The contribution of amino acids, amino acid derivates and short peptides, formed by the heating process and during storage, to the sensory quality of milk and milk products is not well understood and is a fruitful field for further research.

HEAT TREATMENT AND SOME PROPERTIES OF MILK POWDER

The following heat treatments may be considered:
Preheating of the milk immediately prior to entry to the evaporator.
Heating of the concentrate.
Heat treatment during spray drying.
Heat treatment in a fluid bed (two-stage drying).

For the preheating of milk temperatures ranging from 70 to $120^{\circ}C$ for 45 to 60 s may be employed. This rather long holding time causes heat denaturation of the proteins, especially at the higher temperatures. Few details are available concerning preheating temperatures ranging from 100° to 140°C for 3 to 5 s. For the production of low-heat powders with high bacteriological quality and also satisfactory from the viewpoint of viruses, these conditions may provide interesting possibilities. Overall, more information will be required on the influence of pre-heating conditions on the functional properties of milk powder (flowability, mechanical stability, keeping quality) and on the viscosity and the heat stability of reconstituted milks.

It is known that severe preheating of the milk may significantly improve the heat stability of the proteins following reconstitution. Hence the possibility of using high temperatures (up to 125°C) for short times (1 min) should be given further attention. The influence of pre-heating conditions on the heat stability of milk powder can also be affected by seasonal variations in the milk composition. For the production of milk powders with a tailor-made heat stability, more information is needed about the interaction between preheating conditions and seasonal variation.

Recently, two-stage drying involving a fluid-bed has been introduced in the manufacture of milk powder. It is often suggested that a high-quality powder with a low free-fat content may be produced by applying an inlet air temperature of 250°C and an outlet temperature of 80°C, followed by after-drying in the fluid bed. Both the high inlet air temperature and the treatment of the powder in the fluid bed at temperatures sometimes amounting to 100°C will, however, significantly influence the properties of the powder; both the whey proteins and the casein fraction may be denatured and their functional properties may be altered.

Two-stage drying also yields a more or less agglomerated end product with a physical structure markedly different from that of normal milk powder. However the extent to which the heat treatments applied in a fluid bed influence the physical structure of the powder is uncertain. Special attention should be paid to the mechanical stability of the agglomerated product during its pneumatic transport and silo storage.

HEAT TREATMENT AND SENSORY QUALITY OF YOGHURT

Limited heat treatment of yoghurt is increasingly being applied to improve its keeping quality. The question whether such a heat treatment is detrimental to the nutritional value of this fermented product should be examined.

Dairy technology must also have regard for both the keeping and sensory qualities of the product.

The influence of the heat treatment on the keeping quality should be examined with the objective of establishing the appropriate time temperature combinations necessary to attain the desired pH, viscosity, etc. in yoghurt.

A limited heat treatment may be detrimental to the sensory quality resulting in off-flavour, granular structure, whey separation, etc.

Spore-forming bacteria derived from the raw milk or other ingredients are generally believed not to develop in yoghurt. It is, however, possible that some flavour defects in heat-treated yoghurt may arise from the development of spore-forming bacteria.

The consequences of yoghurt thermisation on coagulum contraction depend on several factors. A full examination of the combined influence of the heat treatment and the homogenisation procedure may help to substantially reduce structural defects in the product.

The use of some additives may eliminate rheological defects. It would, however, be of interest to study the heat-treatment conditions under which both a well textured and long-life product can be produced without using additives. This should include both stirred and set yoghurt.

LIST OF POSSIBLE RESEARCH TOPICS

Application of the enzyme - sulfhydryloxidase - for improving the sensory quality of heated milk.

Sediment formation and gelation in long-life milk.

Influence of protein denaturation on the flavour of liquid milk.

Influence of high preheating temperatures for short times on some properties of milk powder.

Influence of two-stage drying on heat denaturation of the proteins and functional properties of milk powder.

Parameters of heat treatment necessary to obtain a long-life yoghurt.

SOME ASPECTS OF THE EFFECT OF HEAT-TREATMENT ON INDUSTRIALLY MANUFACTURED CULTURED MILKS

R. Negri
Direttore del Laboratorio degli Alimenti Istituto Superiore di Sanità, Rome, Italy.

The biological properties of cultured milks are guaranteed by the presence of a high number of specific metabolically active lactic acid bacteria.

The Istituto Superiore di Sanità of Rome has carried out a study on a widely consumed cultured milk, ie yoghurt, in which the lactic acid bacteria have a specific value, in order to correlate the vitamin and enzymatic aspect of this commodity with the microflora which it normally contains.

Particular attention has been focussed on the modifications which certain technological treatments may produce in the 'facies' of the product. To this purpose a comparison has been made between traditionally manufactured yoghurt and yoghurt which has undergone pasteurisation after the normal fermentation process, considering in particular the vitamins of group B and the enzymes Amylase, Cellulase, Proteinase, β-galactosidase.

The methods used for the determination of group B vitamins and of some hydrolases (amylase, cellulase, proteinase and β- galactosidase) were selected after a long series of preliminary investigations. They are based on the microbiological dosage for diffusion on agar for the group B vitamins previously extracted from the product; on specto-photometrical dosage (β-galactosidase) and on titration of the enzymes extracted from the product. (De Felip et al., 1977).

Parallel counts were made of the lactic acid bacteria. The examination of the analytical results obtained on samples of yoghurt (N) from various sources and of known production data, gives the following results:

The count of lactic acid bacteria, which after the manufacturing of yoghurt averages approximately 10^{12} ufc/ml of product, decreases progressively until it reaches, after 45 days from the date of production, average levels of 10^6 ufc/ml of product;

Contemporaneous determination of group B vitamins shows a decrease during storage of approximately 30% and more. Thiamine however, in the conditions of our investigations, did not appear subject to degradation;

The enzymes taken into consideration (Amylase, Cellulase, Proteinase, β-galactosidase) present a significant decrease during storage (Table 1).

TABLE 1

DETERMINATION OF SOME HYDROSOLUBLE VITAMINS AND SOME ENZYMES IN NORMAL (N) AND PASTEURISED (P) YOGHURT DURING STORAGE

Days	0		15		30		45	
Vitamins mcg/g	N	P	N	P	N	P	N	P
Vit B_6	0.3	0.3	0.2	0.1	0.15	0.07	0.14	0.05
Ca Pantothenate	3.5	1.0	3.2	0.5	2.8	traces	2.5	0
Folic acid	0.1	0.05	0.05	traces	traces	traces	traces	traces
Vit. B_1	0.25	0.25	0.25	0.25	0.25	0.25	0.25	0.25
Enzymes mcg/g								
Protease	3.0	1.25	0.7	0.4	0	0	0	0
Cellulase	9.0	6.25	7.5	5.1	7.25	4.9	6.9	4.75
Amylase	5.0	2.5	2.75	1.2	2.0	0.8	1.75	0.75
β-galactosidase	300	0	250	0	225	0	220	0

Parallel investigations on aliquots of the same yoghurt samples which had been heat-treated at 70°C for 10 min. (P), gave the following results:

The count of lactic acid bacteria does not exceed values of 1 000 ufc/ml of product;

Consistently lower levels of group B vitamins as well as of enzymes.

These differences, both in the case of vitamins and of hydrolases, fluctuated between 30 - 70% (and more) less than in the same product which had not been heat-treated.

In particular, in the pasteurised product no presence was found of significant quantities of β-galactosidase, since in heat-treatment of 60°C for 15 min. this enzyme is killed. (Bianchi-Salvadori et al., 1976).

Bianchi-Salvadori suggested that the level of β-galactosidase may be used as in index of the heat treatment to which cultured milks such as yoghurt had been subjected. This suggestion was based on the observation that some commercial yoghurts with low counts of lactic acid bacteria exhibited no β-galactosidase activity.

Another characteristic aspect which differentiates normal cultured milks from those that have been heat-treated, is the structure of the coagulum. In this respect, electromicroscopical researches have been effected on buttermilk containing lactic acid bacteria and on buttermilk which had undergone a pasteurisation process at 70°C for 10 minutes.

Figure 1 shows clearly the ultrastructure of both cultured milks:

> buttermilk (A) presents an almost uniformly perfect distribution of the coagulum over the entire volume of the milk. The fat globules are intact and of the same dimensions as the native ones;

> the heat-treated buttermilk (B) shows strong aggregation of the coagulum, the fat particles are disemulsioned and larger than the original ones. (Portesi, 1977).

From the studies carried out on this subject, the following physiological-nutritional considerations arise: it is known that the lactic acid bacteria administered in a viable state and in great number, as happens with the ingestion of yoghurt,

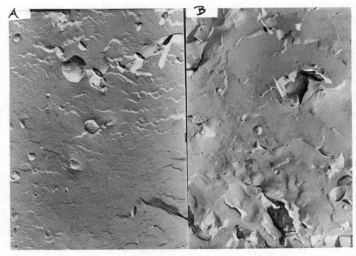

Fig. 1.

are capable of surviving through the gastric barrier (Salvadori and Bianchi-Salvadori,1973), arriving in the intestinal tract where, like the other components of the enteric flora, they are capable of developing significant intestinal lactate activity, as has been demonstrated (Goodenough and Kleyn, 1976) in a study effected on animals fed with 'living' yoghurt (compared to the lactasic activity found in animals fed with heat-treated yoghurt); furthermore a significant reduction in the supply of group B vitamins is observed in yoghurt which has undergone pasteurisation. Heat-treated cultured milks offer inferior digestibility due to the compactness of the casein micelles of merceological nature: apart from the levels of lactic acid bacteria present, yoghurts manufactured traditionally differ markedly in their vitamin and enzyme content from those produced from pasteurised milk. The coagulum masses in cultured milk which have been heat-treated are only slightly hydrated and are heavier, thus resulting in whey separation.

These considerations, together with those of a microbiological nature indicate well-defined differences between heat-treated cultured milks and those manufactured using traditional technologies.

Further research is, therefore, suggested on the contents of vitamins, enzymes and amino acids and on the structure of the coagulum in heat-treated cultured milks.

REFERENCES

Bianchi-Salvadori B. et al., 1976. II Latte 11, 569-573.

De Felip, G. et al., 1977. L'Igiene Moderna 70, 287-296.

Goodenough, E.R., Kleyn D.H. 1976. J. Dairy Sci. 4, 601-606.

Portesi, P. 1977. Graduation thesis - Centro Sperimentale del Latte and Università Statale di Milano.

Salvadori P., Bianchi-Salvadori B. 1973. Minerva Dietologica 13, 8-12.

SOME ASPECTS OF DAIRY CHEMISTRY REQUIRING FURTHER R & D IN RELATION TO THERMAL PROCESSING

G.C. Cheeseman

National Institute for Research in Dairying, Shinfield, Reading RG2 9AT, England.

The requirement of preservation by thermal processing, be it by high or low temperatures, or by complete or partial removal of water, imparts some degree of chemical change on the system. The extent of chemical change and its manifestation is influenced by the severity of the processing and by the pre- and post-processing treatment of the commodity. These changes may have beneficial or adverse effects on quality and nutritive value of the foodstuff.

In the case of milk, the composition, hygienic quality and condition of pre-processing storage would appear to be of prime importance in influencing the subsequent physico-chemical and nutritional properties of the products and these factors must be taken into account when the effect of a thermal process is being considered. The relationship of milk composition, in particular the nitrogen fraction, to behavioural properties during thermal processing is a factor that is becoming more apparent as research unravels the details of milk constituents and the nature of their interactions.

The sensitivity to temperature change of the equilibrium between the caseins and between the caseins and milk salts and between the caseins and whey proteins is well documented. However, the practical importance of some of these changes has to be evaluated. The adverse effect on rennet coagulation of heat treated milk and on subsequent curd formation is considered to be due to interference by heat denatured whey proteins - particularly β-lactoglobulin. A better understanding and control of these heat induced changes might lead to improved yields in cheese production and better control of maturation processes.

Similarly, the changes which occur in colloidal equilibria during chilling (4 - 10°C) of milk need better understanding in order to select the best procedure of subsequent manufacturing treatment. Although temperature controlled equilibria changes, such as the increase in the non-micellar portion of β-casein at low temperature, may be important in affecting functional properties, the action of enzymes both naturally occurring or from microbiological contaminants will also have effect on the properties of constituents.

The activity and survival of enzymes during thermal treatments and possible reactivation after apparent heat inactivation is becoming of more interest, particularly as new methods of milk storage may encourage the selection and growth of psychrotropic organisms, some strains of which produce heat resistant lipases and proteases.

The use of membrane systems for milk and whey concentration must have much attraction in terms of energy saving. Nevertheless such techniques introduce new problems in relation to the use of such concentrates, especially in the manufacture of cheese. Increased effort in research and development in this area must have priority if energy savings are to be maximised. More research is required into the nature and properties of milk concentrates formed by reverse osmosis (RO) and ultra filtration (UF), especially the changes in colloidal properties that the increased solids content bring about.

Consumer acceptance of ultra heat treated (UHT) aseptically packaged products would appear in the main to be related to the convenience of such products. The demand for these products, as packaging improves, is likely to increase in the future and such trends should further stimulate work into improvements in flavour and storage properties. Research into the chemistry of undesirable flavours caused by the UHT process and the chemical changes, particularly protein and carbohydrate interactions occurring both during heat treatment and post-processing storage, is required to improve the quality and nutritive value of these products.

In summary therefore it is suggested that a significant amount of the R & D required for a better knowledge of the physico-chemical and nutritive changes occurring during and after thermal processing is linked with pre-processing treatment and condition of the milk constituents. The suggested research topics which follow therefore reflect this view.

1. The relationship of milk composition to its colloidal stability and the effect that changes in proportions and types of constituents have on the quality of the milk and its products in relation to thermal processing.

2. Factors in the cow's diet which affect milk composition and quality, particularly in relation to keeping quality and off-flavours of concentrated and dried products, eg precursors of oxidative deterioration of milk powders.

3. Details of colloidal micelle structure and the behaviour of the milk colloidal system at various stages of thermal processing, eg cold storage of milk and its effect on cheese manufacture, freezing and thawing of milks and concentrates.

4. Development of low energy procedures for milk and whey concentration. The properties of these concentrates and the quality and nutritive value of their products.

5. Fate and possible formation of toxic factors during thermal processing, eg survival of viruses, biogenic amine development.

6. Control of changes in physico-chemical and nutritive properties occurring during treatment and storage of thermally processed products, eg in UHT aseptically packaged milks and creams, concentrated sterile milks and milk and whey powders.

7. Effects of thermal processing, in particular heat treatment on the functional properties and nutritive

value of whey proteins and of food blends containing milk proteins.

8. Changes in composition and stability of native fat globules and of emulsions formed from milk protein or non-milk protein with various fats and oils which affect quality of thermally processed products.

OPTIMISATION OF QUALITY AND NUTRITIVE PROPERTIES OF STERILISED DAIRY DRINKS FOR USE IN INFANT NUTRITION AND INVALID DIETS

M. Caric
Faculty of Technology, Dairy Technology,
21000 Novi Sad, Yugoslavia.

Nutritional problems arise mainly in two ways: Firstly, lack of food or inadequate diets deficient in certain food components resulting in hypovitaminoses, anaemias. Secondly, an excessive fat intake leading to certain diseases. Inadequate or unbalanced diets are largely composed of carbohydrates and fats, with a deficiency in animal proteins, calcium, iron, vitamins, etc. Since the adverse consequences of such diets are especailly manifested in children characterised, for example, by a high incidence of anaemias, it is particularly important for this category of the population that the nutritional content of the diet be supplemented to perform the correction of nutrition with an adequate amount of animal proteins, minerals and vitamins. On the other hand, the excessive dietary intake, especially of foods rich in fats have other adverse consequences, notably obesity. According to the data published by WHO (1976) in those who are overweight by 25% mortality is due to heart illnesses in 75% of cases. Thus a nutritionally balanced diet with a low fat content is very important in prevention and cure of the aforementioned illnesses.

The data presented here constitute part of a more extensive study which included an investigation of the possibilities of producing nutritionally enriched sterilised dairy drinks suitable for children, sick and aged. With the objective of developing enriched <u>dairy drinks</u> with optimised quality and nutritive properties for infant nutrition, milk was enriched not with fat but also with proteins, sodium caseinate, minerals, iron, vitamins A, D, B_1, B_2 and nicotinamide. The infant drinks were flavoured with coffee, cocoa, chocolate or caramel, sweetened with sugar as required. In those experiments aimed at formulating <u>dietetic dairy drinks</u> suitable for the sick and aged, the fat content was reduced and as already indicated the products were enriched with different nutritive and aromatic additives and vitamins.

For the preparation of <u>enriched dairy drinks</u>, the following formulations were investigated:

1. Sodium-caseinate (0.5, 1.0, 1.5%) with milk fat contents ranging from 3.2, 3.5, 4.0 to 4.5%.

2. Sodium-caseinate and iron from 4, 6, 8, 10, 12 mg/l, Fe expressed as $FeSO_4 \times 7H_2O$.

3. Sodium-caseinate and Fortepan F(iron) vitamins (A, B_1, B_2 and nicotinamide).

4. Sodium-caseinate, Fortepan F, sucrose (3%), and coffee (0.7, 0.8, 0.9, 1.0, 1.1%).

5. Sodium-caseinate, Fortepan F, sucrose (3%), and cocoa powder (1.0, 1.5, 2.0, 2.5 and 3.0%).

6. Sodium-caseinate, iron (12 mg/l), sucrose (5 and 6%) and caramel (0.5, 1.0, 1.5%).

7. Sodium-caseinate and preparation of A, D vitamins (up to 2.1 vitamin A mg/l).

For the preparation of <u>dietetic dairy drinks</u> the following combinations were investigated:

1. Sodium-caseinate (0.5, 1.0%) with milk fat contents ranging from 1.4, 1.6, 1.8, 2.0 to 2.2%.

2. Sodium-caseinate (1.0%) milk fat (1.4%) and Fortepan F.

3. Sodium-caseinate (1.0%) milk fat (1.4%) and preparation of A, D, vitamins.

4. Sodium-caseinate (1.0%) milk fat (1.4%) Fortepan F, sucrose (2.5%) and coffee (0.7, 0.8, 0.9, 1.0, 1.1%).

5. Sodium-caseinate (1.0%) milk fat (1.4%), Fortepan F, coffee (0.9%) and sucrose (2.0, 2.5, 3.0, 3.5, 4.0).

6. Sodium-caseinate (1.0%), Fortepan F, sucrose (1.0, 1.5, 2.0, 2.5, 3.0%) and caramel (0.5, 1.0, 1.5%).

7. Sodium-caseinate (1.0%) and preparation of A, D vitamins with coffee or caramel.

Chemical analysis of dairy drinks with tastes ranging from milk, coffee, cocoa and caramel confirmed that the desired nutritional enrichments had been obtained. The energy value of the modified dairy drinks was also enhanced. Based on organoleptic evaluation of all groups of products, and taking into account the needs of infants, the aged, and sick respectively, the most appropriate formulations were chosen for industrial production.

For the <u>enriched dairy drinks</u> the following milk preparations which may be flavoured with coffee, cocoa, or caramel were selected:

a) Animal proteins (sodium-caseinate), carbohydrates (sucrose) vitamins (A, B_1, B_2, nicotinamide) and iron.

b) Animal proteins (sodium-caseinate), carbohydrates (sucrose) vitamins (A, D_2) and iron.

c) Animal proteins (sodium-caseinate), carbohydrates (sucrose) vitamins (A, B_1, B_2, D_2, and nicotinamide) and iron.

The corresponding <u>dietary dairy drinks</u> contained less carbohydrates and reduced milk fat content.

Further studies involve establishing the technological and nutritive properties of the dietary dairy drinks. In the subsequent phase a pilot spray drying plant will be used to establish the optimum conditions necessary to produce the above products in powder form without undue quality loss and with extended keeping quality.

ENERGY SAVINGS IN EVAPORATION AND DRYING OF MILK AND MILK PRODUCTS

H.G. Kessler
Technische Universität München, Weihenstephan Institut für Milchwissenschaft und Lebensmittelverfahrenstechnik

It is well known that energy costs have virtually trebled in the last few years while costs of plant and equipment have risen only slightly. It is therefore necessary to consider anew how to arrive at an optimum value for running costs and equipment costs. Changes made in processes, in order to decrease heat consumption will usually require additional investment. As prices for energy and for equipment do not change at the same rate, the economic optimum between energy costs and capital outlay is time-dependent.

There are two processes in milk processing which are particularly energy consuming. These are concentration by evaporation and drying by heat. The subject of this talk will be the 'falling film vacuum evaporation' and 'spray drying'. Both methods have been widely accepted in the practice of milk technology for reasons of the quality of the product obtained.

The most widely installed evaporators for milk and milk products of the falling film type have three effects and thermocompressors. The specific steam consumption of these types is about 26%, based on the total amount of water to be evaporated. In the past the cost of adding more effects was higher than the savings which could be obtained by a lower energy use. The new energy costs have changed this economic balance.

At the new energy costs it will most likely be economical for all evaporators to have additional effects, thermal or mechanical compression, heat exchange for liquid flows and condensate recovery for boiler feed water.

Large amounts of heat can be saved by increasing the number of stages. Evaporators with five or six stages are already in operation; specific steam consumption is then decreased to 15%.

A further increase in the number of stages is ineffective because in the evaporation of milk products one is usually restricted to the temperature range of 40 to 70°C. The lower limit is given by the temperature and quantity of condenser water and the upper limit is set because of the problems caused by protein precipitation.

But further savings are possible by using mechanical recompression. The mechanical energy required by the compressor is low compared to the recoverable heat values in the vapour. The equivalent of up to 15 effects can be achieved by mechanical recompression. This means a specific steam consumption of about 6.5%. This favourable energy recovery must be balanced against the high capital costs of the compressor. Steam-turbine driven compressors as well as electricity driven units can be used.

Better heat utilisation can also be achieved in spray drying. Possibilities are an increase in the air inlet temperature or a decrease in the temperature of the exhaust air. Both these measures can give rise to problems. The increased air inlet temperature can cause heat damage to the product if the direction of flow of gas and product in the drying chamber is not accurately controlled. The lowering of the air exit temperature is linked to an increase in the moisture content of the dried product. These measures have therefore limitations if one wants to avoid re-drying the product in a second drying stage.

However, there exist possibilities at the present time by means of which the heat of the exhaust air can be used. Figures 1b and 1c show such an arrangement. Figure 1a depicts a common spray drier with an air heater, drying chamber,

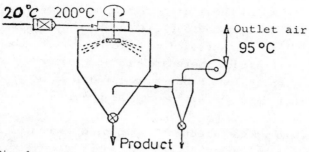

Fig. 1a

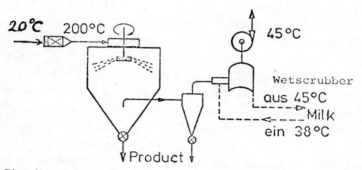

Fig. 1b

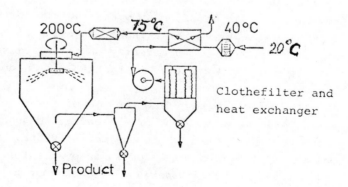

Clothefilter and heat exchanger

Fig. 1c

cyclone and fan. The values quoted for the air inlet and -exit temperatures are those used in many milk drying plants at the present time. It is obvious that enormous amounts of heat are blown into the surrounding atmosphere with the exit air at $95^{\circ}C$. Furthermore there are product losses to consider which can lead, on the one hand, to a large loss of product and, on the other hand, to environmental problems.

Figure 1b shows the frequently employed method of lowering the exhaust air temperature by making use of a wet scrubber. However, the heat can only be usefully exploited if it is used for pre-heating the product and, more particularly, if it is used for pre-concentration. Milk which has not been evaporated previously is used for the purpose. It is let into the wet scrubber at a temperature of about $40^{\circ}C$ which is just below the wet bulb temperature of the air. Both milk and exhaust air leave the scrubber at a temperature of about $45^{\circ}C$, the wet bulb temperature of the air. The milk has thus become pre-concentrated. On no account should cooling be carried out to such an extent that the moisture in the exhaust air condenses and gains access to the product. The scrubber has the further advantage of washing out and thus retaining dust particles contained in the exhaust air. Bacteriological problems are a disadvantage since a temperature of $45^{\circ}C$ offers an ideal climate for the growth of micro-organisms. In practice it is therefore necessary to disconnect the scrubbers from time to time ie after an operating time of about 20 h and to clean them. While the scrubber is being cleaned the exhaust air is, of course, blown directly out to the atmosphere, which mean that the procedure depicted in Figure 1a is in operation.

The energy saving from the use of a wet scrubber is not particularly large as 100% of the energy is used in the pre-concentration of the milk in the wet scrubber. Only a fraction of it would be needed in an evaporator. Consequently only a fraction of the energy of the exhaust air is used effectively. Although the loss of product is reduced, the product recovered from the wet scrubber has to be dried again.

TABLE 1

33 000 kg skim milk per h
27 000 kg water evaporation in the vacuum evaporator 3 000 kg water evaporation in the spray drier 3 000 kg skim milk powder

Evaporator				
spec. steam consumption		Fresh steam needed t/h		Heat needed kJ/h
26 %		~7		$21 \cdot 10^6$
16 %		~4		$12 \cdot 10^6$
6.5 %		~1.75		$5.3 \cdot 10^6$
10^6 kJ/h	$\hat{=}$	$7 \cdot 10^9$ kJ pa.	$\hat{\approx}$	45 000 DM pa.
Saving:				
Spec. Steam consumption 26 % → 15%				DM 420 000 pa.
26 % → 6.5%				DM 735 000 pa.

Drier: Diameter of chamber 10 m	
Amount of air	80 000 kg/h
Inlet temperature	200 °C
Exit temperature	95 °C

	cyclone only	cyclone + wet scrubber	cyclone + bag filter + heat exchanger
Total heat required 10^6 kJ/H	(at 100% efficiency: 8.4)		
Spray drier	14.5	14.5	10
Spray drier + evap. (26%)	(14.5 + 21) 35.5	(14.5 + 19.8) 34.3	(10 + 21) 31
Spray drier + evap. (15%)	(14.5 + 12) 26.5	(14.5 + 11.3) 25.8	(10 + 12) 22
Spray drier + evap. (6.5%)	(14.5 + 5.3) 19.8	(14.5 + 5.0) 19.5	(10 + 5.3) 15.3
Fan capacity 10^6 kJ/h	~0,25	0,5	0,7

Savings on the drier:
about $4 \cdot 10^6$ kJ/h $\hat{=}$ $28 \cdot 10^9$ kJ pa $\hat{\approx}$ 180 000 DM pa

From the point of view of energy saving this is not desirable either.

It is therefore advisable to use the method suggested in Figure 1c. The exhaust air from the cyclone which is at a temperature of $95^{\circ}C$ is led into a bag filter whereby usable product is recovered. From the filter the cleaned exhaust gases are passed into a heat exchanger where filtered incoming fresh air, in countercurrent, can be pre-heated to approximately $75^{\circ}C$. The temperature of the exhaust air as it leaves the pre-heater is just about $40^{\circ}C$. It is recommended for bacteriological reasons to keep the bag filter constantly warm and dry.

An example is shown in Table 1 of what order of magnitude the energy saving can be. To calculate the amounts of heat, a boiler efficiency of 85% and the existence of condensate return was assumed. The cost figures given are those correct at the present time.

Furthermore, by using a wet scrubber or bag filter the usual 0.5% loss of product can be avoided. There is therefore an additional gain of approximately 100 t of milk powder per year which is equal to 200 000 DM pa.

Using the process of milk concentration by means of a falling film vacuum evaporator and the production of milk powder by spray drying as examples it could be shown that substantial savings in energy can be achieved by changing the plant design. It is important to point out that gains in heat have not been obtained at the expense of product quality. Moreover, these energy-saving measures avoid losses of milk powder which is firstly a further substantial gain and, secondly, solves environmental problems.

SOME ASPECTS ON THE EFFECTS OF THERMAL PROCESSING ON QUALITY OF DAIRY PRODUCTS

M. Heikonen* and P. Linko**,
*Valio Laboratory, Kalevankatu 56 B, SF-00180 Helsinki 18,
**Helsinki University of Technology, Department of Chemistry, Kemistintie 1, SF-02150 Espoo 15, Finland.

COOLING AND COLD STORAGE

The natural temperature of raw milk is about $38^\circ C$, and any change in this temperature can be considered to upset the biochemical balance. Yet milk should be able to withstand even prolonged cold storage without any significant losses in quality.

Cooling is known to activate lipolysis in milk. According to the theory of Nordlund and Heikonen (1974), high-melting triglycerides of milk first begin to crystallise as spherical shells, beginning from the membrane of the fat globule. Contraction causes pressure to the molten interior, resulting in a mechanically weakened globule. Such a globule is easily broken by external mechanical stress, forcing out triglycerides susceptible to lipolysis and oxydation, and forming free fat. If not properly controlled, this could result in a decrease in both technical and organoleptic quality.

As milk cools, β-casein is transferred from micelles to milk serum. This is accompanied by an increase in serum proteolytic activity. Such changes may result in complications in cheese manufacture, lowering yield and affecting coagulation (Reimerdes and Klostermeyer, 1976; Reimerdes et al., 1977).

Although cooling is necessary for any prolonged storage of milk, it may also result in a change in microbial flora. The environment may favour psychrophilic micro-organisms over lactic acid bacteria, resulting in proteolysis and lipolysis instead of souring (Nordlund, 1971; DeBeukelaar et al., 1977). Quality assessment of raw milk has traditionally been based on the reductase test, giving no information about potentially detrimental psychrophilic bacteria. Earlier, when cooling methods were often insufficient, lactic acid bacteria were generally dominating. For this reason Valio has recently adopted catalase test for quality control of raw milk in Finland (Nordlund, 1971, 1973, 1976; Nordlund and Kreula, 1974; Nordlund et al., 1972; Junkkarinen and Nordlund, 1977). Thus milk samples containing too many psychrophilic bacteria and/or somatic cells may be easily detected even using automatic analytical techniques.

HEAT TREATMENT

Many new processes recently applied in dairy technology, such as ultrafiltration, electrodialysis, and enzymatic hydrolysis of lactose, require at some stage some form of heat treatment, such as pasteurisation, spray drying, etc. Changing of native condition of milk by processing may also alter resistance towards thermal treatment, although such effects, if any, are likely to be insignificant. It has been shown using rats as test animals that protein coprecipitate obtained by heat and acid treatment from non-fat milk, ultrafiltered skim milk, and whey protein products are excellent sources of lysine and other nutrients (Mäkinen et al., 1977). Lysine, methionine, and cystine were unaffected by heat treatment necessary during calcium coprecipitate preparation (Vattula, 1977). Similarly, spray-drying demineralised whey powder did not adversely affect lysine content (Kalsta, 1975).

There have been some reports of possible formation of lysinoalanine mainly under alkaline conditions in certain milk products (Watanabe and Klostermeyer, 1977; Sternberg et al., 1975). Possible imbalance in essential amino acids due to the Maillard reaction (Renner, 1974) is very unlikely under normal process conditions. Recent results have also indicated that antibodies to milk proteins would not promote atherosclerosis

Toivonen et al., 1975; Scott et al., 1976), as suggested
earlier by Annand (1967, 1972) and others (Davis, 1969; Davis
et al., 1974). It has also been shown that HTST-pasteurisation
and UHT-sterilisation inactivates over 60% of xanthine oxidase
(Greenback and Pallansch, 1962) which is considered to be a
promotor of atherosclerosis (Oster, 1971, 1974).

Heat treatment of milk products may increase the digestibility of proteins and result in desirable functional properties (Kalsta and Kreula, 1972; Kalsta, 1977).

CONCLUSIONS

It may be concluded that although heat treatment may be expected to have both advantages and disadvantages from both a quality and nutritional point of view, thermal processes involving dairy products have been investigated in greater detail and supervised in practice more closely by authorities than almost any other foodstuffs. However, the applications and relationships to quality of new technology requires further research. In particular, the increased use of cooling operations in dairy technology has resulted in technological complications that may, in part, be due to overlooking milk as a biological fluid when applying mechanical engineering in processing. Perhaps the most important problem requiring further research today is the behaviour and subsequent effect on quality of casein components during cooling operations.

REFERENCES

Annand, J.C. 1967. J. Atheroscler, Res. 7, 797-801.
Annand, J.C. 1972. Atherosclerosis 15, 129.
Davis, D.F. 1969. J. Atheroscler. Res. 10, 253-259.
Davis, D.F., Johnson, A.P., Rees, B.W.G. and Elwood, P.C. 1974. Lancet I, 1012-1014.
De Beukelaar, N.J., Cousin, M.A., Bradley, R.L. and Marth, E.H. 1977. J. Dairy Sci. 60, 857-861.
Greenbach, G.R. and Pallansch, M.J. 1962. J. Dairy Sci. 45, 958-961.
Junkkarinen, L. and Nordlund, J. 1977. Nord. Mejeriind. 4, 281-283.
Kalsta, H. 1975. Karjantuote 58 No. 12, 14-15.
Kalsta, H. 1977. Karjantuote 60 No. 6-7, 8-11.
Kalsta, H. and Kreula, M. 1972. Karjantuote 55, 333-337.
Mäkinen, S.M., Horelli, H.T., Heikonen, M.K. and Kreula, M.S. 1977. Ympäristö ja Terveys.
Nordlund, J. 1971. Karjantuote 54, 40-47.
Nordlund, J. 1973. Karjantuoute 56, No. 5, 20-22.
Nordlund, J. 1976. Karjantuote 59, No. 6-7, 6-9.
Nordlund, J., Kreula, M. and Puhakka, M. 1972. Karjantuote 55, 343.
Nordlund, J. and Heikonen, M. 1974. XIX International Dairy Congress Vol. I E, 176.
Nordlund, J. and Kreula, M. 1974. XIX International Dairy Congress Vol. IE, 363.
Oster, K.A. 1971. Amer. J. Clin. Res. 2, 30-35.
Oster, K.A., Oster, J.B. and Ross, D.J. 1974. Intern. Lab. Sept/Oct. 15-21.
Reimerdes, E.H. and Klostermeyer, H. 1976. Kieler Milchwirtsch. Forsch. Ber. 28, 17-25.
Reimerdes, E.H., Perez, S.J. and Ringqvist, B.M. 1977. Milchwissenschaft 32, 154-158.
Renner, E. 1974. Milch und Milchprodukte in der Ernährung des Menschen, 339
Scott, B.B., McGuffin, P., Swinburne, M.L. and Losowsky, M.S. 1976. Lancet II, 125-126.
Sternberg, M., Kim, C.Y. and Schwende, F.J. 1975. Science 190, 992-994.
Toivonen, A., Viljanen, M.K. and Savilahti, E. 1975. Lancet II, 205-207.
Vattula, T. 1977. Thesis, Helsinki University of Technology.
Watanabe, K. and Klostermeyer, H. 1977. Z. Lebensm. Unters-Forsch. 164, 77-79.

MEAT PANEL

PARTICIPANTS

Chairman:	B. Krol *(Netherlands)*	
Rapporteur:	J.V. McLoughlin *(Ireland)*	
Invited Experts:	O. Kvaale *(Norway)*	
	K. Oestlund *(Sweden)*	
	J.R. Norris *(UK)*	
	L. Leistner *(W.Germany)*	
	A.W. Holmes *(UK)*	
Other Participants:	M. Jul *(Denmark)*	Plenary Author Session 1
	T. Ohlsson *(Sweden)*	Plenary Author Session 2
	J. Buckley *(Ireland)*	Plenary Author Session 2
	R. Wirth *(W.Germany)*	Plenary Author Session 3
	A.E. Bender *(UK)*	Plenary Author Session 6
	G.B. Brubacher *(Switzerland)*	Nutrition Coordination Group
	M.E. Paneras *(Greece)*	Observer
Panel Coordinator	A. Cotter *(Ireland)*	

DSC STUDIES ON THE EFFECT OF THERMAL TREATMENT ON MEAT PROTEIN QUALITY

O. Kvåle and H. Martens
Norwegian Food Research Institute,
As-NLH, Norway.

INTRODUCTION

Differential Scanning Calorimetry (DSC) is a technique that can be used for studying the thermal behavior of materials as they undergo physical and chemical changes during heating. For a long time this technique has been applied to analysis of clays, soils and minerals, but with the advent of new instrumentation, the application of DSC has been extended to practically all fields of chemistry, both organic and inorganic (McClain et al., 1968).

In a DSC run, a sample and a reference are both heated at a constant heating rate, through a preset temperature interval. Endothermal or exothermal reactions will be registered as the additional energy required in order to have the sample temperature equal to the linearly increasing reference temperature. This endo- or exothermal pattern with increasing temperature is called a DSC thermogram.

It is generally accepted that the heat-induced unimolecular protein unfolding denaturation is an endothermal process (Martens and Vold, 1976), while multimolecular protein aggregation giving precipitates or gels often shows exothermal patterns and is less well understood.

RECENT WORK IN THIS FIELD

In studying intact meat tissue (beef muscle) Karmos and DiMarco (1970) found a complex endothermal peak starting at $50^{\circ}C$ with maximum at $66^{\circ}C$. A second, more well defined peak started at 73°C with a maximum at 82°C and tailing off about 90°C.

In DSC studies of calf meat scanned at 10°C/min., Martens and Vold (1976) demonstrated three distinct peaks: the first one starting at 49 - 50°C with a maximum at about 57°C, the second one with a maximum at about 65°C and the third one starting at or below 75°C with a maximum near 80°C. The authors suggested the first peak to represent myosin, the second collagen and the third actin.

Later, some of these findings have been verified by Wright et al. (1977). Working with whole rabbit muscle and with its constituent proteins, these authors found that post-rigor muscle yielded a complex thermogram comprising at least three endothermic transitions with maximum values of 60, 67 and 80°C respectively. Comparison with the purified proteins indicated that these peaks correspond to denaturation of myosin, sarcoplasmatic proteins and actin.

By examining pre-rigor muscle, Martens and Vold (1976) and also Wright et al. (1977) found a single large exothermal peak with a maximum at about 54°C superimposed on the post-rigor denaturation peaks. This exothermal peak is probably closely linked with the process of muscle contraction.

In beef, the DSC measured denaturation of myosin, collagen and actin has been related to the changes in texture and appearance by comparative sensoric studies (Martens et al., 1977). By the technique used, both the compression force necessary to chew the meat, as well as the chewing work have been registered. The result of this evaluation is shown in Figure 1.

By raising the cooking temperature from 44°C to about 80°C, the compressing force is steadily increasing. As a native muscle the beef is soft, but is getting increasingly harder as the denaturation of proteins proceeds. The curve for the chewing work, however, shows a steep drop at about the

enaturation temperature of collagen. At higher temperatures,
e with denaturation of actin, chewing work is again increasing.

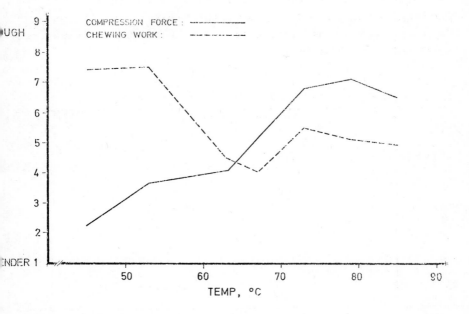

Fig. 1. Sensory Measurements.

According to these findings, an inner temperature of below 60°C would give a soft beef with chewy connective tissue, a temperature in the region of 60 - 70°C would give a juicy, tender beef and a temperature above 75 - 80°C would give a tough and dry product. Temperatures below 50°C would yield beef with a raw colour, 50 - 70°C would yield whitish beef (precipitated sarcoplasmatic proteins), but with red juice. Temperatures above 70°C would yield brown meat and clear juice, probably due to denaturation of myoglobin.

Recently, Martens (1977) has prepared a scheme of temperature data for protein denaturation as shown in Figure 2. It should be stressed that irreversible protein denaturation of this kind is time dependent to a certain extent. The approximate half time thermal denaturation of actin from bovine

M. *semimembranosus* is 30 min at 65°C, 20 min at 67°C and 10 min at 69°C. The figures given represent half time denaturation values from approximately 10 min to 1/10 min. The temperature values represent the beginning of the DSC denaturation peak as opposed to the peak maximum temperatures previously referred to.

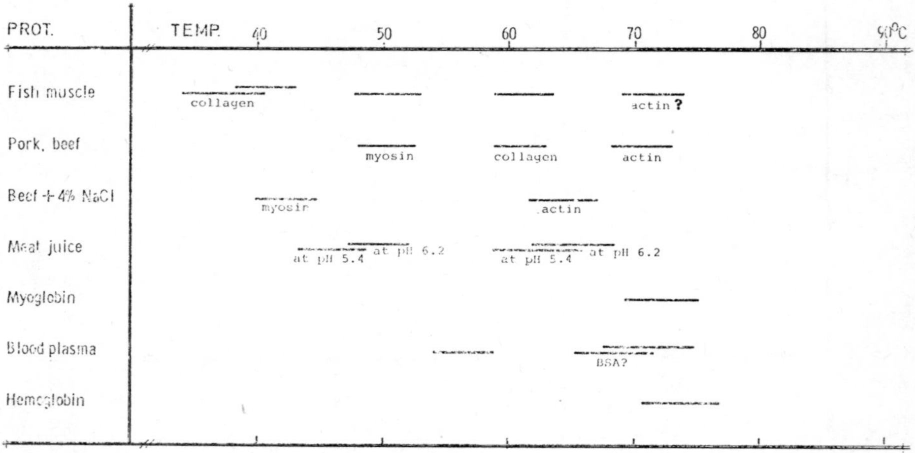

PROT.	TEMP. 40	50	60	70	80	90°C
Fish muscle	collagen			actin ?		
Pork, beef		myosin	collagen	actin		
Beef + 4% NaCl	myosin		actin			
Meat juice		at pH 5.4 at pH 6.2	at pH 5.4 at pH 6.2			
Myoglobin						
Blood plasma			BSA?			
Hemoglobin						

Fig. 2. Denaturation temperature ranges for some muscle proteins.
50% denaturation with 10 to 0.1 minutes. Estimated accuracy: ± 2°C.

Fish muscle is here giving thermograms very similar to those of meat. The addition of sodium chloride to beef is shown to have a drastic destabilising effect on its thermogram, and in meat juice a pH-fall from 6,2 to 5.4 causes a distinct drop in the denaturation temperature.

Denaturation marks without identification represent unidentified meat proteins.

POSSIBLE AREAS FOR FURTHER R & D

So far, introduction of the DSC-technique for studying mechanisms of meat protein denaturation has given interesting results. It has also revealed new fields for further research. Working with conalbumin and ovalbumin, Hegg et al. (1977) have recently demonstrated interesting effects of pH and of the addition of salt and detergent on the protein denaturation and precipitation temperatures, similar to the preliminary results on beef and meat juice earlier referred to (Martens, 1977).

The influence of these parameters on the denaturation temperature of muscle proteins would need a closer study. Here, the effect of polyphosphates and of anionic detergents should also be looked into.

The experiments showing connection between texture and protein denaturation should be extended to include other protein of animal origin as well as minced meat products. More information in this field is needed for an optimisation of the cooking process.

The fact that pre-rigor meat shows an exotherm DSC peak which diminishes as rigor develops and is totally absent in post-rigor meat, needs further substantiation. This might be a novel method for studying the mechanism of rigor, and possibly also the phenomena of cold shortening.

It might also be possible to relate the protein denaturation processes to changes observed when meat is cooked, ie shrinkage, water loss, discoloration etc.

Further, one challenging test lies in applying the DSC technique to study thermal inactivation of amino acids and the mechanism of the Maillard reaction, as well as the study of water binding in meat.

REFERENCES

Hegg, P.O., Martens, H. and Løfqvist, B. 1977. J. Sci. Fd. Agric. In press.
Karmas, E. and DiMarco, G.R. 1970. J. Fd. Sci., 35, 725-727.
Martens, H. 1977. In preparation.
Martens, H. and Vold, E. 1976. 22nd Eur. Meet. Meat Res. Work., Malmø, J.9.
    3-6.
Martens, H., Stabbursvik, E. and Martens, M. 1977. In preparation.
McClain, P.E., Pearson, A.M., Miller, E.R. and Dugan, L.R. Jr. 1968.
    Biochem. Biophys. Acta. 168, 143-149.
Wright, D.J., Leach, I.B. and Wilding, P. 1977. J. Sci. Fd. Agric.,
    28, 557-564.

FOOD DETERIORATING ENZYMES SURVIVING HTST-STERILISATION

K. Östlund

The Swedish Meat Research Centre, S-244 00 Kävlinge, Sweden.

It is well known that a variety of foodstuffs have better sensoric (and even nutritional) properties after HTST-sterilisation than after conventional sterilisation by autoclaving (eg Reichert 1972). This is specially pronounced for vegetables (Giannone and Porretta 1966; Luh et al., 1969; Chen et al., 1970; Paulus 1972) and meat (Persson and von Sydow 1974).

Nonetheless, HTST-sterilisation has been used only to a limited extent in the food industry, and only for a narrow range of products, mainly of dairy origin. An optimal utilisation of the advantages offered by the HTST-sterilisation concept requires the use of continuous production lines and the possibility of aseptic filling of the food directly into consumer packages. These requirements constitute no insurmountable problems today. Modern pumping techniques and the use of, for example, scraped surface heat exchangers enables food preparation and sterilisation continuously and in one step. The developments in the field of packaging have opened the possibility of aseptic filling of foodstuffs in glass containers, sterilised in line.

Thus the technical obstacles for a wider use of HTST-sterilisation in the food industry seem to be eliminated. The remaining hurdle for a promising evolution in this field seems mainly to be the lack of knowledge concerning what role surviving food-deteriorating enzymes may play for the shelf-life of HTST-sterilised food.

Pioneer work in this field has been carried out in investigating heat inactivation kinetics for certain enzymes (for literature, see eg the review by Nehring, 1973 and the works of Clochard and Guern, 1973, and Svensson, 1973). Even if these investigations are limited as concerns temperature range and number of enzymes studied, their results strengthen the assumption that enzymes harmful to keeping quality, may remain active in HTST-sterilised food. This is also in agreement with practical observations from the field of industrial food production.

The detailed knowledge, however, of the mechanisms involved and thus how deleterious effects can be avoided, is too scarce. The picture of which enzymes may be of importance is incomplete, and it is not even known for sure, whether most attention should be paid to enzymes originating from the raw material itself or from micro-organisms active in the material. Nor is it from the present knowledge possible to list the types of food best fitted for high temperature sterilisation, or to outline sterilisation cycles that would be safe as concerns harmful enzymes. It appears that this is the main reason why industrial development in the field of HTST-sterilisation has been retarded. The result has meant a drawback to a positive development within the fields of baby-food, ready-to-eat food and other types of so-called convenience food.

Nor can we expect any rapid development as long as fundamental problems remain unsolved. These problems are too complex and contain too many of the characteristics of basic science to be dealt with as a whole by the food industry and in particular within single development projects.

The knowledge needed can be acquired only from research work of a more basic type. It is not only a question of revealing what types of enzymes that in different types of food may survive sterilisation, but also of exploring their detailed heat inactivation kinetics at temperature ranges not formerly investigated. This work constitutes a wide field of applied science and will, in many cases, require the development of new techniques, for example measuring enzymatic activity or achieving defined short-term heat treatments.

The main part of this type of research programme can hardly be financed by the food industry alone, at least not in small countries like Sweden, where the industrial units are small and their resources limited.

There is therefore every reason to stress the importance of this field of research in order to interest governmental or public research bodies to give a high priority in the funding of this specific research area. An industrial development cannot be expected until we fully recognise the limitations and possibilities offered by the HTST-sterilisation technique.

REFERENCES

Chen, K.C., Luh, B.S. and Seehafer, M.E. 1970. Chemical changes in strained peas canned by the aseptic and retort processes. Food Technol. 24, 821-826.
Clochard, A. and Guern, J. 1973. Revue Générale du Froid, 64, 860-870.
Giannone, L. and Porretta, A. 1966. Ind. Conserve 41, 169-174.
Luh, B.S., Antonakos, J. and Daoud. H.N. 1969. Food Technol. 23, 377-381.
von Nehring, P. 1973. Reaktionskinetische, Betrachtungen zur Hitzesterilisation von Lebensmitteln. Deutsche Levensmittelrundschau 69, 12-20.
Paulus, K. Jahresbericht von Bundesforschungsanstalt fur Lebensmittel-frischhaltung in Karlsruhe. M. 19.
Reichert, J.E. 1972. Neue Erkenntnisse bei der Herstellung von Vollkonserven. Fleischerei 23, 11-12.
Svensson, S. 1973. Thesis, Chalmers University of Technology, Gothenburg, Sweden.

REFRIGERATION STUDIES

C. Bailey
Agricultural Research Council Meat Research Institute,
Langford, Bristol, England.

Studies on the processes of chilling, freezing and thawing of carcass meats and on the effects of these processes on the eating quality of the product have formed a major part of the programme of the Meat Research Institute (MRI) since its foundation ten years ago. Similar work had been carried out at the Low Temperature Research Station prior to the setting up of the Langford laboratory and the opportunity was taken when designing and building the MRI to include a sophisticated wind-tunnel enabling the effects of a wide range of temperatures and air speeds to be studied. A substantial mass of information about the thermal properties of carcasses and of meat in different forms has now been accumulated. These data have been stored in computer-accessible form and have come to provide a background against which new problems can be investigated and advice given to the industry. In recent years the impact of EEC temperature legislation has become important in the United Kingdom and this aspect has received increasing attention from the MRI.

MEAT CHILLING

The principal reason for chilling meat is, of course, to extend its storage life by reducing the activity of the bacteria present both on the surface and in the deep tissues. The more rapidly the temperature is reduced, the longer the subsequent storage life and the smaller the evaporative weight loss with consequent savings which can amount to up to 2% of carcass weight. The more rapidly meat is chilled through the initial temperature range of 40 - 30°C, the smaller will be the drip loss later on cutting. The advantages of rapid cooling rate are obvious but there are disadvantages too; particularly, the phenomenon of cold-shortening leads to irreversible toughness in meat when beef or lamb carcasses are chilled too rapidly before the process of rigor mortis has advanced sufficiently in the muscles. Much of the work of MRI has been aimed at providing the basic data required to enable the plant designer or meat technologist to build equipment and introduce processes which will enable him to chill for maximum economy, maximum tenderness or a reasoned combination of both.

Abattoirs wishing to export meat to the Common Market can also run into problems with cold-shortening, particularly if their operations are geared to a 24 h turnover, as the EEC regulations state that meat cannot be transported or cut until the deepest part of the carcass has attained $7^{\circ}C$. In order to achieve such a temperature in 24 h, beef sides of average weight and fatness would have to be placed in air at $-3^{\circ}C$ at a velocity of 1 m/sec. During this chilling process, considerable portions of the carcass (including the whole of the loin) would fall below $10^{\circ}C$ in 10 h and would consequently be liable to cold-shortening.

The MRI has recently overcome this problem by developing a method which rapidly increases the rate at which a carcass goes into rigor. The technique consists of applying an oscillating electrical current at high voltage through the carcass for 2 mins, after which rapid chilling may be applied without danger of cold-shortening.

MEAT FREEZING

The sole purpose in freezing meat is to extend the storage life beyond that obtained at chilling temperatures. Freezing achieves this by inhibiting the activity of spoilage micro-organisms and by slowing down the rate of biochemical reactions which normally occur in unfrozen foods. Freezing takes place over a range of temperature rather than at an exact point because, as freezing proceeds, the concentration of solutes in the meat fluid steadily increases and progressively lowers the freezing temperature. It is the lack of appreciation by many people in the industry of the time taken for the meat to pass through this latent heat plateau that causes most of the problems experienced. Again, data provided by the MRI are aimed at enabling the industry to design and operate freezing equipment for various purposes.

An important factor to take into consideration when studying freezing is the effect of packaging on freezing rate. Frozen meat is packaged for two reasons:

1) to protect the meat from contamination, particularly during transport and

2) to minimise or prevent evaporative losses during storage.

Such packaging can have a considerable effect on freezing times. Recent work at the Institute has led to the introduction of a novel form of packaging which opens up the possibility of central production of individual sized retail frozen packs for distribution through established cold-chains in the food industry. These are of attractive appearance and seem likely to make a substantial impact on the shape of trade in the coming years.

MEAT THAWING

Thawing, as a process operation, has received much less attention in the literature than either chilling or freezing and in commercial practice there are relatively few controlled thawing systems, even in establishments which specify exact requirements in terms of meat refrigeration. Recent years, however, have seen a marked increase in the usage of frozen meat by the manufacturing industries and a substantial programme at MRI has been aimed at providing information about the effectiveness of the various systems that are available for achieving thawing under controlled conditions.

FUTURE DEVELOPMENTS

The development of practical methods for electrical stimulation of beef and lamb carcasses, coupled with several years of work on the process of hot deboning, opens up the prospect of developing new handling techniques for the meat industry. The MRI has recently reorientated part of its development programme to the exploitation of these opportunities, fundamental to which is a proper understanding of the factors influencing the transfer of heat through meat in various formats and under various packaging systems.

SURVEY WORK

In order to provide the basic information required for identifying and analysing the practical problems of the industry, the MRI has maintained a small survey effort ever since it came into existence. Many studies have been made of refrigeration practice in the UK industry, and indeed on occasions in the Continental industry. This work has provided a clear basis of understanding of the current situation which has formed a useful background for the development of specific research projects in the Institute and has also, incidentally, led the Institute to provide advice to specific parts of the industry.

RESEARCH REQUIREMENTS

Current MRI work is at three levels; fundamental studies of the factors influencing the transfer of heat in carcasses and meat, the application of results from the fundamental work to the solution of industrial problems connected with refrigeration and thawing, and the study of refrigeration in relation to the development of new handling techniques. At all three levels the influence of EEC regulations plays a major role in determining the nature of our experimental programme.

The pressure to work on and solve day-to-day problems

experienced by the industry is considerable and leads to a
distortion in the balance of effort between the three
different phases of the programme. Although facilities at the
institute are good, manpower is short and the more fundamental
aspects of the work and the development of novel processes
suffer from a shortage of staff time. The MRI would welcome
any opportunity to strengthen its existing links with European
research in this field and to apply its facilities, expertise
and information bank to problems of general significance to
the European meat industry.

HURDLE EFFECT AND ENERGY SAVING

L. Leistner
Federal Centre for Meat Research, Kulmbach, Federal Republic of Germany

The primary objective of traditonal and newly developed food preservation processes is the inhibition or inactivation of micro-organisms. In Table 1 these processes and the parameters which govern them are listed. This Table indicates that more processes than parameters are applied in food preservation, and that most processes are based on several parameters or hurdles. In most processes one or two main hurdles are applied. Minor or additional hurdles are, however, needed to accomplish the expected microbial stability achieved by a particular process.

TABLE 1

TRADITIONAL AND NEWLY DEVELOPED PROCESSES USED IN FOOD PRESERVATION AND THE PARAMETERS OR HURDLES THEY ARE BASED ON

Parameters \ Processes	Heating	Chilling	Freezing	Drying	Curing	Salting	Sugar addition	Acidification	Fermentation	Smoking	Oxygen removal	IMF (f)	Radiation
F(a)	X(c)	*	*	*	*	o	*	o	o	*	*	*	o
t(b)	*(d)	X	X	o	*	*	o	*	*	*	*	*	o
a_w	*	*	X	X	X	X	X	o	*	*	o	X	o
pH	*	*	o	*	*	*	X	X	X	*	*	*	o
Eh	*	*	*	o	*	*	*	*	*	*	X	*	o
Preservatives	*	*	o	*	X	*	*	*	*	X	*	X	o
Competitive flora	o(e)	o	o	o	*	o	o	*	X	o	*	*	o
Radiation	*	o	o	o	o	o	o	o	o	o	o	o	X

a = High temperature; b = low temperature; c = main hurdle; d = additional hurdle; e = generally not important for this process; f = intermediate moisture foods.

Since most processes used in food preservation are based on several hurdles, most processed foods also have several inherent hurdles which accomplish the desired microbial stability of the product. Figure 1 illustrates the hurdle effect in foods, using six examples. Example one is a food containing six hurdles, which the micro-organisms present cannot overcome. Therefore, this product has a sufficient microbial stability. In this example all hurdles have the same intensity; however, this is only theoretically possible. A more likely situation is presented in example two. This product could for example be a raw ham. The microbial stability of this product is based on five hurdles of different intensity. The main hurdles are the water activity and preservatives (nitrite and smoke in raw ham), and additional hurdles are the storage temperature, the pH, and the redox potential. These hurdles are sufficent to stop the usual types and numbers of organisms associated with such a product. Example three represents the same product, but a superior hygienic condition, ie only a few micro-organisms are present 'at the start'. Therefore, in this raw ham only two hurdles would be enough. On the other hand, in example four, due to bad hygienic conditions, too many undesirable organisms are initially present. Therefore, the hurdles inherent in this raw ham cannot prevent spoilage. Example five is a food containing two hurdles only. If these hurdles are intensive enough microbial stability will result. There is some indication that the product of the hurdles rather than their number determines the microbial stability of a food. Example six illustrates the synergistic effect that hurdles in a food might have on each other. The synergistic effect of hurdles should be further explored.

The hurdle effect (Leistner and Rödel, 1976; Leistner, 1977) is of fundamental importance for food preservation, since the hurdle concept governs microbial food-poisoning as well as spoilage and fermentation of foods. Table 1 and Figure 1 illustrate the hurdle concept in a simplified fashion. Further research is needed, which should lead to a quantitative understanding of the hurdle concept. As a first step research should

be carried out on the inhibition or inactivation of important
genera and species of bacteria, yeasts and moulds, by a particular hurdle (eg temperature a_w or pH). The effect of a
hurdle should be studied under otherwise optimal conditions
for growth, metabolic activity and survival of the organisms
under investigation. A body of relevant data is already available.
The next step is to study the interaction of hurdles in the
inhibition or inactivation of microbes. Some information of
this kind has been published recently (Roberts et al., 1976).
A further research step could be to computerise quantitative
data on the hurdles in foods. Once this has been accomplished,
much of the present-day food microbiology could become obsolete.
The computerised data could be used to predict what genera and
species of micro-organisms might occur in a food of which only
the physical and chemical hurdles are measured. The initial
microbial population would be a variable more difficult to
predict. However, it could be taken into account by a certain
margin of safety.

Using the hurdle concept, the quantitative data, not only
could the present processes used in food preservation become
better understood but they could be optimised and new processes developed.

The hurdle concept also could be used to save energy in
food processing. Since energy almost certainly will become
more expensive in the future, energy could become a serious
cost factor in the processing of foods. If to-day many products
are tailored to be 'convenient' for the consumer, in the
future the emphasis on 'convenience foods' could be replaced
by an emphasis on 'energy-saving foods'. Particularly energy-
consuming are preservation processes where high temperatures
(eg sterilisation), or low temperatures (freezing, chilling),
or special processes (freeze-drying, radiation) are applied.
These processes should be omitted or modified. Other food
preservation processes are much less energy-consuming, eg those
primarily based on a_w and/or pH adjustment or on preservations.

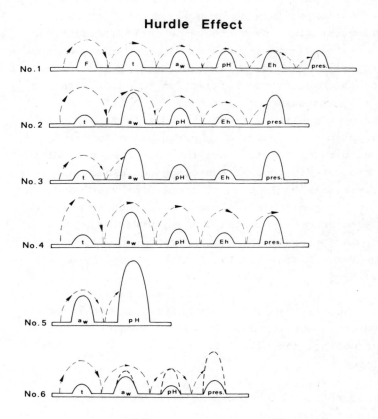

Fig. 1. Six examples illustrating the hurdle effect on which the microbial stability of foods is based.

Thus, research should be conducted to modify energy-consuming preservation methods for foods, such as freezing, chilling and sterilisation. This could be done by the introduction of further hurdles besides temperature. For instance, it is feasible that by increasing the minimal temperatures of micro-organisms by additional hurdles, a comparable microbial stability could be achieved by using +10°C, instead of -1°C as chilling temperature. Storage of foods at -10°C instead of -30°C would probably result in a decreased survival of micro-organisms, including salmonellae. This would be due to the fact that at $-10°C$ the a_w of foods is higher than at $-30°C$. Furthermore, an adjustment of foods to a_w below 0.95 would make temperatures above $+100°C$ unnecessary in canning, since surviving spores of bacteria cannot cause spoilage or food-poisoning below this water activity. In addition energy-saving preservation processes, eg intermediate moisture foods, should be further explored. If energy-saving intermediate moisture foods can be prepared with legally permitted humectans and with superior organoleptic quality, they might gain increasing importance in the future.

REFERENCES

Leistner, L. 1977. In Proceedings IUFoST/CIIA-Symposium held at Karlsruhe, Germany, August 23-24, 1977. In press.

Leistner, L. and Rödel, W. 1976. In Intermediate Moisture Foods (Davies, R., Birch, G.G. and Parker, K.J., eds), pp. 120-137. Applied Science Publishers Ltd. London.

Roberts, T.A., Jarvis, B. and Rhodes, Annette C. 1976. J. Fd. Technol. 11, 25-40.

THE FUTURE USE OF VEGETABLE PROTEINS IN MEAT PRODUCTS

A.W. Holmes
The British Food Manufacturing Industries Research Ass.
Randalls Road, Leatherhead, Surrey, KT22 7RY, UK.

Increasingly, vegetable protein products are being used in meat products. The present state of use is as follows:

1) Full fat flour is not used in meat products.

2) De-fatted flour or grits are used as fillers and binders in sausages, burgers, etc. The use is generally at a fairly low level (less than 3%) being mainly restricted by the flavour problems. Binding properties are limited.

3) Concentrates are used in emulsions, sausages, burgers, meat loaves, etc, at levels of up to 3% of the product. These products have a better flavour than the de-fatted flour, but again have restricted binding properties.

4) Isolates are universally used in a wide range of meat products at levels of up to 6% although the normal use level is about 3 - 4%. The main justification for their use is their ability to emulsify the fat and thus stabilise it and also form a gel which can be heat set. This leads to improved eating characteristics in the meat product with a decreased loss of water and fat on processing. It also allows the increased use of lower grade meat.

5) Textured products. These fall into three main categories, extruded, puffed and spun. Products at present being offered on the market may represent blends of any of these types. They are essentially used as texture-providing fillers or as complete meat substitutes.

The area which has seen the most significant change in the last few years is that of the isolates. The quality of these products is now much improved and a range of products is offered which have various viscosity characteristics. Their handling properties are also much improved - they have less dusting when added to the formulation and they have improved solubility but this solubility still deteriorates significantly with age.

The main vegetable protein used is soya but there is also some gluten being used and in the future we may see an increased use of rapeseed.

The quantities of vegetable protein at present used in meat products represent a balance between cost and functionality and the situation is always bedevilled by the meat product regulations which often fail to recognise the reduced need for meat. Vegetable proteins will be increasingly needed in the future to provide meat-like products with reduced or zero meat content at a reasonable price.

The technology of the existing products has seen rapid advances in the last decade and now appears to be relatively static. Further advances require a more fundamental understanding which will arise from research rather than from simple technological development. The areas of research required may be summarised as follows:

1) An understanding of the factors which are important in the functionality of protein eg heat setting, foaming, emulsification. This may help to explain why the functionality changes with time.

2) Off-flavours continue to present major problems. What is the nature of these off-flavours? Why are they produced? How can they be avoided without seriously affecting the functional properties such as solubility?

3) There is a need for a better understanding of extrusion processes. The questions which remain to be answered are: what phase changes occur during extrusion, how do the proteins align and cross-link, how can products be produced which have a texture more closely in line to that of meat, how can the juicy eating qualities of meat be simulated more satisfactorily?

4) How do vegetable proteins interact with other components in the products eg meat proteins, carbohydrates, etc?

5) How can the useful range of vegetable proteins be extended beyond soya?

6) What is the significance (if any) of anti-nutritional factors in vegetable proteins?

7) How can problems arising from the indigestible carbohydrate in soya be overcome?

The above comments are from the point of view of developed countries with a sophisticated food processing industry. Acceptance within such countries is a prime requirement before one can reasonably expect developing countries to deal with some of their food problems along similar lines.

CEREALS PANEL

PARTICIPANTS

Chairman:	P. Linko *(Finland)*
Rapporteur:	E. Markham *(Ireland)*
Invited Experts:	D.A.T. Southgate *(UK)*
	G. Fabriani *(Italy)*
	J. Olkku *(Finland)*
	W. Seibel *(W. Germany)*
	D. Schlettwein-Gsell *(Switzerland)*
	P.F. Fottrell *(Ireland)*

Other Participants: C. Mercier *(France)* — Plenary Author Session 1

H.A. Leniger *(Netherlands)* — Plenary Author Session 4

A. Dahlqvist *(Sweden)* — Nutrition Coordination Group

B. Spencer *(UK)*) Observers
D. de Ruiter *(Netherlands)*

Panel Coordinator: G. Ni Uid *(Ireland)*

NUTRITIONAL ASPECTS OF PROCESSING WITH SPECIAL REFERENCE TO CEREALS

D.A.T. Southgate

MRC Dunn Nutritional Laboratory, Cambridge, UK.

Although this seminar is primarily concerned with the thermal processing of foods in connection with preservation and freezing where cereals are concerned the most usual thermal treatment involves the application of heat. The nutritional consequences of baking cereal foods have been the concern of the nutritionist for many years with particular regard to the effects of baking in relation to protein quality and losses of some B-vitamins.

The freezing and subsequent storage of prepared cereal products, however, has important technological and commercial implications and the textural changes that accompany this process are important areas for study. The definition of the optimum conditions for the initial freezing, storage and thawing phases are undoubtedly of great interest within the industries concerned.

From the point of view of the nutritionist however, the changes in nutritive value, if any, that accompany the various treatments are, I suspect, largely unknown and this raises an important matter of principle.

This is, that in the consideration of the quality of a foodstuff nutritional factors should form part of the assessment. In a country such as the United Kingdom where the importance of processed foods of all kinds in the national diet has increased considerably over the past few years it is essential that the nutritional effects of processing are well documented. Although it can rightly be argued that foods are not purchased because of their nutritional qualities but because of their appearance and taste, it is potentially dangerous to pursue these ideal attributes in isolation if the health of the community consuming these products is put at risk.

The argument that nutritional quality is difficult and expensive to assess is, in my view, not valid because for each commodity the important nutritional qualities can be assigned a priority by reference to the importance of the commodity in the diet as a whole and as a source of nutrients.

Thus in the case of cereals it is possible to define, for a particular country, the nutritional attributes that the food processor must include in studies of quality.

Cereals deserve particular attention because of their position as the major source of energy-yielding nutrients in most dietaries and in the case of wheat as a source of protein. They are, in many dietaries, important sources of the B-vitamins and some inorganic constituents. In the United Kingdom they have provided foodstuffs that are consumed by almost all sections of the community and have thereby provided a vehicle for fortifying the diet with calcium and other nutrients.

Intuitively one can argue with some confidence that the types of thermal processing that are the concern of this seminar will not greatly affect the particular nutrients which cereal foods provide.

The same could also be said of the major carbohydrates in cereals but processing in the wider context and, in particular, the extraction rate have profound nutritional consequences and it is useful to consider these in relation to the nutritional attributes of cereals as a whole.

In the past few years increasing attention has been directed at the possible links between the consumption of fibre, in particular cereal fibre, and the incidence of many diseases, notably those of the large bowel.

The major proponents of this hypothesis are convinced

that the use of high extraction cereals would be a major factor in protecting the health of the population. This is a nutritional attribute of cereals that has been largely ignored and yet which appears to be of great significance.

The verification of the hypothesis has been hampered by a lack of clarity in the use of terminology and of the appreciation of the chemistry of the constituents of the plant cell wall.

The term 'dietary fibre' has come into use for the sum of lignin and the polysaccharides not digested by the endogenous secretions of the human digestive tract. Values for dietary fibre are not correlated with the conventional crude fibre values. The hypothesis itself rests largely on rather insecure epidemiological evidence which, by its very nature, even when it is carefully collected, can rarely be used to prove casual relationships.

Present research efforts in this field are concerned, firstly with measuring and characterising the components of dietary fibre in human foods, second with assessing the metabolic effects of the components in man and third collecting more epidemiological evidence.

The initial stages in processing cereals are important but there are indications that the availability (that is the susceptibility to enzymatic hydrolysis) of some polysaccharides is decreased during processing particularly baking.

The major role of dietary fibre probably lies in the large intestine where it provides a substrate for the microflora, and increased bulk. The water-binding properties of many polysaccharides are probably also important in the retention of water in the large bowel. There are, however, indications that the texture and physical properties of the polysaccharides in general may be important in some small intestinal effects attributed to dietary fibre.

One can therefore identify a particular nutritional quality of cereals that is affected by processing first by the refining stage and second by thermal processing. The current status of the dietary fibre hypothesis is that the association of low intake with increased incidence of disease and a high intake with protection is not completely established. The evidence does suggest, however, that cereal dietary fibre has special properties and that these may be important in health.

In future work on the processing of cereal foods it may therefore be prudent to consider ways whereby the dietary fibre content can be increased while retaining consumer acceptability and economic processing methods.

THERMIC PROCESSES AND NUTRITIVE VALUE OF CEREALS

G. Fabriani
International Association for Cereal Chemistry,
Rome, Italy.

Cereals have been studied in depth from the standpoint of their nutritive value. If we take into account that these products constitute nowadays the main source of food for the world population, it is obvious why attention continues to be concentrated on them with the aim of both increasing production and obtaining varieties of higher biological value, meaning by this higher protein yield of best possible quality.

Also of interest, however, are all those studies devoted to an evaluation of changes in the nutritive value itself as a result of often necessary technological processing; unfortunately these modifications are almost always negative.

In some of these processes parts of the product which usually contain valuable nutrients for the human being such as vitamins, mineral salts, proteins, enzymes, are removed; this occurs, for instance, in wheat milling or in the preparation of pearled rice. Other processes lead to qualitative or quantitative modifications of some nutrients in that, as has already been mentioned, they reduce the nutritive value of the substances.

In the following we will briefly discuss some of these processes which entail a thermic treatment.

Almost all cereals and derived products meant for human nutrition need to be processed at higher or lower temperatures. Furthermore, the conditions under which such operations are carried out can have remarkable effects independently of the thermic gradient.

The derivatives of cereals, the manufacture of which requires fairly high temperatures include the so-called oven products; among them bread is, of course, the most important one.

Let's briefly indicate in which respect the nutritive value is modified. The most interesting and evident phenomenon is the one known as the Maillard Reaction consisting of a series of modifications which occur, especially in the presence of high temperatures, between reducing sugars and amino acids or proteins

Lysine, by being present in bound form, in proteins, is the most affected amino acid in this reaction; in the case of wheat, although it already is the limiting amino acid, it would be first partly blocked and then destroyed thus leading to a loss in protein value. This effect, however, is not of the same significance in all products: specifically, if we consider bread, we note a destruction of lysine of approximately 70% on the crust and only 10% in the crumb due to the difference in temperature and moisture content of the different parts of the product; further, the phenomenon is reduced in the bread baked in moulds (factory loaf) since the sides of the container act as a shield against the heat.

The Maillard Reaction finds, however, ideal conditions in those oven products which belong to the category of sweets; they are based on wheat flour and also contain in their formulation many other ingredients among which certain quantities of carbohydrates are always present.

Chemical and biological tests carried out in Italy on some of these products, many of which were baby foods, have shown that, even if the formulation proteins of good nutritive value had been included, some 20 - 60% of lysine was unavailable.

It has been equally shown that this phenomenon occurs in dietetic biscuits, rusks, and in all those similar preparations which require intense thermic treatments such as toasted flakes, puffed cereals, etc.

On the other hand, modern baking processes by micro-waves seem to reduce many prejudices of a nutritive character.

Another important area in which thermic processing affects the biological value of proteins is that of alimentary pastes. The modern paste making industry has a tendency to increase the drying temperature with the twin aim of speeding up the operation and obtaining products which are more resistant to the effects of further cooking.

From tests carried out in our country, it has been noted that drying at $18°C$ for 18 hours, which was normal practice until a few years ago, leads to a 22% in nutritive value (blocked or destroyed lysine) if drying is effected at $80°C$ for 6 hours (temperature and time employed by a considerable part of the modern industry) the loss rises to 47%.

Besides a reduction in lysine, a reduction in methionine and threonine has also been observed, due to the paste making process.

On the other hand, alimentary pastes, once produced, have to undergo proper boiling before being consumed; under certain conditions this boiling must be considered as a genuine technological treatment since it is an essential operation to make the product available and ready for use by the consumer.

Also, boiling can produce losses due both to the influence of heat and the fact that some nutrients are dissolved in the water - which is in most cases, thrown away.

Tests have been carried out on the subject in order to evaluate these losses. It has been found that on boiling, losses in cystine reach values of up to 16%; lower losses have been recorded with regard to aspartic acid, glutamic acid, leucine, proline and serin. Losses in lysine are not very significant in this case.

On boiling alimentary pastes, when the water is then thrown away, a certain loss in the vitamin content has also been recorded especially with regard to those of the B Group. There is a reduction of about 40% in thiamine, 60% in riboflavin, and 30% in niacine.

It should be pointed out, however, that these losses are only of relative importance with regard to nutritive effects since alimentary pastes, and this also holds true for pearled rice, are not considered an important source of these vitamins.

Another point of interest with regard to the correlation between thermic process and nutritive value is the possible formation, as a result of rather high temperatures, of an aromatic polycyclic hydrocarbon, benzo(α)pyrene, which has been recognised as carcinogenic if present beyond certain limits.

This phenomenon mainly concerns other types of foods, different from those with which we are dealing, such as smoked or roasted meats; it must not, however, be underestimated in the case of some derivatives from cereals such as toasted bread, rusks, water biscuits and sweet biscuits. In these products the toasting process leads to remarkable increases in benzo(α)pyrene content compared to the quantity present in the original product. However, this content must be still considered modest and in any case inferior to the limits of tolerance recommended by PAG and IUPAC (5 ppb).

On the other hand, attention should be paid to the danger involved in the very frequent use of toasted and heated products such as those mentioned earlier.

We have tried to describe briefly some aspects of the inter-relationship between thermic processes and nutritive value of cereals; but we have not intended to diminish the important role which they represent for human nutrition.

An in-depth knowledge of these phenomena can be of interest to the experts in the study and implementation of technological processes which would not impair the nutritive value of the products to be processed.

EFFECT OF THERMAL PROCESSING ON CEREAL BASED FOOD SYSTEMS

J. Olkku* and P. Linko **

*Technical Research Centre of Finland, Food Research Laboratory
Biologinkuja 1, SF-02150 Espoo 15, Finland.
** Helsinki University of Technology, Department of Chemistry
Kemistintie 1, SF-02150 Espoo 15, Finland.

INTRODUCTION

Thermal processing is involved in the manufacture of most cereal based foods. The most important operations during processing normally include modification of physical properties of raw materials by heat application to obtain desired product quality. The food processing system may be described as a general block diagram as shown in Figure 1.

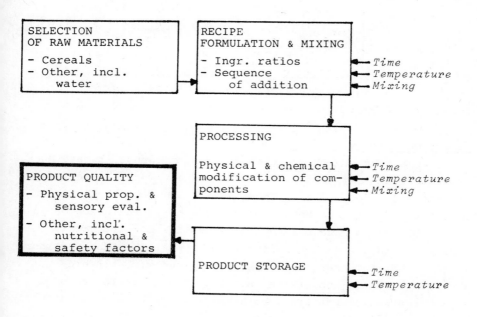

Fig. 1. The food processing system.

Typical key phenomena during heat treatment include starch gelatinisation and denaturation of proteins, both normally water dependent. Such physico-chemical changes tend to be modified by other food system components, thus affecting physcial properties of product. Such ingredients include sugars, proteins, lipids, mineral salts, etc. (Osman, 1975; Donovan, 1977; Olkku and Rha, 1977). The intensity and duration of heat treatment may also have a considerable effect (Olkku, 1975; Suzuki et al., 1976). The diagram in Figure 2 demonstrates the interactions of chemical and physical descriptions during passage through the process.

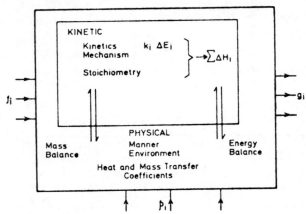

Fig. 2. Interaction of chemical and physical descriptions (Branch, 1975)

Starch gelatinisation - a theory

In native starch granules starch components are arranged in crystalline, micellar form, embedded in amorphous regions (Schoch and Elder, 1955). Upon heating in aqueous solutions the amorphous regions begin to absorb water. The reaction is reversible up to a temperature at which the birefrigence typical of native starch begins to disappear (Schoch and Elder, 1955). This temperature is also associated with a rapid increase in water and starch mobility (Jaska, 1971), loss of X-ray crystalline pattern (Charbonniere, 1977), and an increase in viscosity leading to a paste (Miller et al., 1973). The viscosity increase has been attributed to an exudate released from starch granules to surrounding medium to form a three-dimensional network that binds swollen granules, and to fragments due to mechanical breakdown of starch particles under shear.

Donovan (1977) recently described the endothermic reaction taking place both in proteins and in starch during heating in aqueous solutions as denaturation. Both phenomena involve a change in the tertiary molecular structure formed by hydrogen bonding. Water is considered essential, especially in the reaction involving starch. Berry (1966) and Olkku et al. (1977) observed that the initial temperature of starch gelatinisation in solutions of sugar that has a delaying effect on swelling and gelatinisation (Dean and Yamazaki, 1973; Bean et al., 1974) is a linear function of sugar concentration. Olkku et al. (1977) also reported that water vapour pressure at gelatinisation temperature may be expressed as a simple function of sugar concentration:

$P_1 = P_o (10 \ x_s)$, where

P_1 = water vapour pressure at gelatinisation temperature in sugar solution

P_o = water vapour pressure at gelatinisation temperature in pure water

x_s = weight fraction of sugar in sugar-water mixture.

It was also found that the gelatinisation/mechanical breakdown phenomenon proceeds uninterrupted, even after the heating applied to increase the temperature is discontinued at the onset of gelatinisation.

It may be suggested that water acts as the necessary energy transfer agent to break down hydrogen bonds of the tertiary structure of native starch. After reaching a certain level of water reactivity, as indicated by the vapour pressure, the energy level is high enough to rupture hydrogen bonds to initiate gelatinisation. Further gelatinisation is a typical time-temperature relationship at a certain minimum level of water reactivity. Continued heating is required to break down more firmly associated areas of starch. The reaction of starch with water follows first order kinetics (Suzuki et al., 1976), and the limiting factor at low temperatures is the reaction of starch with water and at high temperatures the diffusion of water into the granules. The authors have recently observed that such changes may take place in cereal starches due to an increased water reactivity during HTST extrusion cooking even at much lower water levels that are generally believed to be necessary to initiate gelatinisation. Other ingredients, such as sugars, suppress the reactivity of water. Starch complexing agents limit the reactivity of starch and/or interfere with the formation of a network, thus affecting macrostructure (Miller et al., 1973).

CONCLUSIONS

High temperature short time extrusion cooking is rapidly becoming an important means of heat processing and texturisation of cereal based food systems (De la Gueriviere, 1977). One interesting possibility is the production of protein enriched cereals by extrusion (Smith, 1976). Further intensive research is needed to understand the phenomena involving macromolecules during extrusion cooking. Our programme will aim in developing response surface models of extrusion (Aquilera and Kosikowsi, 1976), involving technical, technological, and quality variables. It is also intended to perform kinetic studies involving physical quality factors (Suzuki et al., 1976; Hayakawa and Timbers, 1977; Lenz and Lund, 1977) as indicators of extrusion product quality. A part of the study will deal with extrusion system models necessary for automatic control of the process, and with quality prediction through models for thermally treated foods. Related problems exist in processing other commodities, such as dairy products, and a comprehensive study of changes in biopolymers during thermal treatment and their effects on quality and nutritional value needs international effort.

REFERENCES

Aquilera, J.M. and Kosikowski, F.V. 1976. J. Food Sci. 41: 647-651.
Bean, M.M. and Yamazaki, W.T. 1973. 58th Annual Meeting of AACC, St. Louis, Abstr. Cereal. Sci. Today 18 (9): 308 No. 194.
Bean, M.M., Yamazaki, W.T. and Donelson, D.H. 1974. 59th Annual Meeting of AACC, Montreal, Abstr. Cereal Sci. Today 19 (9): 44 No. 114.
Berry, G.K. and White, G.W. 1966. J. Food Technol. 1: 249-256.
Branch, J. 1975. Chem. and Ind. 1975. No. 19: 832-835.
Charbonniere, M.R. 1977. Extrusion Cooking in Food, Cucle CPCIA Seminar E6, Paris, 1977.
De la Gueriviere, J.-F. 1977. Ibid.
Donovan, J.W. 1977. J. Sci. Food Agric. 28: 571-578.
Hayakawa, K.I. and Timbers, G.E. 1977. J. Food Sci. 42: 778-781.
Jaska, E. 1971. Cereal Chem. 48: 437-444.
Lenz, M.K. and Lund, D.B. 1977. J. Food Sci. 42: 997-1001.
Miller, B.S., Derby, R.I. and Trimbo, H.B. 1-73. Cereal Chem. 50: 271-280.
Olkku, J.E. 1975. A model for starch-sugar paste texture prediction, Doctoral Dissert. U. Mass, Amherst, Mass.
Olkku, J.E. and Rha, C. 1977. Food Chem. (In press).
Olkku, J.E., Rha, C. and Fletcher, S.W. 1977. (In press).
Osman, E.M. 1975. Food Technol. 29 (4): 30-32, 34-35, 44.
Schoch, T.J., and Elder, A.L. 1955. Advances in chemistry series 12: 21-34.
Smith, O.B. 1976. In: Altschul, A.M. (ed) New protein foods. Vol. 2, Acad. Press. Inc., pp. 86-121.
Suzuki, K., Kubota, K., Omichi, M. and Hosaka, H. 1976. J. Food Sci. 41: 1180-1183.

FREEZING PROBLEMS WITH GERMAN ROLLS
BREAD VARIETIES AND OTHER BAKED GOODS

W. Seibel

Federal Research Centre of Cereal and Potato Processing,
Detmold, W. Germany

1. SITUATION OF FREEZING IN WEST GERMAN BAKERIES

In West Germany, bread, inclusive of rolls and other baked goods, may be specifically defined (Seibel et al., 1977). Bread, including rolls, is produced from cereals or cereal products - milled or ground - by the preparation of dough, followed by scaling, moulding and baking. In the production of bread (including rolls) besides the cereal products already mentioned, water, leavening agents, salt, fat, sugar, milk (or milk products), leguminoses, potato products, spices, raising, oilseeds, and cereal germs are also used. The fat and sugar content accounts for less than 10% of the total, the remaining 90% is composed of cereal products (average moisture content 15%). The fat and sugar content of milk and milk products should also be taken into consideration. This percentage is approximately 11% based on the cereal products. Rolls have the same definition as bread; the only difference arises from their shape and weight.

The minimum bread weight for a whole loaf is 500 g. All other weights for whole loaves must be divisible by 250. The minimum weight for sliced bread is 125 g - the only permissible weights apart from this weight being 250 and 500 g. The weight of rolls may vary up to 250 g. It is apparent therefore that no loaf of bread or a piece of roll can be produced and sold with a weight between 250 and 500 g.

Products with higher fat and/or sugar contents are called 'Feine Backwaren', or other baked goods (eg, sponge cake, cake, biscuits, etc.)

Wheat bread and wheat rolls have a particularly short shelf life, although this could, of course, be extended by the use of freezing. For economic reasons, very little bread is stored at freezing temperatures (-18°C). Some bakeries use freezing in order to maintain a small stock of bread specialities. Freezing is, however, used to a considerable extent to increase the shelf life of rolls as well as confectionery goods.

2. STORAGE OF ROLLS BY FREEZING

Almost all bakeries have facilities to freeze rolls. Freezing of rolls should start 15 - 30 minutes after baking when the crumb temperature of the roll is lower than 50°C (Stephan, 1973). After 30 - 60 minutes the crumb temperature should have reached -7°C.

The storage condition entails a temperature of 18 - 20°C with slowly moving air (1 m/sec).

In these conditions rolls can be stored without quality deterioration for the following periods:
- rolls with 0.5% fat: 1 - 2 days
- rolls with 1 - 2% fat: 2 - 4 days

The use of fat and milk powder in the recipe and the addition of sugar and malt increases shelf life because these ingredients improve the browning reactions in the crust and form only a thin crust. The wrapping of rolls increases the shelf life by one to two days.

Thawing of the frozen rolls needs to be undertaken very carefully. This can be done in a baking oven at a temperature of 200 - 220°C for 3 - 4 minutes. Thawing with room temperature needs 35 - 60 minutes. The main defect with the freezing of rolls is the flaking off of the roll-crust. This defect is caused during freezing and is particularly serious in bread with a thick crust.

3.   STORAGE OF BREAD BY FREEZING

It was already mentioned that for economic reasons there is very little freezing of bread in West Germany. Wheat and wheat mixed bread should be frozen directly after baking. For rye mixed bread a cooling period of 1 - 2 hours has some advantages (Stephan, 1973; de Vries, 1973).

Using the aformentioned storage conditions bread can be stored over the following periods:
- wheat and wheat mixed bread: 5 - 8 days
- rye and rye mixed bread: 7 - 10 days.

In addition, it is always advantageous to wrap the bread.

Thawing at room temperature yields somewhat unsatisfactory results. It is therefore suggested that bread loaves be thawed for 2 - 3 hours at a temperature of 30 - 50°C, following a 10 minute period in a baking oven at 220°C.

4.   STORAGE OF OTHER BAKED GOODS

Almost all confectionery shops have freezing facilities since a large number of different products are capable of being frozen. Yeast-raised products can be stored for 5 - 10 days, and baked products with cream for 2 - 5 days. It should also be noted that coated tarts should be wrapped prior to freezing (Belderok et al., 1971; Bretschneider, 1973: Jünger, 1973).

5.   RESEARCH REQUIREMENTS

The length of time taken to achieve the desired temperature decrease during freezing and the temperature increase during thawing, is the most important factor affecting the conservation of bread quality. Within the temperature range from +5°C to -5°C, the bread quality alters rapidly. Therefore it is necessary to find a freezing system with a short freezing time and a thawing system with a shorter thawing time. To overcome the insulation effect on the bread, particularly the roll crumb itself, it is necessary that the low freezing temperature or the high thawing temperature, penetrate directly into the crumb.

The storage time over which quality does not deteriorate is still very short, especially for confectionery goods. Basic research should be conducted to determine all biochemical changes taking place at these low temperatures, and which cause quality changes.

In regard to roll and bread varieties the relationship between the optimum bread recipe and storage stability is still unknown. Certain compounds in the recipes effect a reduction of the freezing point and the freezing time. Therefore it is necessary to identify the optimum recipe for freezing.

Freezing of rolls, bread varieties and other baked goods, is still very expensive. Economics research should be conducted in order to identify the cheapest energy source and the most economical system for freezing.

REFERENCES

Belderok, B., van't Root, M.J.M., de Vries, L.W.B.M. and Wiebols, W.H.G.
    1971 Diepfriezen in de Brood en Banketbakkerij. Agon Elsevier,
    Amsterdam/Brussels
Bretscyneider, F. 1973 Getreide Mehl und Brot 3, 110-112
Jünger, H. 1973 Getreide Mehl und Brot 3, 112-114
Seibel, W., Brümmer, J.-M., Menger, A. and Ludewig, H.-G. 1977 Brot und
    Feine Backwared, Arbeiten de DLG, Band 152, DLG-Verlag, Frankfurt
Stephan, H. 1973 Getreide Mehl und Brot 3, 108-110
de Vries, L.W.B.M. 1973 Getreide Mehl und Brot 3, 104-108

INVESTIGATIONS INTO THE INFLUENCE OF THERMAL PROCESSING ON CEREALS

Daniela Schlettwein-Gsell
Institute of Experimental Gerontology,
Basel, Switzerland

Investigations into the influence of thermal processing on cereals should give priority to two aspects:

1) Influences of thermal processing on the nutritional content of cereals.

2) Possible relationship between the influence of thermal processing and clinical or epidemiological findings.

1. <u>Influence of thermal processing on nutritional content</u>

In the course of this seminar, Brubacher has pointed out that in industrialised countries like Switzerland, the supply of thiamine, vitamin B_6 and iron is suboptimal in substantial segments of the population.

Bread - the most common cereal consumed in industrialised countries - could be a valuable and ample source for all three of those nutrients. Even with the present low consumption level of only 150 g/day, bread's contribution to the average daily requirement is important

TABLE 1

BREAD AS A SOURCE OF THIAMINE, VITAMIN B_6 AND IRON. (Compiled from common food tables).

	Percentage of daily requirement supplied	
	150 g of unenriched white bread 70% extract	150 g of whole wheat bread 95% extract or revitaminised bread
Thiamine	12%	35%
Vitamin B_6	10%	40%
Iron	13%	22%

Bread supplies more thiamine than any other single food; only meat is higher as a source of iron and only potatoes as a source of vitamin B_6.

Thermal processing may influence the micronutrient content in different ways:

a) Influences on thiamine content

Thiamine content of bread compared with the dough is influenced by temperature and duration of the baking process and the moisture of the dough.

TABLE 2

THIAMINE LOSSES DURING THE BAKING PROCESS (Compiled from Marks, 1975, Seibel, 1976 and Stransky, 1974).

Rolls	20 - 30 %
White bread	10 - 20%
Crust	30%
Crumb	7%
Whole grain bread	30%
Rusk	40 - 50%
Knäckebrot	10 - 50%
Pumpernickel	> 90%

Prolonged baking times as are necessary in special types of whole grain bread and which also occur in domestic baking, lead to losses of up to 100%. Differences between the loss in the crust and the loss in the crumb are crucial because an increasing preference is observed for breads with a high proportio of crust (Seibel, 1975).

b) Influences on vitamin B_6 content

Vitamin B_6 content is entirely destroyed in whole grain breads like Pumpernickel (Seibel, 1975). The mechanism of this decrease is not clearly understood.

c) Influences on iron content

We have recently shown (Seiler et al., 1977) that the tota mineral content of potatoes increases or decreases during the cooking process in various utensils. Such variations can be explained by deposits or by corrosion. A striking decrease of the iron content (up to 73%) was observed when the potatoes were prepared in utensils with a non-metallic surface.

We do not know of such investigations in cereals but we have to assume that similar processes can take place.

We do not know either in which way the bioavailability of iron is influenced by such interactions. Considering the worldwide shortage of iron supply, intensive research on this problem seems necessary.

These examples show that our knowledge of the actual nutrient content of ready-to-serve cereals is lacking. It has been claimed that nutritional surveys cannot be evaluated because of this lack of data (Wirths, 1970). The increasing variety of breads, breakfast cereals and baked foods offered to the consumer calls for more general and systematic investigation into the influence of thermal processing and domestic cooking on the micronutrient content of cereals.

2. Possible relationships with epidemiological findings

Two examples may underline the importance of further investigations in this field:

a) The browning of the crust of the bread has been reported to decrease protein quality. Rats fed the crust of the bread gained less weight than rats fed the crumb. Long-term investigations into reduction diets have been initiated (Seibel, 1976). In view of the widespread problems posed by obesity, the actual changes arising from browning need better understanding.

b) Epidemiological data suggest associations of the incidence of Crohn's disease (granulomatous jejunoileitis) with the intake of corn-flakes (James, 1977) and sweetened baked foods (Martini et al., 1976). Substances which may be formed during thermal processing have to be reviewed as possible causes for this frustrating disease.

SUMMARY

Investigations should give priority to:

1) Influences on nutritional content of nutrients, for which cereals are an important source and among these to nutrients which have been reported to be deficient in certain populations - ie, thiamine, vitamin B_6 and iron.

2) Relationships between the effect of thermal processing and epidemiological data including data on manifestations which are either frequent (eg, obesity) or severe (eg, Crohn's disease

REFERENCES

James, A.H. 1977 Brit. Med. J. $\underline{1}$, 943-945

Marks, J. 1975 A guide to the vitamins. MTP Ltd., Lancaster, UK, p 176

Martini, G.A. and Brandes, J.W. 1976 Klin. W.schrift $\underline{54}$, 367-370

Seibel, W. 1976 Verbraucherdienst 21, 107-113.

Seiler, H., Schlettwein-Gsell, D., Brubacher, G. and Titzel, G. 1977 Mitt. Gebiete Lebensm. Hyg. $\underline{68}$, 213-224.

Stransky, M. 1974 Alimenta $\underline{13}$, 131-132.

POSSIBLE APPROACHES FOR REMOVING THE TOXICITY OF CEREAL GLUTEN TO PATIENTS WITH COELIAC DISEASE (GLUTEN ENTEROPATHY)

P.F. Fottrell, J. Phelan, F. Stevens, B. McNicholl and C.F. McCarthy,
University College, Galway, Ireland.

INTRODUCTION

(a) Definition of Coeliac Disease

Coeliac disease is a human malabsorption condition where patients are sensitive to gluten (or alcohol soluble gliadin) from wheat, rye, barley and probably oats. In patients with this disease the villi of the small intestine become atrophied and general malabsorption occurs accompanied usually by diarrhoea. The disease which affects both children and adults may be treated with a gluten-free diet after which the structure of the villi return to normal.

Normal Small-Intestinal Villi

Coeliac Villi

gluten-free diet

(b) Incidence of Coeliac Disease

Accurate figures for the incidence of the disease are only available for a few European countries. It is obvious from these figures that the incidence is very high in some countries. In Austria for instance the incidence is about 1 in 500 and in Ireland 1 in 300 of the population suffer from the disease.

(c) Inheritance

The mode of inheritance is not known but there is a strong familial incidence and studies from University College Galway have shown an incidence among first degree relatives of 1 in 10.

(d) Associated Diseases

One of the most interesting and disturbing associated diseases is the very high incidence of cancer especially cancer of the digestive tract which occurs in patients with coeliac disease. A very important aspect of this association is whether the incidence of cancer is lower among patients diagnosed with coeliac disease early in life and who strictly adhere to a gluten-free diet compared with patients diagnosed later in life. The latter patients may have been ingesting gluten for many years. For this reason it is very important to diagnose the disease as early in life as possible.

Other associated diseases are certain forms of dermatitis (*Dermatitis herpetiformis*) which improve on a gluten-free diet. Suggestions have been made of an association between coeliac disease and schizophrenia but this has not been substantiated in studies at Galway.

(e) Screening Tests

There is no accurate test for the disease. However, recent studies from the USA suggest that the presence of two antigens, $HLA-DW_3$ and a specific B cell antigen are indicative of sensitivity to cereal gluten and susceptibility to coeliac disease.

(f) Diagnosis

The disease is diagnosed following several clinical and biochemical tests which include morphological and enzymic studies on a per-oral intestinal biopsy.

BIOCHEMICAL NATURE OF GLUTEN TOXICITY TO COELIAC PATIENTS

One of the most urgent problems in coeliac disease is the chemical identification of that part of the gluten molecule responsible for its toxicity to coeliac intestine. That this has so far not been achieved is partly due to the chemical complexity of gluten and to the lack of suitable *in vitro* assays for monitoring toxicity. The only reliable test system to date is the feeding of gluten fractions to coeliac volunteers. Despite these drawbacks a number of approaches aimed at identifying the nature of gluten toxicity are showing considerable promise. The following are examples of these approaches

a) At Oxford (UK) the approach is to prepare by chromatographic fractionation the smallest fragments of wheat gluten that are toxic to coeliac patients. The rationale behind this is that small fragments are more readily purified to chemical homogeneity and subsequent molecular studies more easily accomplished. Although this approach has difficulties it is yielding small peptide fragments which may provide some insight into the nature of the toxic fraction.

b) At Galway the stability of gluten toxicity to a wide variety of proteolytic enzymes has been interpreted as evidence that the toxicity resides not in a specific region of the primary structure of the gluten molecules but in a non-protein side chain substituent closely associated with the molecules. Because the most common side-chain substituents on proteins are carbohydrates an enzyme mixture was prepared from *Aspergillus niger* which cleaved the bound carboyhdrates (xylose, arabinose, glucose, galactose) from wheat gliadin. When this enzyme-treated gliadin was fed to voluntary coeliac patients no evidence of toxicity was detectable by clinical and laboratory tests. An important aspect of this enzyme treatment was that the gliadin protein molecules were not altered and only the side chain substituents were removed. Encouraging evidence has been obtained to date that the chemical nature of the toxicity may be identified by monitoring alterations in chromatographic profiles of gliadin peptides following treatment with the enzyme mixture which detoxifies the molecule

c) An interesting approach for eliminating gluten toxicity to coeliacs by using genetic variants of 'Chinese Spring' wheat is being studied in California. Nullisomic 6A-tetrasomic 6B Chinese Spring is a genetic variant of normal Chinese Spring wheat in which the 6A chromosome is completely missing but has been compensated for by two extra doses of the 6B chromosome. It is known that the 6A chromosome codes for most of a particular gliadin fraction called α-gliadin is toxic to coeliac patients. When this chromosome is absent α-gliadins are not present but the amounts of other gliadins remain unchanged. Bread made from this wheat has excellent texture and taste. Although results to date are very preliminary and are unpublished the bread made from the genetic variant had decreased toxicity when fed to two coeliac patients. If this finding is repeated it raises some very interesting possibilities for selecting wheat variants that would be without or have reduced toxicity to coeliacs.

POSSIBLE AREAS FOR FUTURE RESEARCH

Progress is being made towards identifying the factor in gluten from wheat and other cereals harmful to coeliac patients Future research should be concerned with

a) continued studies on chemical nature of gluten toxicity

b) when the toxic factor has been identified, different varieties of wheat should be screened for toxicity to coeliacs. This will greatly increase the possibility of producing cereal variants non-toxic or coeliac patients.

c) better techniques should be developed for *in vitro* testing of toxicity with minimum involvement of patient volunteers.

d) screeing tests on the type mentioned in Section 1 (e) for patients who might be at risk for coeliac disease should be established at a European Centre.

REFERENCES

Hekkens, W. Th. M. and Pena, A.S. Eds. 1974. Coeliac Disease (A.S. Stenfert, Publ).

McNicholl, B., Fottrell, P.F. and McCarthy, C.F. Eds. 1978. Perspectives in Coeliac Disease (M.T.P. Lancaster, Publ.). (In Press).

FRUIT AND VEGETABLES PANEL

PARTICIPANTS

Chairman:	W. Spiess *(W.Germany)*	
Rapporteur:	A. Hunter *(Ireland)*	
Invited Experts:	M.J. Woods *(Ireland)*	
	V. Wenner *(Switzerland)*	
	S.D. Holdsworth *(UK)*	
	J. Solms *(Switzerland)*	
	F. Escher *(Switzerland)*	Plenary Author Session 4
	Mrs. J. Ryley *(UK)*	
	K. Gierschner *(W.Germany)*	
Other Participants:	C. Cantarelli *(Italy)*	Plenary Author Session 1
	B. Blanc *(Switzerland)*	Plenary Author Session 4
	J.A. Munoż-Delgado *(Spain)*	Plenary Author Session 5
	R. Zacharias *(W.Germany)*	Plenary Author Session 6
	G. Tomassi *(Italy)*	Nutrition Coordination Group
	S.M. Passmore *(UK)*	} Observers
	M. Sp.Georgakis *(Greece)*	
Panel Coordinator:	D. O'Doherty *(Ireland)*	

INDUSTRIAL R & D PRIORITIES IN FRUIT AND VEGETABLES

M. Woods
Associated Producer Groups Limited, Dublin, Ireland.

INTRODUCTION

With the advance of science and technology and the development of the European Nations mankind in Europe is facing dramatic changes in his environment and style of living. The choice of lifestyle is one which will ultimately emerge from the intermingling of the people of Europe. But once it has been made then this will indicate the likely priority areas in the industrial treatment of fruit and vegetables. There can be little doubt that life in Europe will be increasingly competitive. It will for the average individual involve less physical toil and a reduction in the amount of exercise gained in the course of one's work. The widespread incidence of obesity and of heart attacks especially in men are evidence of the changing trend in our mode of living. Families are not as large as they used to be and it is increasingly commonplace for women to be working. The weaker sectors of the community suffer at least in the short term in the race for higher living and income standards. Consequently, their freedom and ability to eat properly is often impaired. Accordingly those who wish to set priorities for R and D in fruit and vegetables on a European basis must take into account these and other relative factors. The objectives of Food Research for the European Community must include the ultimate provision of nutritious food for all the people of Europe. This will include balanced diets for the rich and poor and the supply of suitable foods for use in the home and in institutions.

When we eat food it is either in the fresh or cooked form. In some cases we may eat processed foods which have not been cooked in a conventional way but processing itself can be considered a form of cooking. Food is cooked either in the home or by caterers in hotels, canteens, airlines or institutions. In some cases the food is cooked by food processors and is then frozen and sold to the purchaser in this form.

Such food is simply reheated at the point of consumption and is then served. Whichever food is presented there is today a recurring need for new forms of processing which meet the needs of the eventual consumer.

RAW MATERIAL

The quality of thermally processed food is only as good as the quality of the raw material. Cooking does not generally improve food quality. It is a common mistake to assume that food of poor quality in the raw state can be improved by thermal processing or cooking. To ensure that the food products supplied to the consumer are of good quality then the processor and caterer must have rigorous quality control procedures to select suitable fruit and vegetable raw material. Processors generally have such control procedures in combination with contract growing. The caterer in contrast does not generally have such control over the supply of raw material and may often be at a disadvantage in this respect. Practical methods of quality control at the catering level would be of considerable benefit to the consumer.

FOOD QUALITY

When food is thermally processed the cooking of the food must be such that the colour, flavour and texture, are preserved to the best possible degree. This may be done by the addition of sauces, gravies, spices or other materials which contribute to the appearance or flavour of a cooked product. The colour, flavour and texture of most fruit and vegetables can be preserve by cooking only to the minimum degree necessary. This applies particularly to fruits and vegetables which rapidly lose quality if overcooked. Research is needed into the optimum temperatures

and quickest methods of cooking foods to preserve their
texture and flavour. In addition the optimum conditions for
the storage of cooked foods are becoming increasingly
important.

Nevertheless, it is the nutritive value of cooked foods
that is one of the most important areas now requiring research.
Fruits and vegetables contain numerous vitamins and minerals
which are very important to our diet and the correct thermal
treatment will go a long way towards preserving high levels of
these important constituents. In general it is desirable to
slightly undercook all green vegetables as this results in
better retention of vitamins and minerals. In certain meats
and dairy products overcooking results in serious protein denaturation. The advent of the deep freeze has brought with it
its own problems and questions for further investigation.
More research is now needed on the nutritive value of cooked
foods which have then been frozen and stored in a deep freeze
for long periods of time prior to reheating and consumption.
It is also a common practice both in the home and in the
catering institution to keep foods warm in ovens before serving.
The loss of nutritive value in such foods which are kept warm
in ovens before serving should also be investigated in the
interests of the consumer. The importance of the dietary
fibre content of fruit and vegetables is receiving increasing
attention in recent times. Little research has been done on
the effect of thermal processing on the properties of these
fibres and investigation of this area could also be very
beneficial to the health and wellbeing of the consumer.

Research is in progress under Dr. Ronan Gormley at
Kinsealy Research Centre in Dublin on the properties of dietary
fibres in foods which have been cooked. There is also
considerable medical interest in the provision of dietary fibre.
It has been shown that dietary fibre has many beneficial
effects on humans eg it acts as a laxative agent by binding
water, it may bind bile acids which can reduce cholesterol; it

also binds minerals and may have application for diabetics. All fruits, vegetables and cereals contain dietary fibre. Little work has so far been done to establish whether cooking has an adverse effect on these beneficial properties of fruit and vegetable fibre.

MICRO-BIOLOGICAL AND SAFETY FACTORS

Since cooking is the final process before the food is eaten it affords an opportunity to ensure that the food is safe from a microbiological point of view. For example, freshly cooked chips should be sterile while in contrast meat which has been grossly undercooked could have a very high content of micro-organisms. In view of the trend towards pre-cooked foods special attention must be given to such pre-cooked foods as these may well become re-contaminated with micro-organisms during the subsequent stages of storage and/or distribution and/or packaging. This same risk of re-contamination would also apply to foods processed by the cook-freeze system as there is an opportunity for contamination between cooking and freezing.

HOME COOK SPECIALISTS

One of the features of our modern society is the availability of cooking advisers to groups of consumers. These advisers can play an important part in the dissemination of information about thermal processing. It is common to have home economics advisers, domestic economy instructresses, home farm advisers and demonstrators in such organisations as the Electricity Supply Board, the Gas Company and the Deep Freeze Suppliers. It is important that existing knowledge and information scattered throughout Europe be co-ordinated and made available to these people in particular and consumers in general to ensure that they have available the most recent information on food preparation, blanching techniques, home freezing in deep freeze units, bottling and a clear understanding of the possible microbiological hazards and nutritive deficiencies likely to be encountered in all aspects of food preparation and cooking.

INSTITUTIONAL FEEDING

This is a very important area and includes school food programmes, catering in hospitals and in many other institutions. The quality and nutritional value of cooked foods needs to be very carefully monitored in these instances because the people in these institutions will normally be wholly dependent on the feeding programme provided. In this connection there is clearly a need for research on short term surveillance systems for monitoring the nutritional status of populations especially those on institutional feeding programmes.

BULK CATERING AND THE COOK-FREEZE SYSTEM

In the case of bulk catering the scale of operation of the thermal processing can in itself cause problems. For example in a system with a large and continuous throughput it is not difficult to find a batch of units such as chickens being either overcooked or undercooked with consequent poor quality and/or health hazards for the consumer. In addition the correct storage of pre-cooked foods can present major problems for bulk caterers. This is why we occasionally hear of outbreaks of food poisoning caused by the contaminated cold meats.

The cook-freeze system has a great potential in bulk catering. However, further research is required on elaborate quality control systems to ensure that the quality of each component in a frozen meal meets an acceptable standard. Each component of a cook-freeze meal must be cooked to the correct degree and portion control must be exact to ensure that each unit meal contains a minimum amount of meat, vegetable and potato etc. Freezing and subsequent packaging must be such that the product is preserved in top quality. The outlets at which the cooked frozen products are consumed will also have to be controlled very carefully to ensure that all personnel

are fully familiar with the reheating treatment required.
Otherwise all the correct practices which have gone before may
be negated at the final stage when reheating of the product
takes place for example in a convection or micro-wave oven.
Date stamping meals prepared by the cook-freeze system will be
essential to ensure that the product is consumed before the
top quality shelf life deterioration point of the most
perishable component in the meal has been reached.

ENERGY REQUIREMENTS OF COOKING

There has already been considerable research and
discussion about the energy requirements of different food
processing methods at industrial level for example comparisons
have been made of the energy content and cost of canning versus
freezing versus dehydrating. In contrast little attention has
been given to the energy requirements of thermally processing
foods in the home or at institutional level. The present system
whereby the food is cooked in half-one pound lots in the home is
very uneconomical from an energy point of view and should be
looked at when the national conservation of energy is being
considered. In contrast if foods are largely cooked at
industrial level this would mean a much lower energy input at
the home and consequently a more efficient overall use of
energy from a national point of view. The possible savings in
such systems should be considered in terms of the national and
European energy cost benefit. It would be particularly important to study the energy requirements of an increasing use of
the cook-freeze system since with this method the foods require
energy when they are cooked, they are then cooled requiring
further energy and frozen which requires yet a further input of
energy. They are then reheated and this process requires a
considerable energy input as the food has to be thawed and then
warmed to the serving temperature. Accordingly the trend
towards an increasing use of the cook-freeze system throughout
Europe could have serious implications in terms of energy
requirements.

SOME EFFECTS OF THERMAL PROCESS CONDITIONS ON FOOD PROTEINS FROM VEGETABLE ORIGIN

V. Wenner

Nestlé Products Technical Assistance Co. Ltd. Research Department, CH-1814 La Tour-de-Peilz, Switzerland.

The heat induced changes on vegetable proteins from pulses have been studied, particularly in soya beans as a food raw material. The heat effects on proteins are numerous, some are beneficial others are undesirable or detrimental in respect to the nutritional value or general quality considerations.

1) **The beneficial effects of heat treatments are:**

  better hygienic quality

  better texture (softening, gelation)

  improved digestibility

  better taste and smell

  improved colour

  destruction of antinutritional factors (trypsin inhibitors)

  elimination of undesirable components by extraction (flatulent sugars) or by evaporation or steam distillation (crude, beany flavours)

  inactivation of enzymes for better keeping quality (hydrolases, oxidases)

2) **The detrimental effects of heat treatments**

can be classified according to Mauron (1974) in five categories:

  First type: Mild heat treatment   - With reducing sugar
  Second type: Severe heat treatment - Without sugar
                                      - with some sugar ⟶ Reducing
                                                        ↘ Non reducing
                                      - With oxidised lipids

Third type: Alkali treatment — Carbonates
 — Ammonia
 — Strong bases
Fourth type: Oxidations — H_2O_2
 — Photoxidation
 — Presence of SO_2

All these types of damages are encountered in the field of vegetable protein processing, generally combined with heat treatments at levels from 50°C (mild extraction in alkaline solution) to 150°C (sterilisation or frying temperatures). Some of the detrimental effects due to heat treatments of protein materials (see Mauron, 1970) are:

The Maillard reaction of proteins with aldehydes (reducing sugars or thermal degradation products of sugars) blocking the biological availability of lysine or inducing its chemical destruction.

The hydrolytic cleavage and racemisation with strong bases.

The oxidation of sulfur amino acids (methionine and cystine) and other amino acids of lesser importance.

The splitting off of sulfur from cystine and the formation of lysine-alanine under normal cooking conditions (Sternberg et al., 1975) or at alkaline pH (de Rham et al., 1976).

Polymerisation with oxidised lipids as described by Karel (1973).

Reduced digestibility if heated at low water activity, due to crosslinkages of terminal amine (lysine) and carboxylic groups (aspartic and glutamic acids).

It has immediately to be said that the Maillard reaction is considered as detrimental in respect to the diminished biological availability of the essential amino acid lysine but on the other hand the Maillard reaction is at a certain level desirable to produce the caramel flavour which is one of the most attractive flavours in sweet foods (see also Mills et al., 1969). This caramel flavour is not equal to the burnt sugar flavour often called 'caramelised' and which is not desirable, as found in sterilised, concentrated fruit juices.

3) The case of the heat induced generation of lysino-alanine (LAL) as potential toxic substance in proteins

The formation of LAL at neutral or alkaline pH is a well known fact (Sternberg et al., 1975). LAL is formed during soya protein processing by the extraction operation to produce isolates or if the protein is dissolved in alkaline solution for the spinning operation (de Rham et al., 1977). The formation of LAL is a consequence of the destruction of cystine/cysteine by heat and/or alkaline pH. The production of LAL is not only undesirable as a potential toxic substance but also because of the decrease of the nutritional value of the proteins, especially in the case of the soya protein where the sulfur amino acids are already deficient.

The results published by de Rham et al. (1977) may be summarised by the following remarks:

At pH 12.5 more than 50% of the cystine/cysteine of soya protein isolates are destroyed after 3 min at $65^{\circ}C$.

At $65^{\circ}C$ the pH of 11 is critical since after 3 min cystine losses are noticeable.

At temperatures higher than $65^{\circ}C$ the critical pH is lowered and the destruction is faster.

Comparatively, a high cystine containing animal protein, like milk whey proteins, reacts the same way as soya protein but if the % destruction is less important the absolute cystine losses are equal or even larger depending on the time-temperature relations.

The LAL found was equivalent to roughly 50% of the cystine destroyed.

4) Conclusions

4.1) As shown by Sterberg et al. (1975) LAL is normally found in protein foods at concentrations from a few ppm to several thousand ppm.

4.2) The more sophisticated analytical methods facilitate the detection of more undesirable substances in foods, generated by heat treatments.

4.3) Traditional heat treatments in the kitchen are generally safer than industrial heat processing conditions provided they are correctly applied.

4.4) However, in the industrial food processes the possibility of repeated and excessive (time and temperature) heat treatments exists generating eventually new and unknown reaction products.

4.5) Toxic substances in foods are inevitably present in variable amounts and legal requirements on toxic substances can not eliminate unavoidable reaction products like LAL. However process conditions generating unreasonable amounts of those reaction products, or certain processing conditions like elevated pH should not be allowed.

5) Proposed COST programme on heat treatments of protein foods with special emphasis on vegetable proteins (soya, cereals, peanuts, rape seed, etc.)

5.1) Elaborate guide lines for generally acceptable heat treatments of protein foods taking into account:

protein concentration (time-temperature and water activity)

pH limits

limits of lysine blockages due to reducing substances

minimum heat treatments for bacteriological quality, together with the minimum water activity levels required

maximum limits for heat induced unavoidable but undesirable reaction products (example lysino-alanine).

5.2) Analytical survey on undesirable heat reaction products in traditional cooked foods and industrially prepared foods, including cooking and frying operations in large kitchens for mass feeding programmes.

5.3) Elaborate guide lines for nutritional testing (animals and human beings) of protein foods with emphasis on vegetable proteins in respect to:

acceptable amounts of unavoidable toxic reaction products (animals)

optimum biological value of vegetable proteins heat treated to destroy antinutritional factors (animals and humans)

biological value of vegetable proteins foods complemented with animal proteins for child feeding (humans).

5.4) Propose new food legislation based either on available data or emanating from the proposed studies.

REFERENCES

De Rham, O. Van de Rovaart, P., Bujard, E., Mottu, F. and Hidalgo, J. 1977. Cereal Chem. 54, 238-245.

Mauron, J. 1970. J. Internat. Vitaminol. 40, 209-227.

Mauron, J. 1974. Proc. IV Int. Congress Food Sci. and Technol. Vol. I, 564-577.

Mills, F.D., Baker, B.G. and Hodge, J.E. 1969. J. Agr. Food Chem. 17, 723-727.

Karel, M. 1973. J. Food Sci. 38, 756-763.

Sternberg, M., Kim, C.Y. and Schwende, F.J. 1975. Science, 190, 992-994.

RESEARCH AND DEVELOPMENT PROBLEMS FOR THE FRUIT AND VEGETABLE PRESERVATION INDUSTRY

S.D. Holdsworth
Campden Food Preservation Research Association, Chipping Campden, Gloucestershire, UK.

INTRODUCTION

The future research and development priorities for the fruit and vegetable preservation industry will be determined by the main requirements for producing high quality products viz. suitable raw materials of the highest possible quality, processes which will maintain the quality attributes of the raw materials and packaging systems which cause least deterioration of the products during storage. The main quality factors which will be considered in this communication are organoleptic properties, microbiological factors and chemical composition including nutrients.

RAW MATERIALS

The importance of carrying out research to ensure that the best possible raw materials are grown cannot be overstressed. There is a constant need to investigate high yielding <u>varieties</u> both for quality attributes and also handling characteristics in relation to ability to withstand mechanical harvesting procedures, as well as processing machinery. This is particularly relevant to fruit processing, eg strawberries and other berry fruits. Many quality factors are directly related to the <u>maturity</u> of the crop and more effort is required to develop methods of <u>maturity measurement</u> in order that raw material quality may be adequately controlled prior to processing. An area of specific interest to raw material production is the fate of <u>pesticide residues</u> (and possibly the degradation products which may be formed during the processing operations especially heating).

The effect of heating, cooling and freezing on the flavour, texture and colour needs far more detailed <u>chemical investigation</u> with regard to the breakdown of the natural constituents and 'processed' flavours. This requires intensive chemical and biochemical studies of the raw materials and products, and changes in the key components occurring during post-harvesting operations and processing. This type of information is most useful when investigating defects accentuated by processing.

Similar types of studies are required in connection with all types of <u>additives</u> or proposed new additives especially sugar substitutes, natural and artificial colouring matters, flavouring substances and inhibitors.

An important area of research, which has been neglected, is the determination of <u>physical properties</u>, including <u>thermo-physical properties</u>, of plant produce before and after processing. There is an increasing need for this type of data especially with regard to maturity and variety for food processing engineers.

POST-HARVEST PREPARATION

Mechanical harvesting techniques are developing rapidly and being used more extensively for crops for processing. Although these may be economically viable, there is no doubt that there is considerable wastage due to damage as well as an overall loss in quality due to bruising accompanied by enzymatic and other biochemical changes. Development of new <u>harvesting techniques</u> to minimise these problems requires the cooperation of engineers and food scientists especially when the crop is to be processed. More research is required into the effect of <u>chilling</u> immediately after harvesting and prior to transport, on the quality of processed materials.

Cleaning and washing techniques require investigation especially with regard to <u>water usage</u> and <u>leaching</u> of soluble components. In cases where the problems are serious research should be carried out on alternative methods of cleaning produce. <u>Hygienic</u> handling of raw materials and <u>sanitation</u> of preparation equipment both require continuous assessment in relation to possible contamination and tainting of the product by sanitising agents used for cleaning equipment or present in the water supply. Alternative sanitising agents should be investigated eg chlorine dioxide, in order to reduce corrosion problems with equipment.

With regard to the progressive movement towards <u>safer equipment</u>, it is essential that research and development work is carried out to ensure that safety requirements are sufficiently simple in design to facilitate <u>cleaning operations</u>.

Much of the existing preparation equipment used in fruit and vegetable processing requires modification in order to prevent wastage and damage of product during <u>transfer</u> from one unit to another.

<u>Blanching</u> of vegetables in relation to quality of canned and frozen products is at present being actively studied in the UK at CFPRA. Detailed studies of enzyme systems and their inactivation by heat are being carried out in order to assess the degree of blanching required for a limited number of products, peas, green beans and Brussels sprouts. This work is now being extended to other products in order to determine the optimum conditions required for thermal treatment in relation to quality of the stored product. There are several other important areas of research in relation to blanching and these include the study of different methods of thermal inactivation, the possibility of chemical methods and the prevention of leaching losses during blanching.

Current techniques of <u>sorting</u> prior to processing require development with respect to removal of defective product, colour defective material, extraneous matter and hygiene.

THERMAL PROCESSING

The major area of research required in <u>thermal processing</u> is the detailed study of the effects of various combinations of time and temperature on the <u>physico-chemical behaviour</u> of the constituents of raw materials. This type of work requires chemical and biochemical analyses before and after processing and during storage at different temperatures.

With regard to both flexible pouches and lacquered containers the <u>transfer of components</u> from the package to the food by diffusion requires consideration in relation, not only to present or proposed legislation, but also in relation to tainting of the product.

No discussion of thermal processing would be complete without referring to the safety aspects of processing. A more critical appraisal of the <u>instrumentation</u> used in sterilising equipment and equipment for carrying out <u>heat penetration</u> is required. Alternative methods of measuring temperatures require careful appraisal especially platinum resistance thermometers, thermocouples and thermistors.

With the development of new packaging systems eg flexible packages, it is necessary to use <u>heat transfer media</u> which are less efficient than steam. The use of air/steam mixtures and pressurised hot water as heat transfer media requires investigation with regard to velocity of flow and geometrical arrangement of packages in order to ensure that a safe process has been used.

Microbiological data is also required on the thermal death characteristics of spores of *Clostridium botulinum* in various vegetable packs. This requires special facilities and expertise and should receive high priority. Similar investigations with other spoilage <u>micro-organisms</u> is also necessary.

The usefulness of <u>mould counts</u> for fruit products in relation to specifications of raw materials such as tomato products could also be evaluated further.

QUICK-FREEZING

The deterioration reactions which proceed during long term storage require more detailed study especially with regard to freezing conditions and changes in the biochemical structure of the products. In view of the progressive expansion of the freezing industry improvements in the quality characteristics of frozen fruit and vegetables will only take place by obtaining more information about the basic processes which accompany freezing eg role of residual fluids in products during and after freezing, nature and shape of ice crystals, rate of growth of ice crystals, concentration of flavour constituents etc.

Control procedures also require development at all stages of the frozen food line especially reliable <u>time-temperature indicators and controllers</u> to prevent abuse of the product.

Adequate <u>microbiological research</u> is also required to ensure that any proposed standards of microbiological quality or codes of practice are realistic both with regard to the suitability of sampling procedures and analytical techniques. The development of the methodology of microbiological assessment of frozen food samples requires considerable research because standards can be so unrealistic.

DEHYDRATION

The main purpose for dehydration research is to develop processes for the production of <u>rapidly reconstituted products.</u> This involves establishing the dehydration behaviour during drying under conditions which will simulate explosion puffing. Processes which might do this involve the short-time exposure of the product to high temperatures. The main objective of these processes in relation to the drying of solid particles is to cause expansion of the porous structure which will then facilitate rapid reconstitution.

In relation to <u>processing and storage</u> changes in the <u>chemical components</u> require to be investigated in detail in order to improve the general quality attributes. In particular methods of preventing non-enzymic browning and other biochemical reactions should be studied. Detailed investigation of the relationship between water adsorption isotherms and storage stability is required. There is also a need for development of <u>analytical techniques</u> for assessing the quality of dehydrated products.

WASTE MATERIAL UTILISATION

A major area for active research is <u>waste material utilisation.</u> Although considerable effort has been put into the disposal of both liquid and solid wastes relatively little effort has been expended on the problems of economic utilisation of waste materials. The problem may be divided into two parts viz. the conversion of waste materials into useful chemicals and the retexturing of low quality raw material for human and animal consumption.

In the fruit and vegetable processing the utilisation of <u>energy</u> must also be an area for considerable research effort. This topic is related to measures which have to be taken to prevent losses within existing processing plant and also to aid the choice and design of equipment for new processes.

COOPERATIVE RESEARCH REQUIREMENTS ON FRUIT AND VEGETABLES

J. Solms and F. Escher,
Department of Food Science, Swiss Federal Institute of Technology, Universitätstr. 2, 8092 Zurich, Switzerland.

A large variety of industrially processed fruits and vegetables are produced throughout Europe. Production and distribution methods vary from one region to another due to different raw materials available, different eating habits and quality judgments. There are, however, few data available to define the relationships between the quality of the processed foods, the raw materials and the various processes, and production and distribution. In general, data are available on raw materials, the 'unit operations' applied in processing, and the properties of finished products. But there is a need for additional cooperative research efforts.

Processing of fruits and vegetables and its impact on quality should be investigated no longer from the point of isolated 'unit operations', but of 'unit processes', including all steps from the raw material to the finished food to be consumed. Investigations of the general relationships between processes and quality changes will be important. For example: If we measure the quality of air-dried vegetables by analysing for nonenzymatic browning and pigment stability, is the drying process really the most important step influencing the quality of the product to be consumed?

Or, what is the exact influence of processing and storage conditions on the quality of frozen vegetables, evaluated by measuring vitamin C content, pigment stability and fat rancidity?

In this connection there is also a need to look for processing methods in which raw materials of lower quality can be utilised, since sufficient raw material of high quality may not always be available, and thus low quality raw material must be used, eg in potato and apple processing. There is certainly also a need to look into better utilisation for human consumption of fruit and vegetable wastes obtained in food processing. As an example: In transferring fresh potatoes into french-fries, only 80 - 85% of the raw material goes into the product and everything else is used as waste for animal feeding.

With increased interest in 'biological' food products combined with 'simple home operated' processing methods, a re-investigation of sophisticated, large scale food operations would be most interesting. This should also comprise economic aspects, especially energy consumption. For example: Simple air drying of carrots, combined with a short salt treatment of the raw material, results in products with rather high stability and quality.

Quality aspects comprise nutritional content, sensory properties and convenience. Data on the content of nutritional components of many foods have to be completed and analytical methods have to be improved. Just one example: Present methods for the determination of vitamin C contents in foods are problematic and have to be revised.

There are practically no criteria on variations in sensory properties and procedures for their determination. For example, how does optimal quality of processed potatoes vary between countries; what are the rheological properties of mashed potatoes produced and consumed in Germany and in Switzerland? What is the difference in composition between an apple juice produced in Holland and in Switzerland? A first step would be an evaluation of the physical and chemical properties of different vegetable and fruit products from different regions. It would then be possible to 'translate' one country's opinion of high quality into that of another country and to establish vocabularies and scales for measuring sensorial qualities and food acceptances in different countries. This would then give a better understanding of regional differences.

RETENTION OF ASCORBIC ACID, THIAMINE AND FOLIC ACID IN VEGETABLES IN MASS FEEDING - THE ROLE OF SYSTEMS AND EQUIPMENT DESIGN

J. Ryley and G. Glew
Catering Research Unit, Procter Department (Food Science), University of Leeds, England.

It is well documented that the main causes of loss of labile nutrients such as ascorbic acid, thiamine and folic acid from vegetables are enzyme activity during preparation, leaching during washing and cooking and prolonged heating during cooking and subsequent storage prior to service. The processes of preparation, cooking and storage between cooking and service are necessary to provide a palatable product and some nutrient loss is inevitable. One experiment in an institution using a cook-freeze process found that retention of ascorbic acid greater than 37% in white cabbage was associated with products judged to be undercooked by a taste panel (Millross et al., 1973). However, in large scale situations, massive losses can occur as a direct result of lack of control in food service systems which are fundamentally scaled-up domestic techniques.

In the past decade considerable advances have been made in systems design, the main feature being the separation of production schedules from consumption times, achieved by utilising a period of frozen, chilled or frozen and thawed storage in place of hot storage, followed by a reheating stage.

The replacement of the conventional cooking process by a two stage 'blanching' and 'reheating process' combined with the contribution made by freezing to texture change in vegetables means that control over time and temperature of heating processes is essential for both high nutrient retention and optimum quality.

'Blanching' prior to freezing is a critical operation for which no specific equipment is available to the institutional caterer. The need for 'blanching' arises when fresh vegetables are used and when it is desired to use frozen vegetables and standardise reheating times for all meal components. The loss of free folic acid from white cabbage during blanching on a domestic scale has been shown to be 40% in boiling water but only 20% in steam blanching (Hill, 1975).

Retention of ascorbic acid, thiamine and folic acid during reheating by convection oven or microwave after chilling, freezing or from the thawed state has been shown to be very high. In a hospital situation a level of 12.5 mg of ascorbic acid/100 g edible portion was achieved when frozen peas were reheated in a convection oven compared with a level of 1.8 mg/100g edible portion when a boiling pan was used (United Leeds Hospitals/University of Leeds, 1970). This increase is attributed to the use of minimum quantities of water and controlled heating times and temperatures. In a laboratory situation, losses of thiamine from frozen peas were of the order of 20% when heating took place in a boiling pan but between 5% and 10% when microwave and convection ovens were used. Losses of free folic acid from white cabbage during reheating in a convection oven after freezing were very low relative to losses during the initial blanching stage (Hill, 1975). However, both convection and microwave ovens require the use of relatively thin layers of food causing handling difficulties and high capital costs.

Food service systems utilising frozen or chilled food replace hot storage by storage at $-18^{o}C$ or between $0-5^{o}C$. The effects of storage at $-18^{o}C$ on nutrient retention have received considerable attention. In the institutional situation, storage at $-18^{o}C$ is likely to be very short because of high costs and is therefore unlikely to be of importance for nutritional quality. However, storage between $0-5^{o}C$ is more likely to be a cause of nutrient loss and has received less attention. Although there are no regulations or guidelines in the UK governing the duration of storage between $0-5^{o}C$, the French regulations state that the period must not exceed 6 days

(J. Official de la Republique Francaise, 1974). However, our investigations into the loss of ascorbic acid in pre-cooked cauliflower, peas and white cabbage show significant losses during the first 24 h (15-25%, 10-20% and 25-40% respectively). Losses of thiamine from peas were between 5 and 10% (Daly, 1977).

Thus, recent developments in systems design facilitate the elimination of those areas of traditional vegetable processing which result in low vitamin retention, but require a greater sophistication of equipment design. Storage between $0-5^{\circ}C$ prior to reheating may result in greater nutrient loss than reheating from the frozen state.

REFERENCES

Daly, L. 1977. M. Sc. Thesis. University of Leeds.

Hill, M.A. 1975. Personal communication.

Journal Official de la Republique Francaise. Plats Cuisines. a l'Avance. July 16, 1974.

Millross, J., Speht, A., Holdsworth, K., Glew, G. 1973. The utilisation of the cook-freeze process for school meals. University of Leeds. 145-149.

University of Leeds and United Leeds Hospitals. 1970. An experiment in hospital catering using the cook-freeze system.

THE QUALITY OF FRUIT AND VEGETABLE JUICES. INFLUENCE OF PROCESSING METHODS ON FINAL PRODUCT QUALITY

K. Gierschner
Universität Hohenheim (LH) Institut für Spezielle,
Lebensmitteltechnologie, Stuttgart, W. Germany.

Manufacture of liquid products from fruit and vegetables requires the size reduction of the raw materials and destruction of the cells and membranes. This enables enzymes and substrates to react. Other substances, however, can also enter into reactions. Thus oxygen either from the intercellular material or from outside plays an important role. When producing stable clear or turbid products, many factors exercise an influence on the changes which occur in the constituent substances of the original raw material. Conditions of storage, filling, preservation and despatch all affect the product.

Of particular technological significance are:

1) Changes brought about by enzymes inherent in the raw material. (Enzymes include oxidases, hydrolases, lyases and transferases. The production of secondary aromas and flavours from unsaturated fatty acids of the cell membranes and from carotenoids is an example of an enzyme reaction. Another example is the liberation of volatile nitrogen or sulphur containing precursors.

2) Changes brought about by other raw materials
Especially important are the proton catalysed reactions such as the Maillard reaction; the production of bitter limonidilactons and the apparent loss of acids through lactonisation.

3) Changes brought about by Micro-organisms
For example, the formation of volatile acids, diacetyl and acetoin.

4) Changes brought about by added enzymes.
(Example pectin enzymes)

5) <u>Changes brought about by other added materials or impurities</u>

(Example; reactions resulting from sulphiting or heavy metals).

6) <u>Changes brought about by the Heat Process necessary for Preservation</u>.

The aim should be to maximise the desirable reactions and minimise undesirable reactions.

FISH PANEL

PARTICIPANTS

Chairman:	A. Hansen *(Norway)*	
Rapporteur:	J. Somers *(Ireland)*	
Invited Experts:	G. Londahl *(Sweden)*	
	F. Bramsnaes *(Denmark)*	
	T. Strom *(Norway)*	
	G. Varela *(Spain)*	
	J.C. O'Connor *(Ireland)*	
	W. Vyncke *(Belgium)*	Plenary Author Session 5
Other Participants:	J. Connell *(UK)*	Plenary Author Session 1
	D.H. Buss *(UK)*	Nutrition Coordination Group
	C.J. McGrath *(Ireland)*	Observer
Panel Coordinator	K. Partridge *(Ireland)*	

SOME PROBLEM AREAS IN FREEZING AND FROZEN STORAGE OF FISH

G. Löndahl
Frigoscandia AB, Helsingborg, Sweden

Fish and other seafoood products have been of great importance for a number of countries for years. Today fisheries are of increasing importance in our efforts to feed the growing population of the world. The annual catches during the last five years have been between 65 and 70 million tons. According to FAO estimates the catch could yield some 130 million tons by year 2 000. Today some 25% of the world fish catch is frozen and the utilisation by freezing techniques in fishing is expanding fast and has trebled since 1960.

Along with the commercial success a number of quality problems have arisen, mainly due to changes during storage and distribution; though in some cases caused by inadequate processing techniques. The mechanism of the changes is in most cases known and the problem areas have been and are tackled to improve the overall quality. The quality of frozen seafood is, as for any other frozen food, determined by a number of factors like:

- The quality of the raw material
- Handling and preparation prior to freezing
- Freezing methods
- Packaging
- Storage
- Defrosting methods
- Cooking methods.

Intensive research related to those factors has been carried out in recent years and the understanding of changes during the handling from catch to consumption is fairly good as long as we deal with traditional or conventional species and processing methods. Even if more knowledge is needed, scientists the world over are aware of what is missing and working on those problems. In the following I will therefore limit my views to some likely problem areas related to one major change and one major goal in the modern fishing industry.

FISHING LIMITS

The new 200 mile fishing limit does not only cause political discussions but also technological problems. Some traditional fishing waters, in which a country has caught traditional species as the base for the fishing industry, will be closed. On the other hand some countries will be facing large resources of fish which they normally have not, or to a limited extent, processed.

The potential catches are sometimes very big, eg Baltic herring, and processing will not be possible directly following the landings. Large quantities will therefore be frozen pending further processing. The freezing method and measures during storage are of course related to the final product frozen, canned or preserved by other means. The measures to be taken are not necessarily the ones, which are normally taken, as the 'new' resource may have other characteristics as compared with species normally used in the traditional way.

It is also likely that the quantities of frozen raw material for later processing by the European fish food industry will increase further. One of the technical bottle-necks in this processing is the thawing operation. Sometimes the fish is thawed and refrozen twice before the product reaches the consumer. The quality from a sensory, nutritional as well as from a technological point of view is influenced by the thawing method as well as by the handling and refreezing methods. The latter are, however, in most cases streamlined fulfilling most requirements, whereas the thawing normally is done in a less controlled way.

Thawing is often done in the same way for any type of fish, while it is probable that each species requires eg a special temperature programme. It is recognised that only 25% frozen haddock can be used in canned fishballs without loss of texture, to take an example. By changing the time temperature programme during thawing by forced air it was possible to use 100% raw material without changing the significant texture of the final product. Similar experience has been obtained in lab. -scale tests on plaice frozen in blocks, thawed for filleting and refrozen.

Currently, thawing procedures are significantly influenced by the fish species and the nature of the processing and are likely to become even more so in the future.

IMPROVED UTILISATION

Improved utilisation of the first catches and of the resources in total for direct human consumption is the main goal in the world fisheries. This means that we will be dealing with unconventional species, or at least species, which normally are utilised for reduction into fish meal only. One of the main problem areas is that some of these species deteriorate very rapidly after catch, eg Blue Whiting and Krill. It is obvious that this calls for a very fast processing or freezing. As the catches as a rule are very big, the necessary freezing installation on board the ships has to be large or research has to be carried out to find ways of handling and freezing ashore still maintaining the quality of the product. In the case of Blue Whiting the thawing of frozen round fish prior to filleting is likely to become a problem.

Many of the new species are very small and filleting is not possible. Mincing by using deboning machines can be an answer to an optimum utilisation of these species. Deboning equipment is also a way to increase the yield from traditional species.

Minced fish either from whole fish or from waste from the filleting operation of traditional species has a very limited storage life in the frozen state. For example it may be mentioned that only small parts of liver in the minced product will decrease the possible storage life to a few weeks at $-25^{\circ}C$. Further studies on the deterioration mechanism and alternative measures are necessary before the benefit of deboning and mincing can be fully utilised.

SUMMARY

The aim of this communication has been to indicate new or known problem areas which are likely to be accentuated when trying to improve the utilisation of conventional as well as unconventional fishes. The new fishing limit will in some instances cause a change in the raw material used for further processing. This change may well call for changed processing, freezing as well as thawing techniques. Thawing is the main problem area in the industry of today.

TWO FROZEN FISH PRODUCTS REQUIRING SPECIAL CARE

F. Bramsnaes

Food Technology Laboratory, Lyngby, Denmark.

The aim of the present brief communication is to draw attention to two fish products, frozen peeled shrimp and frozen trout, retail-packed, which sometimes are offered to the consumer in an unsatisfactory quality condition.

The quality draw backs are dehydration, loss of colour and loss of flavour and now and then off-flavours occur because of fat changes.

Both items are sold at a fairly high price level. It is very likely, therefore, that the consumer uses these products for special occasions and consequently may be looking for products of prime quality.

From the available literature one would deem it possible for the packer to get acquainted at least with the greater part of the technological advice necessary for maintaining a good quality of these products up to the time of final sale. Some earlier experiments in this field are outlined below.

There may, however, still be research needed. In addition it may well be that the scientific and technological information available is not widespread and sufficiently known or forgotten or perhaps the technical procedures which can be deduced from the available information are not adaptable to or practical for use in commercial operations.

SHRIMP

Canning of shrimps had in many years been the predominating method of preservation when freezing of shrimp was taken up after the last world war. The art of canning had been brought to a stage where good quality products were and are still made. There were several problems earlier. One part of the canning process which had to be developed was the packaging side. Special care in constructing the can and in using it turned out to be essential to good quality of the canned shrimps.

When freezing of shrimp started it soon became obvious that the quality of frozen peeled shrimp was difficult to assure during cold storage even at low temperatures. The same was true with frozen lobster or crab meat. Experiments with various treatments like glazing, antioxidants, etc failed to achieve essential progress in this respect, and it was not until removal of air was tried (Anon, 1955) that the thing began to look better. Later it was shown with lobster meat, that the yellow discolouration arising in frozen storage was due to oxidation of the red astacene colour to a yellow pigment by fat peroxides (Bligh et al., 1957).

When vacuum-packing in the late fifties became a possibility in the food industry it was natural to start using this technique commercially. The differences in storage life between loose-packed and vacuum-packed peeled frozen shrimps can be illustrated with the results from a storage test made where the amount of quality deterioration which was easily detectable after 2 months storage at $-20^{\circ}C$ in the loose-packed product had not appeared after 9 months in the vacuum-packed shrimps.

During the years following the start of vacuum-packing research work was mostly concerned with other parts of the process ie the handling and processing before freezing and the cold storage of whole shrimps (Aagaard, 1965 and Anon, Torry Advisory Note No. 54). Among other things it was shown that when fresh shrimps are bulk frozen well glazed and cold stored at about $-25^{\circ}C$ this raw material can be stored for 6 months and after thawing and processing can lead to fully satisfactory frozen peeled shrimps. Another important step was the introduction of pasteurisation of shrimp packages before freezing in order to kill contaminating organisms introduced during the hand peeling of shrimps.

About ten years ago new freezing techniques and among them the so called IQF-freezing began to spread. The advantages by using such methods are a free flowing product and a fast freezing operation. As an example, IQF meats of North Atlantic deep water shrimp require a freezing time of about 10 minutes in an air blast freezer at $-30^{\circ}C$.

Individual quick frozen peeled shrimps for sale to caterers and retailers are normally packed loose in plastic bags which are sealed and packed in outer cartons for storage and distribution. In the retail freezer cabinets the shrimps are displayed in these plastic bags. Some packers glaze the shrimps to provide for better protection against dehydration and rancidity. However, glazing has various disadvantages and it is doubtful whether these are compensated for by the advantages gained (Anon., Torry Advisory Note No. 54).

Danish investigations were recently carried out on the methods of packing for IQF shrimps. The shrimps were caught in Greenland, machine-peeled, pasteurised briefly at $85^{\circ}C$ and immediately cooled in brine to get 1.5% salt in the meat, flow frozen at -25 to $-30^{\circ}C$, glazed and packed in 1 lb bags, partly loose and partly vacuum-packed. At $-20^{\circ}C$ the HQL[+] time was found to be 2 months for the loose-packed shrimps and 4 months for the vacuum-packed product. The acceptable storage life was 3 - 4 months and 7 months respectively at $-20^{\circ}C$. At $-30^{\circ}C$ the HQL was 4 months and the acceptable storage life 6 months for the loosely packed products. With vacuum-packing at $-30^{\circ}C$ the shrimps were still fully acceptable after 12 months.

[+] High quality life.

In the shrimp market to-day some production series are vacuum-packed but quite a large proportion is packed free flowing, in order to give convenience to the consumer. It appears therefore that from a quality stand point we are partly back to the earlier times of poor storage life. Furthermore fairly heavy salting seems to have crept in in some productions. It is well known from several earlier investigations that salt may enhance toughness and flavour changes of frozen shrimp.

In conclusion, there seems to be a need for further research in this field.

TROUT

Many years ago trout raised in pond farms always used to be frozen whole, that is not-eviscerated, in blocks or as single fish, heavily glazed and packed in large cartons or wooden boxes. The fish would not be fed during the last week before catching in order to improve the storage life in frozen as well as in iced condition. In the frozen form this product has a satisfactory storage life at low storage temperatures, and is suitable for hotels, institutions, etc.

However, as time progressed, the market increasingly requires consumer packs and the bulk of the production to-day consists of such packs. A common type is two eviscerated fish packed in a carton, each fish in a small plastic pouch. The eviscerating of the fish results in the belly cavity becoming exposed to air and the fish which is normally frozen in the cartons in a plate freezer cannot easily be glazed. A much shorter storage life than for the un-eviscerated glazed fish is therefore to be expected. This has also been the general experience in the trade.

Being aware that the storage life of this product in some commercial applications was too short experiments were carried out with antioxidant treatment of the fish before freezing. It was found that ascorbic acid dips gave a reasonably good improvement and as this antioxidant dip is easily accepted by the authorities besides being easy to apply, this procedure was recommended to the Danish packers. From storage life tests it was found that at $-10^\circ C$ the HQL was extended from about 2 months to about 5 months by using a 0.5% ascorbic acid dip (Bramsnaes et al., 1960). Later vacuum-packing was investigated and this in fact proved to be by far the best method tried. While the control packs passed the HQL stage at $-20^\circ C$ after about 2 months, the vacuum-packed product retained its high quality after 9 months at $-20^\circ C$. Unfortunately samples were only stored at one temperature. There is a need for repeating these experiments in order to get TTT figures.

Vaccum-packing, however, is only used to a small extent in commercial practice and the same is the case with ascorbic acid treatment. Thus, there seems to be room for improvement. From later experiments with ice packing of fresh trout it was found that there is a variation, so far unexplained, in the resistance to rancidity and similar fat changes in different fish lots. This field also requires further research (Hansen, 1971).

REFERENCES

Anon. 1955. Technology Laboratory of the Ministry of Fisheries, Copenhagen, Annual Report for 1954. p. 27.

Bligh, E.G. et al. 1957. J. Fish. Res. Bd. Canada $\underline{14}$, 637.

Aagaard, J. 1965. Konserves og Dybfrost (Copenhagen) No.4.

Anon. Torry Advisory Note No. 54. Torry Research Station, Aberdeen.

Bramsnaes, F. et al. 1960 Proceedings of the Xth International Congress of Refrigeration, Vol. 3, page 247.

Bramsnaes, F. and Sørensen, H.C. 1960. Annexe 1960-3 au Bulletin de l'Institut International du Froid, p. 281.

Hansen, P. 1971. Internal report, Technology Laboratory of the Ministry of Fisheries, Copenhagen.

PROCESSING OF UNDER- AND NON-UTILISED FISH SPECIES FOR HUMAN
CONSUMPTION: NEEDS FOR R AND D.

Terje Strøm
Institute of Fishery Technology Research, Tromsø, Norway

In recent years the fish technologist throughout the world
have been paying more attention to the problems of processing
under- and non-utilised fish species for human consumption,
as seen in Europe in the research projects on Blue Whiting and
Capelin. A number of different products may be produced
(traditional and non-traditional) like 1) fresh, frozen, 2)
salted, pickled, 3) dried, smoked, 4) sausages, hamburgers, and
5) fish protein concentrate (FPC) with functional properties.
The success of products like these will depend on the quality and
the nutritional value of the products, as well as the efforts
used in marketing the products.

NUTRITIONAL ASPECTS

Fish is traditionally regarded as an excellent food, being
a source of top quality proteins and containing important
nutrients like minerals, vitamins and fat. An increasing demand
for new protein sources and emphasis on the utilisation of proteins
as a source of essential amino acids and not as energy in the
diet demands more detailed knowledge of protein and protein
quality, and the factors affecting this. A better understanding
of the protein requirements of humans is needed, and existing
biochemical and biological methods for assessing the protein
quality must be improved.

In fish processing the chemical reactions affecting the
protein quality must be characterised, and of special importance
are: 1) reaction between formaldehyde and L-lysine (and other
amino acids) following the possible formation of formaldehyde
during frozen storage, 2) reaction between amino acids and
products from lipid oxidation (intermediates and end products)
formed during storage and processing of fish, 3) loss of amino

acids and possible formation of harmful compounds (eg L-lysinoalinine) caused by alkali and heat treatment in eg extrusion, production of FPC, 4) loss of amino acids and reduced nutritional value in processes of fat-removal and dehydration, eg production of FPC type A and B. These reactions must be studied under conditions similar to those applied during processing and cooking, frying at home and the final product.

Factors affecting the digestibility of the different proteins are important. Very little is known of how the processing conditions will affect the nutritional value of sarcoplasmic protein, and connective tissue, eg denaturation during frozen storage and heating.

Utilisation of controlled enzymatic hydrolysis of the proteins may in this respect improve the nutritional value of the fish.

Fish oil is an important nutrient, and the production of fish oil an important process. Factors affecting the nutritional quality of fish oil must be better understood eg seasonal changes in fatty acid composition, reactions of fatty acids during storage and processing, importance of controlled hydrogenation in order to reduce the content of unwanted fatty acids.

During storage of fish postmortem, changes take place as a result of microbial activity. Harmful results of the bacterial growth may be: 1) loss of nutrients, eg essential amino acids, 2) production of toxic compounds (eg botulin) 3) formation of compounds (eg SH-compounds) that during the following production will react with amino acids (eg L-lysine) and thereby reduce the nutritional value of the fish.

A few different analytical methods are used to-day to assess the overall quality of the fish. With emphasis on more specific nutritional changes that may take place during storage and processing of fish the validity of these methods must be investigated, and new methods may have to be developed.

FUNCTIONAL PROPERTIES

In the food industry the problems of functional properties like water binding capacity, emulsifying properties, flavour and off-flavour of the different ingredients used, and possible reactions in the final product are important. The effect of these ingredients on the final product must be evaluated for each individual product. Some problems of relevance to the fish industry may be listed: 1) denaturation of fish proteins during frozen storage and heating/drying will result in products with poor texture quality and less water-binding capacity. 2) The properties of the different fish proteins and the native fish muscle itself in processed food like sausages are poorly understood. 3) Preparation of fish proteins with good functional properties is possible. However, with fatty fish the problems of fat removal (eg using organic solvents, water, pre-enzymatic treatment), protein solubilisation (eg alkali, enzymes), dehydration (eg complete or partially dehydrated), and stabilisation of partially dehydrated products, are not too clearly understood. 4) Enzymatic hydrolysis of proteins may produce peptides with a bitter flavour. 5) Lipid oxidation may cause off-flavour by the formation of aldehydes and ketones, and by reaction between lipid oxidation products (end-products and intermediates) and other components like proteins and peptides. These reactions are in many cases only poorly understood, and well-known chemical analyses are not always as sensitive as required to detect changes.

Knowledge of the underlying causes of the problems listed above will be of importance to the fish technologists, in order to develop methods to minimise the effect of these reactions.

CONCLUDING REMARKS

The problem and challenge to the fish technologist, now and in the future, will be to increase the total food output from the sea and to utilise the fish in the best possible way. The utilisation of non- and under-utilised species like small pelagic fish demonstrates this problem and challenge clearly. The underlying problems for this utilisation are facing the technologist internationally, and a joint international research effort to solve these fundamental problems is called for.

AREA OF R AND D

1) Nutrition

- Evaluate protein requirements of humans, and improve existing biochemical and biological methods for assessing protein quality.
- Protein digestion in vivo of different fish proteins.
- Nutritional value of fish oil.

2) Protein modifications

- Reactions with formaldehyde, alkali, products from lipid oxidation, heat treatment; changes in nutritional and functional properties.
- Protein denaturation; changes in nutritional and functional properties.
- Application of enzymes to modify proteins; changes in nutritional and functional properties.

3) Research yet to be identified.

NUTRITIONAL ASPECTS OF FROZEN FISH

G. Varela, Olga Moreiras-Varela, and M. de la Higuera.
Instituto de Nutricion, Consejo Superior
de Investigaciones Cientificas, Facultad de Farmacia,
Ciudad Universitaria, Madrid-3, Spain.

Spain is one of the countries with the highest level of fish consumption, especially certain types such as hake (*Merlucius merlucius*), which is very popular. Consequently, the consumption of this food group has increased, and we find that the present price of some kinds of fish, such as hake, exceeds that of the best meat. This may seem strange, given the importance of our fishing industry and unsuitability of this country for the production of high priced meats such as beef.

These considerations also justify the extraordinary growth of our frozen fish industry which in a very short time, as is known, has made Spain one of the leaders in this field. Naturally, the appearance of a new technology of processing such a widely consumed food presented several problems, problems which our laboratory helped to solve. We believe our experience may be of interest to other areas where frozen fish is beginning to be consumed or where there are great possibilities of making it a basic source of protein in the diet.

Our experience will be set forth in the following three points:

a) The influence of the freezing process on the nutritional value of fish

For our study we have compared the hake, a high priced fish, and the drum (*Sciaena regia* Assol), of similar nutritional qualities and having great potential for preservation by freezing since it is a seasonal catch.

The quantification of the nutritional quality of these two species, fresh and frozen, was studied in rats according to the Thomas-Mitchell technique. The results are shown in

Table 1.

TABLE 1
INFLUENCE OF FREEZING ON NUTRITIONAL PROTEIN UTILISATION (%)

	ADC	TDC	BV	NPU
Fresh hake	87.7 ± 0.5	97.7 ± 0.3	88.8 ± 0.3	86.7
Frozen hake	90.1 ± 0.9	98.3 ± 0.2	88.3 ± 0.2	87.0
Fresh drum	84.6 ± 1.4	96.5 ± 0.8	87.9 ± 0.4	84.8
Frozen drum	86.4 ± 0.9	97.4 ± 1.1	85.4 ± 0.9	83.2

The percentage of protein absorbed expressed as an apparent digestibility coefficient (ADF) or true digestibility coefficent (TDF), shows the excellent quality of the protein of both kinds of fish, either fresh or frozen, which indicates that the technological processes of freezing do not affect the digestion and absorption of protein.

The quality of this absorbed protein will be reflected by the degree to which it is retained or eliminated. The greater the quality of this protein, the better it will be retained by the organism. This capacity for retention is given as the biological value (BV). From the Table we can conclude that the utilisation of the proteins tested was not significantly affected by freezing, and even though the values for drum are somewhat less, the difference is not statistically significant Similar percentages for the biological values of these and other fresh marine species have also been published (Pujol et al., 1958; Varela et al., 1977).

b) Nutritional bases for the evaluation of a protein containing food

By net protein utilisation we mean the percentage of that protein ingested by the animal which is utilised, and we believe this should be the basis for economic evaluation of a food product, as we have indicated above. The NPU's obtained (Table 1) are similar to the BV's and also in this case there are no significant differences either between fresh and frozen or between the kinds of fish.

A nutritionist is immediately tempted to make an economic evaluation of a food rich in protein, in accordance with the objectives he seeks. That is, a food of this sort should be valued according to the grams of protein really utilised.

If the value of a food were a function of its nutritional utilisation (Varela et al., 1957), on acquiring a gram of protein we should know how much of it the organism really utilises.

In this regard, we consider net protein utilisation of fresh and frozen fish and their muscular protein content and wastes. We can observe in Table 2 that, according to the market prices for fresh and frozen hake, with the price paid for 1 kg of hake we are buying only 79.2 grams. This would suppose a cost of 4 pesetas per gram of protein utilised.

Given that the nutritional utilisation of the protein is approximately the same in fresh and frozen fish and taking into account the differences in price, one can conclude that the nutritional return for the same price is much greater in the case of frozen fish, since for the same amount of money, one can buy three times more utilisable protein in frozen than in fresh fish.

c) Problems of consumer acceptance of frozen fish

Man does not consume foods for nourishment alone. Eating is a right and a pleasure and, even when a food has an excellent nutritional value and a very good economic value in relation to its nutritional quality, as is the case with frozen fish, it is the consumers who in the end determine whether a new food, or a food treated with a new technology, will be successful on the market. When we say this, we must not forget that in the process of accepting a new food or technology there are two different stages: first, the encounter with the new situation and second, the process of accepting or rejecting it.

TABLE 2

COMPOSITION AND COST OF FRESH AND FROZEN HAKE

	Muscular protein content (%)	Waste (%)	Price/kg (pesetas)	Price/g proteins utilised
Fresh	16.5	45	500	4.0
Frozen	17.8	45	170	1.2

Even when the nutritional and economic value of frozen fish is excellent its appearance as a new food or technology may make it difficult for consumers to accept.

In this way we made a study, in 1967, sponsored by the OECD and the University of Granada, about the attitudes and behaviour of the Andalusian population with respect to different problems regarding foods (Varela, 1969). That survey was made among two thousand families.

According to the results (Table 3) approximately half of the population eat frozen fish while half does not. It is curious that in spite of its being a food of excellent nutritional quality and having a rather low price, the upper class uses it in greater proportion than the middle and working-class population

TABLE 3

SURVEY REGARDING CONSUMPTION OF FROZEN FISH. QUESTION: DO YOU OFTEN EAT FROZEN FISH IN YOUR HOUSEHOLD? RESULTS IN % OF TOTAL POPULATION OF SURVEY

	Yes	No	No answer	Totals
All Andalusia	47.2	50.7	2.2	100 (2042)
Upper class	69.0	31.0	-	100 (29)
Middle class	56.1	41.9	2.0	100 (754)
Working class	41.2	56.4	2.3	100 (1229)
No answer	42.8	57.1	-	100 (21)

The main reason for the consumption of frozen fish (Table 4) lies in its price, a fact which is reflected in the stratification by social classes. Next in importance, is the lack of fresh fish in local markets. Only 10% believe it is more nutritious. The last two reasons seem to have no relation to social class.

TABLE 4

REASON FOR THE CONSUMPTION OF FROZEN FISH. RESULTS EXPRESSED AS % OF TOTAL AFFIRMATIVE ANSWERS.

	Price	Fresh unavailable	More nutritious	Other	No. ans.	Totals
All Andalusia	46.7	20.8	10.6	18.3	3.6	100 (1003)
Upper class	28.6	23.8	9.5	33.3	4.8	100 (21)
Middle class	44.4	19.9	12.7	19.0	4.0	100 (426)
Working class	49.3	21.2	8.7	17.3	3.3	100 (537)
No answer	44.4	33.3	11.1	11.1	-	100 (9)

The principal reason why half of the population does not eat frozen fish (Table 5) is taste (48.8%). On first impression, it would seem understandable that those members of the upper class who do not like it do not buy it, and that those of the lower class would find this sufficient reason not to buy it. However, we should also take into account that this could be simply an attitude of reservation regarding any change in the presentation of food products. It is important to mention that the lower percentages of families who do not buy frozen fish because there isn't any or because they believe fresh fish is more nutritional show no variation among the different social classes studied. However, it is not satisfactory that in all Andalusia, 15% of the families do not consume frozen fish because it is unavailable. Also, it is interesting to note the high percentage of the population which gives reasons other than those mentioned.

From these studies it can be concluded that the consumption of frozen fish is largely a function of its price.

TABLE 5

REASONS FOR NOT EATING FROZEN FISH, GIVEN AS % OF TOTAL NEGATIVE ANSWERS.

	Unavailable	Don't like	Less nutritious	Other	No. ans.	Totals
All Andalusia	15.0	48.8	7.6	24.3	4.2	100 (1039)
Upper class	22.2	66.6	-	-	11.1	100 (9)
Middle class	12.8	53.6	9.1	20.4	4.0	100 (319)
Working class	16.3	47.7	7.0	26.9	2.1	100 (673)
No answer	7.1	35.7	14.3	42.8	-	100 (14)

The most frequent reason for its non-consumption seems to be its taste, aside from the influence of the novelty of its appearance on the market. What is needed here is not a comparison of the flavours of fresh and frozen fish, since it is evident that freezing influences the flavour, but rather a different culinary preparation of each.

In summary, we should remember here that these comments refer to the period when this study was done. Consequently, we believe it would be valuable to undertake a new study similar to the one made ten years ago. We believe it would show that in this lapse of time, frozen fish has come to occupy an important place in the Spanish diet, both nutritionally and economically.

Although it has not been included in this paper it would be very interesting to study how freezing and thawing processes affect nutritive value and acceptance of other fish. This research line should be completed by trying to find which culinary process is more appropriate for frozen fish.

REFERENCES

Pujol, Amparo and Varela, G. 1958. Valor biológico de la proteina de algunos pescados de consumo en Espana. Anal. Bromatol. X. 437-478.

Varela, G. 1969. Actitudes y comportamiento de la población andaluza ante algunos problemas alimentarios. An. Sociol. 4-5: 3.

Varela, G., Pujol, Amparo and Moreiras-Varela, Olga. 1957. Bases para una posible valoración económica del pescado congelado en función del valor nutritivo del mismo. II Asamblea General de Frio. Vigo.

Varela, G., Moreiras-Varela, Olga and de la Higuera, M. 1977. Valeur nutritive et problematique de la consommation de poisson congele. Cahiers de Ceneca. 5217. Paris.

PROBLEMS ASSOCIATED WITH THERMAL PROCESSING OF FISHERY PRODUCTS

J.C. O'Connor

Food Technology Department, Institute for Industrial Research and Standards, Dublin 9, Ireland.

INTRODUCTION

Research cannot be conducted in isolation and it is becoming more and more important that it reflects the needs of the industry it serves. The fish processing industry is now facing severe constraints and it seems inevitable that the needs of the industry in terms of R and D will be governed by increasing pressure on availability of supply. This pressure will continue and there will be a greater emphasis on the development of new products and on the re-education of the consumer with regard to utilisation of more readily available species, as in the case of Blue Whiting as a replacement for cod in the UK and elsewhere. However, it may not be possible to convince the consumer to accept less well known species on the fresh market and the prudent approach would undoubtedly be to work on the development of new products in the formulated field. R and D activity aimed at utilising these raw materials as ingredients in processed products therefore will be vital.

There is an increasing interest in canned fishery products at present both from the point of view of increasing the value of certain species and also in taking up spare canning capacity which is available as herring supply dwindles and more and more conservation measures are implemented.

This communication will consider only one area of thermal processing, viz. heat sterilisation problems.

RAW MATERIALS

It is important that research is carried out to ensure that suitable raw materials can be selected and that procedures for quality control at the processing plants can be more adequately established. Work is needed to reveal more precise information on the economics of some processes and to help optimise yields from valuable and increasingly scarce resources. Indeed, the economic basis for some operations can only be termed precarious.

In particular, the microbiological condition of raw materials needs examination. The occurrence of *Clostridium botulinum* in fish is well documented but more information is required. In what species is this organism most frequently found and in what geographical locations? A co-ordinated review of the distribution, types and heat resistance characteristics would be valuable. Indeed, work is needed in the whole area of microbiology of the fish processing plants. The increased use of thawing equipment followed by handling the fish requires investigation.

As newer species are introduced to the industry further questions are raised. How do the handling and pre-processing methods affect the final product quality?

PRE-PROCESS PROBLEMS

The operations prior to the heat sterilisation itself require investigation in a number of areas. Perhaps the most significant problem in heat sterilisation of fishery products is the deterioration in quality which occurs during the period between sealing the containers and the actual heat sterilisation itself. Even under the most ideal conditions some delay is inevitable and under normal commercial practice delays can be severe. Quite important changes in colour, flavour and odour are noticeable and product quality is measurably altered. In my view not sufficient attention has been paid to these problems. While some of these changes are well understood, eg microbial and enzymatic activity, non enzymatic browning (Maillard reaction) etc detailed studies of these products are rare. Yet the effect of these changes on the quality and nutritional value of the final product may be severe.

In the absence of any co-ordinated microbiological studies it is surely a disturbing situation. I submit that these phenomena require quite intensive biochemical and microbiological studies, which would be of immense value to the technologists working in the industry and not least to the workers responsible for the investigation of faults or defects in the process.

The effects of pre-process handling and in particular machinery on the yield is an area for investigation. While there may be some advantages in using high speed machinery perhaps the losses through wastage are too severe?

THERMAL PROCESS

I would suggest that more work is required in this area than in any other. Quite apart from the need for a study of the effect of various heat treatments on the nutritional and quality attributes of the products the most significant gap in research concerns the actual thermal process and in particular the parameters on which it is based. The safety aspect of any thermal process is obviously the most important factor; yet it seems that little co-ordinated work is being carried out in this area. If one examines the practice in various countries one is amazed by the different Fo values suggested for particular products. For example a 500 g can of mackerel in tomato sauce receives Fo 6.0 in one plant and Fo 12.0 in another. While there may be good reasons for these differences no authority appears willing to give precise information which will assist the processor. An urgent need exists for a co-ordinated survey which would establish the existing practice in the various industries for selected products. This could form the basis of a future study which could define the parameters and enable the industry to operate in confidence. Minimum Fo values could be established and standards would then be more easily applied. Undoubtedly this would be a large project but it should not be put off because of its complexity or size.

A related area of study concerns equipment used in the industry. Instruments used for retorts vary considerably in type and quality and, in particular, the equipment used in the heat penetration studies for process determination could be subjected to scrutiny. Many homemade thermocouples are in use and some are of doubtful value. The development of a 12 point or 20 point slip ring contact unit for rotary retorts would be a welcome advance.

Cannery retorts also vary considerably in design and construction and one could be forgiven for assuming that some are manufactured by engineers who are unaware of the function of a retort. Heat distribution characteristics in retorts are but vaguely known and work could usefully be undertaken in this area also. Indeed, a comparative study of retort characteristics using steam, steam/water and steam/air heating media would be most interesting. Having regard to the current high cost of energy, the heat losses from retorts become a significant factor in any energy balance studies in fish canneries.

RESEARCH PROPOSALS ON FISH TECHNOLOGY
(TECHNOLOGICAL, QUALITATIVE AND NUTRITIONAL ASPECTS)

(WEFTA-meeting, 4-5th May 1977, Brussels)
Presented by W. Vyncke, member of working group

INTRODUCTION

During a number of years the West-European institutes involved in fish technology research have organised annual conferences in order to promote the exchange of knowledge and results of their research. Moreover, to a limited extent the institutes have undertaken co-operative research with regard to the development of grading schemes for fresh and frozen fish.

For both purposes, the organisation of the conferences and the combined research, the institutes have formed the Western European Fish Technology Association (WEFTA).

As a result of visits of Mr. J.J. Rateau from the Commission to two WEFTA conferences the representatives of WEFTA were invited to Brussels, 8th December 1976, in order to facilitate further the flow of information between the Commission and WEFTA and together look into ways of helping the fish technology association to continue its activities.

According to one of the conclusions of the forementioned meeting a working group would be set up and meet in 1977 in Brussels to develop a joint research programme. The resulting research proposals, prepared during such meeting - held the 4th and 5th May - are given below.

PROJECT PROPOSALS

1) Background

In a period that European fisheries are confronted with serious problems - resulting from overfishing and strongly increased costs for fuel and auxiliary materials - maximum attention should be given to achieve an optimal valorisation of the limited amounts of fishery products to be harvested. For this purpose an intensivation and co-ordination of fish technological research is considered to be of great help to increase utilisation possibilities of these products. In order to promote the means for co-ordination of research and exchange of knowledge between WEFTA members the help of the Commission is sought.

According to the opinion of WEFTA members the form of assistance which could best be provided by the Commission is that which will further the co-ordination of the activities by all institutes by allowing for a separate fish technology programme for the pooling of information and experience and for the consequent acceleration of co-operative work on those lines of research disclosing greatest promise. In connection to the postulated minimum requirements as mentioned uner point(7) it can be concluded that a relatively small financial contribution is asked to foster the fruitful co-operation with regard to a vast and costly amount of research paid from different national resources.

The number of personnel working in fish technology research (excluding general servies, administration, liaison and informat services) in the nine WEFTA research centres in the European Economic Community is about 230, being 130 professional staff and 100 assistant personnel. For the five WEFTA centres outside the Community these numbers are in total 70, about 50% professional or assistant staff.

The general outlines of the field of research needed for the improved utilisation of fishery products for food purposes are given under point (3) in this paper. The more specific research proposals to be made within this general farmework are mentioned under (6).

These projects are at present or will in the near future form part of the day to day activities of all or several of the WEFTA institutes. Though at this moment the exact research

costs can not be specified, from the general character and
importance of the research proposed it can be easily understood
that the greater part of the total research capacity of WEFTA
institutes will be involved in this research.

The projects presented under (6) do form together the
present realisation of the general outlines for the utilisation
research given under (3). The specific projects (6) are
listed according to the technological aspects of various phases
of utilisation of fishery products, going from raw materials
to final products. These projects do show a strong inter-
relationship; therefore it should be said that numbers I-VII
have about the same priority, whereas the urgency of project
numbers VIII-IX may be somewhat less. Moreover it should be
mentioned that the results of projects II-VII are considered
to be of great importance with regard to the fishing industry
in developing countries.

2) Past and present research activities

In addition to the initial aim of WEFTA - mainly the
exchange of knowledge and research results - it was felt very
soon to be necessary to embark upon co-operative work with
regard to quality assessment of fishery products which asks
for an approach on an international level.

A working group, formed by WEFTA members and related
organisations in EEC-countries paid considerable attention to
these problems during six separate workshops. The subjects
studied were the quality grading schemes in the present EEC-
regulation for fresh fish and the proposed regulation for
frozen fish. Moreover some aspects of the proposed Draft
Standard for Quick Frozen Fillets of Flat Fish *(Codex Alimentarius)*
were studied. The main result of this work was the development
of a new grading scheme for frozen fish.

Similar successful research co-operation can be achieved
on other aspects of fish technology. This may include further
work on quality assessment of raw materials and their derivatives,

but also the improvement of methods for handling, processing and preservation, the (bio)chemistry and microbiology of these products and their nutritive value.

As a matter of fact WEFTA members are actively engaged in one or more of these fields of fish technology which have been discussed during the annual WEFTA conferences.

3) Purpose of project

In the present situation of the European fisheries it is strongly felt that more co-operative research on various aspects of fish technology is needed to reach the final goal: to maintain and to improve an efficent and economic supply of good quality fish products to the consumers.

According to this the following title and sub-titles are given as an indication of the function of various more specific research projects:

Improved utilisation for food of fishery products:

a) The influence of biological and technological factors on the yield and quality of edible parts and how to offset adverse effects on quality.

b) To investigate to what extent those fish species or materials at present not used as food could be utilised as such in the near future.

c) Investigation of spoilage, effects of pollutants and ancillary hygiene and sanitary aspects of fishery products.

d) Nutritive aspects of fishery products.

4) Participating research centres

The studies will be based on the activities of WEFTA members which are government, semi-government and university research centres in the countries of the Community (except Luxemburg) and Iceland, Norway, Portugal, Spain and Sweden.

5) Scope of project

Because of the diversity of the problems with regard to improvement of utilisation and valorisation of fishery products it will be necessary to investigate different fish species, processes, quality and nutritive factors.

Each of the participating research centres will select research areas of major importance with regard to the present situation.

The engagement of the research centres is in accordance with national needs and is dictated in most cases by the requirements of the national fishing industry of the respective countries.

In general it will not be feasible or practicable for the research centres to engage in activities outside their approved programmes or to take directions from external sources except perhaps where such requests arise from programmes which are completely funded by those requesting the particular research work. Moreover in most cases such engagement should be approved by the national authority as meeting a need that otherwise cannot be provided for at the particular time.

As a matter of fact it has been shown that there is sufficient agreement between greater parts of programmes of the participating research centres to allow for the formulation of the specific research proposals under (6), which cover most important parts of the present principles for technological utilisation research on fishery products as listed under (3).

6) Specific research proposals

At present the following specific research projects are presented.

1) Influence of the biological condition on the utilisation of fish.

Many fish species, both lean and fatty, show important variations in their composition, texture, flavour, odour, surface appearance and colour of the flesh due to changes in their biological condition, which is mainly influenced by the sexual cycle and the food supply.

For these reasons the characteristics of fish intended for processing may vary from season to season, from different fishing ground, even from batch to batch. Therefore it is difficult to be sure in advance that a particular consignment can be processed in a particular way.

The proposal is directed to gathering more systematic information on seasonal and ground to ground variation of the most important fish species, including analytical methods to assess the condition of the fish.

Systematic information on quality and processing yields related to biological condition and fishing grounds is needed to provide objective information on the rational exploitation of fish stocks.

Further technological research should be carried out to establish the relation between the condition of the raw material and its suitability for processing including new end products in order to promote the valorisation of fish with a low value under present conditions.

11) Quality of wet fish.

Quality changes during wet storage of raw materials especially of under-utilised species in order to provide basic information on which to design practical systems of handling, storage and processing of food fish.

111) Research on quality of frozen products.

The project is aimed at giving better information on the effect of quality of raw materials and other factors on the greatly varying shelflife and quality of frozen products. In connection with this it is necessary to improve methods for quality assessment of frozen products.

Expansion and co-ordination of research between laboratorie is needed with regard to a number of parameters such as fis species, quality of raw materials, product type, temperatur and time of frozen storage.

IV) Research programme for minced fish.

In the 1970's mechanical deboning has come to use in the western world mainly to improve the yield of edible flesh from traditional white fish used for filleting with 25 - 50%. Several under-utilised fish species have been taken into consideration also.

The resulting production of minced fish was followed by a backlash caused by an unexpectedly fast decline of the quality during frozen storage.

The project embraces both fundamental research (identification of the species used, methods for texture grading, causes of quality changes) and development (influence of manufacturing processes and additives on quality, utilisation of small fish and improvement of the recovered minced flesh, development of new products based on minced fish, development of equipment for the utilisation of the minced fish).

V) New uses for fish.

Development of new types of food products from species and raw materials not previously used or underexploited in Western Europe, eg new types of canned sprat, mackerel, squid and crab; soups, pastes, spreads from flesh recovered from small Crustaceans; reformed large pieces from small fragments of flesh. New food products to be made from fish which at present is regarded as suitable only for animal feeds, eg hydrolysates, ingredients with functional properties, fillers, spun or extruded forms.

VI) Mechanisation in fish processing.

The fish processing industry is labour intensive and the further exploitation of certain species depends on the development of new equipment. Mechanisation should not however lead to deterioration of working conditions.

The general aims of the proposed R and D should be:
- The design of equipment to facilitate the exploitation

of fish especially small bony fish (eg capelin, Norway pout, sprat, blue whiting).
- Improvement of working conditions, including equipment design.
- Improvement of fish utilisation.

Other more specific areas are listed below:
- Mechanical preparation and cutting of fish.
- Automated fillet defect detection and sorting (removal of bones, blood spots, skin parts, etc).

VII) Utilisation of by-products from fish processing.

Typical by-products from fish processing include intestines (digestive tract, roe, milt, liver), heads, skin after filleting, bones and fillet-trimmings, fish constituents in waste water.

The project is aiming at the improved utilisation of by-products for human food (including roe, liver and minced meat to be separated from heads, bones and fillet-trimmings), animal feed (by silaging or freezing) or manufacture of special products (sterols from milt, fish glue).

VIII) Irradiation of seafood.

Earlier experiments, partially carried out in co-operation with the Commission (Bureau Eurisotop) showed the good prospects of the application of irradition with regard to prolongation of shelflife of brown shrimps and white fish and fillets thereof. Moreover, in contrast with foregoing periods, there is reason to be more optimistic on the acceptance of the irradiation process by the health authorities in the near future. For these reasons the application for commercial use should be developed. Test marketing will be carried out.

IX) Functions of fish oils in the human diet.

Former experience has shown that there is often a great need for a sufficient number of objective data on the occurrence of particular constituents of food commodities which may affect their nutritional value in a positive or negative manner. For these reasons such type of research cannot be omitted in a programme of research on food utilisation and may be combined effectively with technological research for which often similar data are needed.

The following projects on fish oils are at present parts of the research programmes of some WEFTA institutes.

Fish oils have attracted attention because of reports in which their ability to lower blood cholesterol levels in both man and experimental animals has been claimed. The effect appears to be largely due to their abundant supplies of polyunsaturated fatty acids (PUFA's). Apart from the effect on blood cholesterols PUFA's are said to have depressing activity on the triglyceride level in blood as well.

It is proposed to carry out more systematic studies in order to elucidate the role of fish oils and fish lipids as a preventive agent in the control of cardiovascular diseases.

There is also some concern about the presence in fish oils of fatty acids similar to erucic acid which are alleged to cause heart lesions.

7) General considerations, working methods and estimate of costs

The Commission is not being called upon to finance specific research projects as those carried out by the individual research centres, but it is asked to further the speedy arrival at adequate solutions by enabling greater collaboration between fish technology research centres by financing travel and subsistence expenses of officers with regard to the research purposes explained before. Existing financial constraints under which the various institutes operate at present will not allow these possibilities.

Travelling and subsistence allowances are asked for:

a) The members of a Co-ordinating Committee, comprising the directors or leaders of fish technology departments from the research centres combined in WEFTA;

b) The members of some working groups to be set up for exchange of knowledge and experience on some very relevant parts in the co-operative work, consisting of some research workers engaged in the same line of work;

c) The exchange of research officers, allowing some persons to work for some period in another WEFTA research centre for co-operative research.

The formation of working groups and exchange of research officers will be a subject for the discussion in the Co-ordinating Committee, which will advise on the disbursement of funds for these purposes to the Commission. Moreover the Co-ordinating Committee will report on the progress made with regard to the research programme.

The meetings of the Co-ordinating Committee could take place best at the place and time of the Annual Meeting of WEFTA, to be held in one of the research centres. In case other consideration from the Commission's point of view should demand such a meeting to take place at Brussels (eg to allow for matters under examination to be discussed with officers of the Commission or to keep the relevant industry informed) this meeting could be arranged also, eventually as an additional meeting or as the main meeting as the case may be considered.

The expenditure involved in these various arrangements mentioned before is estimated to be of the order to 20,000 units of account per year, divided as follows:

Meeting Co-ordinating Committee (about 10 persons once per year)	6,000
Meeting of Working Groups (about 10 persons once per year)	6,000
Exchange of research officers (2 man years once per year)	6,000
sub-total	18,000
Secretarial expenses	2,000
total	20,000 units of account.

The approval is asked for a period of three years.

APPENDIX

ADDRESSES OF PERSONS AND RESEARCH CENTRES AND ORGANISATIONS INVOLVED IN WEFTA-ACTIVITIES

A. In the European Economic Community

Belgium:　　　　　　　　　　　Dr. W. Vyncke
　　　　　　　　　　　　　　　　Rijksstation voor Zeevisserij
　　　　　　　　　　　　　　　　Ankerstraat 1
　　　　　　　　　　　　　　　　8400 OOSTENDE

Denmark:　　　　　　　　　　　Mr. L. Herborg
　　　　　　　　　　　　　　　　Fiskeriministeriets Forsøgslaboratorium
　　　　　　　　　　　　　　　　Building 221, Technical University
　　　　　　　　　　　　　　　　2800 LYNGBY

France:　　　　　　　　　　　　Madame F. Soudan
　　　　　　　　　　　　　　　　Institut Scientifique et Technique
　　　　　　　　　　　　　　　　　　　des Pêches Maritimes
　　　　　　　　　　　　　　　　La Noë, Route de la Jonelière 44
　　　　　　　　　　　　　　　　B.P. 1049
　　　　　　　　　　　　　　　　NANTES

Germany:　　　　　　　　　　　Professor Dr. W. Schreiber
　　　　　　　　　　　　　　　　Bundesforschungsanstalt für Fischerei
　　　　　　　　　　　　　　　　Palmaille 9
　　　　　　　　　　　　　　　　2 HAMBURG 50

Ireland:　　　　　　　　　　　Mr. C.J. McGrath
　　　　　　　　　　　　　　　　Department of Agriculture and Fisheries
　　　　　　　　　　　　　　　　Fisheries Division
　　　　　　　　　　　　　　　　Agriculture House
　　　　　　　　　　　　　　　　Kildare Street
　　　　　　　　　　　　　　　　DUBLIN 2

Italy:　　　　　　　　　　　　Professor L. Mancini
　　　　　　　　　　　　　　　　Centro Universitario di Studi e Ricerche
　　　　　　　　　　　　　　　　sulle Risorse Biologiche Marine di
　　　　　　　　　　　　　　　　　　　　　　　　　　　　　Cesenatico
　　　　　　　　　　　　　　　　Piazza Saffi, 36
　　　　　　　　　　　　　　　　47100-FORLI

Italy:　　　　　　　　　　　　Dr. G. Baldrati
　　　　　　　　　　　　　　　　Stazione Sperimentale per l'Industria
　　　　　　　　　　　　　　　　delle Conserve Alimentari
　　　　　　　　　　　　　　　　Viale Faustino Tanara, 33
　　　　　　　　　　　　　　　　43100.PARMA

Netherlands:　　　　　　　　　Ir. J.J. Doesburg (also responsible for
　　　　　　　　　　　　　　　　the liaison between the Commission and
　　　　　　　　　　　　　　　　　　　　　　　　　　　　　　　　WEFTA)
　　　　　　　　　　　　　　　　Instituut voor visserijprodukten TNO
　　　　　　　　　　　　　　　　Postbus 183
　　　　　　　　　　　　　　　　IJMUIDEN

APPENDIX (Cont.)

United Kingdom: Dr. G. Burgess
  Torry Research Station
  P.O. Box 31
  ABERDEEN AB9 8DG

B. Outside the European Economic Community

Iceland: Dr. B. Dagbjartsson
  Icelandic Fisheries Laboratories,
  Skugulata 4,
  REYKJAVIK

Norway Mr. A. Hansen
  Fiskeriteknologisk Forskningsinstitutt
  P.O. Box 1159
  9001 TROMSØ

Spain: Dr. M. Lopez-Benito
  Instituto de Investigaciones Pesqueras
  Muelle de Bouzas
  VIGO

Portugal: Dr. L. Torres
  Institute Portuges de Conservas de Peixe,
  Avenido 24 de Julho, 76
  LISSABON

Sweden: Professor E. von Sydow
  Swedish Institute for Food Preservation
  Research
  Kallebäck Fack
  S-40021 GÖTEBORG

FINAL SESSION

SEMINAR CONCLUDING REPORTS

MEAT PANEL FINAL REPORT

Chairman:
B. Krol

Rapporteur:
J.V. McLoughlin

Research into the thermal processing of both red and poultry meat, involving the removal or addition of heat, requires the consideration of many interacting factors. These include the effects of the processes on sensory properties, ie colour, texture, waterholding capacity, taste and flavour; microbiological stability; the presence or absence of toxic substances; and the nutritive value of the food. Research must also include an examination of the efficiency of the processes themselves, and the design of plant, transport and storage facilities as well as increasing the value of the carcass.

Since the use of vegetable protein in meat products will most probably become more widespread in the future, the possibility of adverse interactions occurring between meat and non-meat proteins during thermal processing must not be overlooked. Special attention might be given to the texture, flavour and nutritive value of these products and also to their benefits to the consumer in terms of price and acceptability.

Research and development in the thermal processing of food will no doubt produce traditional products of better quality and improve the economy and efficiency of the process. They will produce new products and possibly at a competitive price but these must be acceptable to the consumer. There is need for an investigation into the psychological basis of food preferences and an attempt should be made to assess the influence of advertising and food promotion campaigns on preferences basically derived from a complex of cultural, religious, traditional and socio-economic factors. Such studies might in particular have a long-term value since changes in world patterns of food production and distribution may require changes in eating habits if all sectors of the world population are to receive a diet which is nutritionally adequate. The members of the Panel discussed food preference and, recognising its importance in relation to the nutritional status of a population, considered that attention should be drawn to this matter.

The members of the Panel agree that the following are important areas for research and development and appear to be suitable for international co-operative research:

1. CHILLING AND FREEZING OF MEAT

Current EEC regulations require that a temperature of $7^{\circ}C$ attained in the deepest part of the carcass before cutting or transportation is permitted. The rate of chilling required to reduce the temperature in a time interval suitable for commercial operation can produce a serious problem of meat quality. Such a rapid fall in the temperature of the carcass before the onset of *rigor mortis* causes the contraction of the skeletal muscles (cold-shortening) which toughens meat. This is an example of a commercial problem created by the use of a single criterion for legislative purposes. It is a problem which is being resolved by electrical stimulation of the carcass so that *rigor* sets in before the critical temperature which initiates cold shortening is reached. The use of electrical stimulation has also opened up possibilities for hot de-boning operations so that primal cuts can be produced in a chilled, packaged form with no detrimental influence on meat quality. Work on this topic is presently being carried out in Denmark, Holland, the UK and Germany.

2. THAWING OF MEAT

Thawing of meat presents problems which are not encountered in the freezing of this commodity. In particular, there is a risk of microbial contamination and subsequent spoilage, development of undesirable sensory characteristics and loss of nutritive value. The development of standardised procedures for thawing is necessary in order to alleviate the problems often associated with this process. Work on this topic is in progress in the UK, Denmark and the Netherlands.

3. MAXIMUM USE OF THE ANIMAL CARCASS

Carcass materials which at present are often wasted might be converted to edible materials. For instance, approximately 45% of the carcass weight constitutes connective tissue, including bone, so the use of collagen-based preparations in heat-treated products and as meat extenders is worthy of investigation. Preparations from the viscera might be used in a similar way. There is scope for extension of current work on the use of fractionated proteins from blood in meat products. The use of whole blood to prepare foods which have a high nutritive value and a relatively low cost should be further developed. Research into aspects of this topic is being done in the UK, Germany, the Netherlands, Denmark and Ireland.

4. THERMAL PROCESSES

Research into the heat transfer processes themselves and the development if necessary of new processes is desirable. Research and development in this should include the study of flexible packaging materials and aseptic closing procedures.

5. ADDITION OF NON-MEAT PROTEINS TO MEAT PRODUCTS

Many meat products are likely in the future to contain non-meat proteins. The effects of thermal processing on such mixtures and in particular the nutritive consequences, require not only the study of the meat and non-meat proteins themselves but also their possible interactions. These considerations invariably lead to the question of the source of non-meat proteins. In this context, and although it may not relate directly to the thermal processing of food, the Panel considers that the use of protein from plants indigenous to Europe should be investigated. The use of vegetable protein, in particular soya, in meat products was considered in

detail by an EEC Study Group (under the Chairmanship of Professor
A.G. Ward) and the report of the Study Group is in preparation.

Caseinates and whey proteins have certain technological
advantages in the stabilisation of high-heat-treated meat products.
The potential for the use of these proteins in meat products
should be further investigated. An additional advantage of research and development in this area would be the use of a food
commodity which at present is produced in surplus within the EEC.
Cooperation between meat and dairy research institutions might be
appropriate in this area.

All countries represented on the Panel are involved to some
extent in research into the use of non-meat protein in meat products.

6. CURING INGREDIENTS AND THEIR INTERACTIONS

Curing ingredients have been used for centuries mainly to
maintain or improve the colour, flavour and microbiological
stability of meat. It is well known that these ingredients have
individual and synergistic effects. Their optimum use might improve the sensory properties and microbial stability of meat products but methods of analysis and organoleptical evaluation must
be developed to ensure that this objective is reached. Furthermore, an optimal combination of curing ingredients might lead to
the use of lower temperatures in various heat treatment processes
This is a particularly suitable subject for international cooperation since many countries will almost certainly change their
legislation on this matter in the near future. Work is already
being done in all countries represented on the Panel.

7. MONITORING THE NUTRITIONAL VALUE OF HEAT TREATED FOODS

There is need to develop methods of routinely monitoring
the effects which thermal processes may have on the nutritive
value of foods, particularly those consisting of mixtures of meat
and non-meat proteins. Such changes in thermally processed foods

will, of course, have to be related to the total nutritional value of the diet.

Research on items 1 to 7 is already in progress in many institutions in several of the countries represented on the Panel. All members of the Panel expressed interest in the idea of cooperative research and there appears to be no reason why such research might not be undertaken. Therefore members would welcome moves to initiate such co-operation.

DAIRY PANEL FINAL REPORT

Chairman:
J. Moore

Rapporteur:
P.F. Fox

Nowadays, some form of heat treatment is used in preparing practically all milk products for use by the consumer. Quite apart from making milk bacteriologically safe, heat treatment assists in the preservation of milk by inactivating certain enzymes that might otherwise cause spoilage. An example of such a milk enzyme is the lipase that is derived from the mammary gland; under certain circumstances this enzyme can cause hydrolytic rancidity. However, most heat treatments do have some disadvantageous effects on the nutrients in milk, several of which are heat labile. Nevertheless, those losses in nutrients must be set against two other factors.

1. Milk production throughout the world is uneven in the extreme; there are large supplies available in some areas and hardly any in others. Without processing with some form of heat treatment, large populations would have no milk at all. Surely it is better for a population to have sterilised milk, even though the biological value of the protein has been reduced some 10% by the heat treatment, than to have no milk at all.

2. The loss in nutritive value caused by heat treatment must be considered in relation to the contribution that milk and milk products make to the total intake of nutrients by any given population. For example during the preparation of condensed milk there can be a loss of up to 60% of the vitamin C originally present in the raw starting material. In isolation, this loss seems sizeable, but in the UK for example milk and milk products contribute less than 10% of the total vitamin C intake.

Losses in nutrients caused by heat treatment must, of course, be guarded against: this is particularly important in the artificial feeding of infants of whom milk may be the major or sole food for a short but vital period during development. For adults generally, the part played by milk and its products is to contribute to the total value of a mixed diet.

In general, the members of the panel considered that the effects of heat treatment on the losses of nutrients from milk and its products were relatively small and, in any event, were fairly well documented and understood. However, a product containing large amounts of high quality nutrients is of no value if it has an objectionable taste or is unacceptable to the consumer for some other reason. Therefore it was agreed that the panel should concentrate their discussions on the effects of heat treatment on the acceptability of dairy products to the consumer and on the efficiency of subsequent processing. A large number of problems were discussed during the two meetings of the panel and the following topics received the most support.

1. Objectionable flavours in UHT products

The consumption of UHT products is increasing rapidly, both in Europe and in some non-European countries. This suggests that these 'off-flavours' merit special attention in investigations of high priority.

2. Psychrotropic micro-organisms in refrigerated milk, and the significance of their enzymes in dairy processing

The enzymes produced by these micro-organisms appear to be relatively heat stable, and certain of these enzymes may cause quality defects, eg the lipases elaborated by psychrotropic micro-organisms can result in lipolytic rancidity.

3. Effect of cooling on the structure and properties of the casein micelle

Certain enzymes (eg lipases, proteinases) that originate in the mammary gland appear in milk as a complex within the casein micelle. Cooling of milk tends to cause dissociation of the enzyme from the micellar system, and the liberated lipase, for example, can result in the hydrolysis of milk fat and the production of off-flavours.

4. Heat stability of milk

There are widespread problems caused by the coagulation of milk proteins during the processing of milk by heat treatment. More research is required into the mechanisms of, and factors that control coagulation in standard and concentrated milk.

5. Thermal denaturation of indigenous milk enzymes

Kinetic parameters should be established for the denaturation of a number of indigenous milk enzymes (eg proteolytic enzymes, acid phosphatase) under laboratory and commercial conditions.

6. Development of 'Prediction Tests'

There was widespread support for the development of tests that could result in the selection of milk for certain processes. For example, can the heat stability of standard milk be used as an index of the stability of the concentrated or dehydrated product derived from it?

7. Methods for determining characteristics related to thermal processing

Correlations should be established between subjective (organoleptic) tests and objective (chemical and physical) tests.

8. Comparative studies on sheep and goats milk

The projects listed from 1-7 should be extended to include sheep and goats milk.

Two other aspects should receive special attention:

a) Some of the projects listed above should also be extended to new, thermally processed dairy foods, so that the technology of producing dietary-enriched and modified dairy foods can be established. The factors affecting the stability, storage characteristics and other parameters of product quality of these dietary-modified dairy foods are somewhat different from those of standard milk and milk products.

b) Some projects may not have been recommended because they are of only moderate priority for any commodity group. However, they may be important to several groups and so accumulate combined priority. This may mean that potentially important areas are missed because of the commodity group structure for the selection. It is suggested that under this heading come:

    i Problems of thermobacteriology (relevant to meat, fruit and vegetable, milk and milk products, and possibly fish commodity areas).

    ii Relationship between container properties, eg oxygen permeability, light transmission, or the loss of product quality during storage (referred to in papers for most of the commodities).

FISH PANEL FINAL REPORT

Chairman:
A. Hansen

Rapporteur:
J. Somers

BACKGROUND

Fish resources hitherto not exploited, or used for fish meal and oil production, represent great potential for the future increase of food supplies.

European fisheries are presently confronted with serious problems resulting from the over-fishing of traditional species and greatly increased costs. Maximum attention should therefore be paid to achieving optimum utilisation of all old and new resources with respect to quality.

In common with research workers in other food commodity sectors, fish technologists are interested in knowing more about the raw material which they are using, its biological condition and the quality changes which occur between harvesting and consumption, through solving problems associated with deterioration, the objective being to ensure that the consumer has the most organoleptically and nutritionally acceptable products available.

TOPICS

The study panel has identified the following topics as being of major interest to 'COST' countries and where useful collaboration can be achieved:

- Under and non-utilised species
- Biological condition and end product quality
- Upgrading of waste material

- Methodology for quality assessment
- Microbiological safety
- Nutritional considerations.

These were among topics suggested by the West European Fish Technologists Association (WEFTA - an informal annual forum for discussion between fish technology laboratories) in a submission to the European Commission at a meeting in Brussels (May 1977).

UNDER AND NON-UTILISED SPECIES

As already stated almost all traditional resources available to COST countries are over exploited. As a result of this and also due to the introduction of new extended national fishery limits, supplies in many cases have declined radically. Many countries are therefore looking for alternative species like, for instance, blue whiting, horse mackerel, capelin, and others which hitherto have not formed the basis for direct exploitation for human food. Consequently, data regarding the quality and nutritive properties of these species in their various biological conditions are scarce, and very little is also known about quality changes during handling and processing. The proposed study comprises an investigation of the deterioration of little or non-utilised species during chilling, freezing, storage and heat processing, to provide information for the design of efficient systems of handling and processing.

INFLUENCE OF BIOLOGICAL CONDITION AND END PRODUCT QUALITY

Many fish species, both lean and fatty, show important variations in their composition, texture, flavour, odour, surface appearance and colour of the flesh due to changes in their biological condition, which is mainly influenced by the sexual cycle and the food supply.

For these reasons the characteristics of fish intended for processing may vary from season to season, between different fishing grounds, and even from batch to batch. Therefore it is difficult to be sure in advance that a particular consignment can be processed in a particular way.

The proposal is directed to gathering more systematic information on seasonal and ground to ground variation of the most important fish species.

Systematic information on quality, nutritional value and processing yields related to biological conditions is needed to provide objective information on the rational exploitation of fish stocks.

Further technological research should be carried out to establish the relationship between the condition of the raw materials and their suitability for processing (including new end products) in order to promote the upgrading of fish with a low value under present conditions.

UPGRADING OF WATER MATERIAL

With the decreasing supply of traditional species it has become essential in most countries to recover as much material as possible. An important development here has been the introduction of machines capable of stripping flesh in the form of mince from parts of the fish previously discarded. The quality of the recovered material has often been poor; colour tends to be too dark; texture and water binding capacity is sometimes poor; spoilage rates tend to be high; the material may contain unacceptably high proportions of connective tissue, bone or other non-fleshy parts. If the fullest possible use of this recovered material is to be achieved it is necessary, therefore, to ascertain either conditions of machine operation under which products of different acceptably qualities can be obtained or the means by which quality can be improved after recovery. A good deal remains to be found out about the sensory and microbiological quality of these materials and how they limit storage under different conditions. It has been observed, for example, that the texture of certain kinds of recovered flesh deteriorates very rapidly under normal commercial conditions of frozen storage.

METHODOLOGY FOR QUALITY ASSESSMENT

International collaboration on quality and nutritive value pre-supposes that collectively agreed methods for assessing these are available. There are growing moves both nationally and internationally towards the introduction of standards and grading schemes which partly rely on agreed methods of assessment. This movement would be greatly facilitated by further work on methodology.

The West European Fish Technologists Association have already organised successful collaboration which resulted in an agreed scheme for the sensory grading of whole frozen fish of main commercial interest. WEFTA have proposed a similar collaboration on fillets. There is also a need for agreed methods of quality assessment for Codex Alimentarius Standards. There is a need for the extension of this collaboration to cover all industrially important products. Other non-sensory methods are needed for assessing:

a) The fish flesh, bone and parasite contents of products,

b) The proportions of glaze and of minced flesh in products containing a mixture of mince and fillets,

c) The identity of species in products containing mixtures of species, and

d) The degree of oxidation of fish lipids.

MICROBIOLOGICAL SAFETY

While the incidence to damage of health caused by micro-organisms in fish and fishery products is relatively low, products are being marketed where limited information is availabl on their micro-biological quality. The major agents of fish spoilage are micro-organisms. During process and storage conditions the microbial population changes in character and in numbers. Examples of these processes are the pasteurisation of shellfish and vacuum packed, smoked fish products. Further study could give the information necessary for shelf life extension as well as product improvement. Information on these microbial changes is essential in order to define the exact parameters of the process and storage conditions, which will ensure consistently good and wholesome products.

NUTRITIONAL CONSIDERATIONS

Fish is traditionally regarded as an excellent food, being a source of high quality proteins and containing important nutrients such as vitamins, trace elements and fat.

Limited information, however, is available regarding nutrients in fish, especially species hitherto not used as a food source. Very little is known as to what extent different processing methods and conditions affect the nutritional value in the final product.

Research is therefore required to determine the nutritional value of traditional species and, especially, those species which have been recently identified as having commercial potential. The stability of nutrients during processing and storage also merits attention.

The panel notes that this research topic is related to earlier proposals and while we conclude with it, we do not infer that is is low in priority.

CEREALS PANEL FINAL REPORT

Chairman:
P. Linko

Rapporteur:
E. Markham

Cereal grains have, since ancient times, played a central role in human nutrition and still today provide some 25 - 50% of man's energy requirements. In a properly balanced diet they are an important source of protein. In some areas they provide a significant amount of mineral nutrients, especially of iron, but also of magnesium, manganese, etc. Cereal products are also important as a source of certain B-vitamins, and whole grain or high extraction rate products are a major source of dietary fibre. This key role of cereal products in human nutrition justifies a closer look into the interrelationships of processing and quality.

Thermal processing is involved in the manufacture of nearly all cereal based foods intended for human consumption. Yet, it is important to view food manufacture as a total food processing system, composed of a number of unit operations and processes.

Most important operations during the processing of cereal grains normally include some degree of modification of physical properties of the raw material by the application of thermal energy to obtain desired product quality and to improve digestibility and, thus, nutritional value. Simultaneously, adverse affects may be realised, and in optimising a thermal process a compromise has to be made.

Typical key phenomena during heat treatment of cereal grains include starch gelatinisation and denaturation of protein, both normally water dependent. Related changes that affect functional properties of importance in further processing and quality of final product take place during freezing and thawing.

Recently, development of new technologies has made possible both an improvement in existing processing, and the application of completely new techniques in food manufacture to improve quality, to reduce production costs and to better utilise by-products and waste materials for human consumption. Little is yet known of the interrelationships of these new techniques and the quality of final product. It should be emphasised that application of such new techniques in food processing systems frequently involves thermal treatment in some form.

One should also recognise that the availability of raw materials suitable for industrial processing, pretreatment techniques, and storage conditions are very important in determining final quality and nutritional value. Here changes in and interrelationships of enzyme activity, temperature and water activity play a central role. Thus, various dehydration techniques may be used to decrease water reactivity or thermal treatments to reduce actions of biocatalysts. However, application of most of these techniques may simultaneously result in deleterious changes.

Main types of thermal processes involved in cereal technology, their interrelationships, and effects they cause during processing are indicated in Figure 1.

Related processes are also general to many other commodities and it is useful to recognise these areas of common ground. A number of key research areas common to most commodities are shown in Figure 2.

In general, it should be the goal of food engineers, technologists and nutritionists together to strive towards process optimisation to minimise losses of nutrients and other adverse effects, and to maximise quality, with least costs. This seminar is recognised as the beginning of bringing together

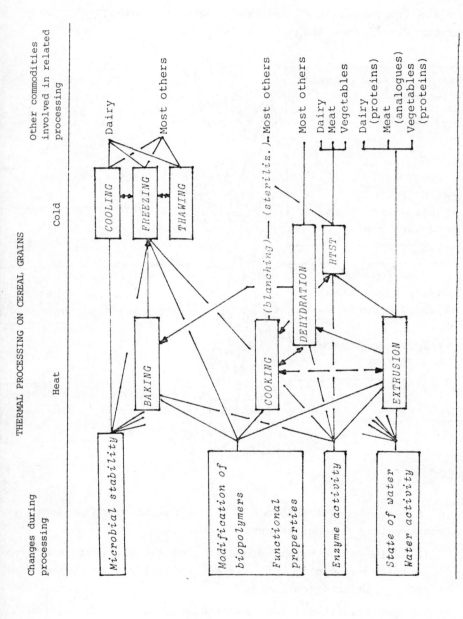

Fig. 1.

for the first time in Europe, such experts who represent these various disciplines and various commodities.

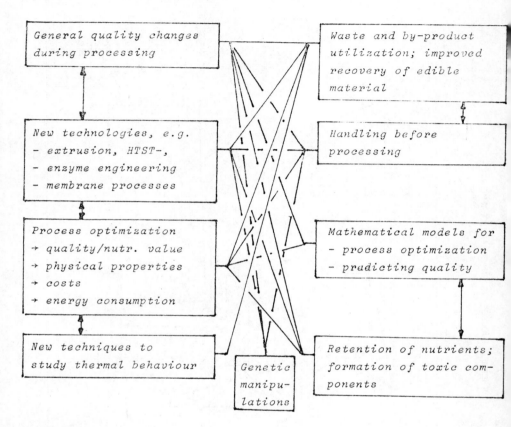

Fig. 2.

The cereal panel recommends the following three projects as most needing further investigation:

1. Effects of thermal processing on the biopolymers of cereals, especially with respect to lipid, starch, and protein interaction.

2. Non traditional thermal treatments in the production of cereal based foods.

3. Effects of freezing and thawing on the quality of cereal based foods.

EFFECTS ON THERMAL PROCESSING ON THE BIOPOLYMERS OF CEREALS, ESPECIALLY WITH RESPECT TO LIPID, STARCH, AND PROTEIN INTERACTION

The biopolymers of cereals are of importance in several respects. An understanding of the basic phenomena involved can lead to substantial importance in product quality and development of more effective processing techniques. It would also provide better understanding of the role of dietary fibre and other carbohydrates.

The objectives of the research are to study (a) functional properties of biopolymers as affected by thermal processing, with special reference to starch structure and gelatinisation, and (b) nutritive value, with special reference to starch denaturation and dietary fibre.

The topic is of special interest to developed countries. Work in this area is currently being undertaken in many research centres of COST countries which would provide an excellent basis for collaboration.

NON TRADITIONAL THERMAL TREATMENTS IN THE PRODUCTION OF CEREAL BASED FOODS

A number of new technologies has recently been introduced in food processing. Research on extrusion has been initiated in many COST countries, and it is necessary to investigate the effects of processing on product quality and nutritional value.

The objectives of the research are to (a) explore new technologies, eg extrusion and microwave cooking techniques, to (b) improve the utilisation of cereal grains important to COST countries, to (c) minimise total costs whilst maintaining quality, and to (d) increase the range of available cereal foods in order to increase consumption which would be beneficial to health and the economy.

Because of increasing interest in and utilisation of such new technologies, research in this field has been initiated in a number of research centres in several COST countries. The panel considered international collaboration in this field to be of high priority, in order to fully understand the effects of these new processes.

EFFECTS OF FREEZING AND THAWING ON THE QUALITY OF CEREAL BASED FOODS

Freezing and thawing are of crucial importance to the quality of final product, yet notably thawing has been little studied. The objective of the research is to ensure that commercial and consumer benefits of freezing and thawing are consistent with the maintenance of quality. The panel felt that special attention should be given to the thawing under domestic conditions.

Whilst the project is of immediate interest to all developed COST countries, it will also be increasingly important to developing countries. There is already considerable interest in this area in eg Finland, France, Germany, Italy, The Netherlands, and The United Kingdom. Work in this area could possibly be linked to research on other commodities in this field.

FOLLOW-UP SEMINAR

Recognising the importance of this seminar the cereal panel unanimously recommends that a follow-up seminar would be held in about three years time.

FRUIT AND VEGETABLES PANEL FINAL REPORT

Chairman:
W. Spiess

Rapporteur:
A. Hunter

INTRODUCTION

Within Europe a rich variety of processed fruit and vegetables is available. Production and distribution methods vary from one region to another depending on availability of different raw materials, eating habits and quality judgements.

The most important operations in fruit and vegetable processing are the heat treatments necessary for thermal sterilisation or for freezing. By these preservation techniques it is possible to provide the consumer with products of high nutritional value and sensory appeal but during the process important changes occur in the raw material. The choice of processing condition can also have important economic implications.

The high quality standard of fruit and vegetable products in Europe is a result of extensive and successful research by Government-sponsored and Industry-sponsored research units.

There is still scope for research in many areas which could lead to the further development of the industry and facilitate the wider distribution of high quality fruit and vegetable products at an economic price.

These research efforts succeed only as a result of the application of highly developed analytical methods and new approaches, (eg the study of complete unit processes rather than of unit operations in the process chain) and the close co-operation of scientists of different disciplines from plant breeders to nutritionists. The better appreciation of the interactions within the products and between products and their environment so obtained must be supplemented by an improved understanding of the reaction kinetics and transfer phenomena taking place at various stages of the process to enable the manufacturer to establish more controlled and more profitable conditions.

Results of research of this type will be applicable not only in Europe but also in countries which lack the necessary funds or scientific background (eg countries in tropical regions)

The main areas where joint efforts within the framework of COST would be of most value are outlined briefly below. These are the areas where research is possible and necessary if results are to be obtained which can usefully be applied in the near future. The emphasis placed here on certain unit processes in heat sterilisation and freezing does not mean that these are the only fields where research would be of value. The Fruit and Vegetable Panel felt, however, that by restricting its observations to certain unit processes a more manageable programme could be formulated.

RAW MATERIAL

The importance of research to develop the best possible raw material, the best growing conditions and optimum degree of maturity for given processes cannot be overestimated.

Processors must work with plant breeders in the preparation of raw material specifications which will ensure that high quality processed foods can be produced. It will be necessary therefore to establish the relationship between the final product quality and the quality of the raw material. The effects of harvesting and handling up to the processing line must also be monitored.

PREPARATION

Sorting and washing are processing steps which ensure that the raw material reaches the processing line in required condition. The effects of washing and of additives such as alkalis used in washing have not been thoroughly investigated.

As heat and mass transfer processes are dependent on particle size and configuration, the size reduction of the raw material after washing has a marked effect on heat transfer in subsequent thermal treatments.

The mechanical cell damage which occurs during size reduction further influences the mass transfer between the product and its environment. The leaching out of minerals, during blanching and subsequent cooling and the accumulation of calcium from hard blanching or cooling water, are examples. Size reduction processes and the consequences of size reduction in later processes have received little scientific attention. (An important example of a reaction resulting from a size reduction process is the interaction of oxidised plant phenolic compounds and proteins.)

BLANCHING AND PRECOOKING

Blanching, which achieves a number of results including the reduction of microbial contaminants, the breaking of cell turgidity and the inactivation of enzymes is widely used in fruit and vegetable processing and mass feeding. The process has been studied extensively but many questions remain unanswered - particularly in relation to the combined effect of blanching and subsequent heat treatments or freezing processes.

The inactivation of enzymes is especially important in blanching operations, as enzymes play an important role in the processes of deterioration which occur in heat treated and frozen material. Blanching is also important in that it improves the palatability of food for final consumption.

A better understanding of the chemical reactions which take place in blanching is required so that conditions may be adapted to maintain, as far as possible, the original and desirable qualities of the food. In addition, knowledge of trading problems, household cooking methods and many other factors is required to arrive at the best systems to be employed.

A very important area requiring investigation concerns the blanching of plant material in closed systems such as in a sealed can or, in the case of chopped or finely comminuted materials such as spinach or vegetable purees in closed liquid, liquid heat exchangers.

Better process control and heat recovery are important features of this blanching system. Problems in this area have not been examined to any extent but would appear to be of importance in quality optimisation and energy saving.

HEAT STERILISATION

The exposure of plant materials to elevated temperatures of heat sterilisation (and of blanching) results in the occurrence of major structural and chemical changes in the product which are not well understood. Both desirable and undesirable substances may be either produced or destroyed in such conditions The understanding of these chemical reactions and their pathways is of prime importance (eg, the reaction of protein and reducing sugar, the destruction of amino acids and the formation of 'off' flavours).

One important result of such research may be the identification of substances which will serve as better quality indicators than are available at present.

The complete understanding of the kinetics of desirable and undesirable reactions will also contribute to the formulation of optimal process conditions and improved equipment design.

Microbiological problems, such as behaviour of *Clostridium botulinum* in fruit and vegetable products, were considered but were thought to be outside the scope of this study.

HOME AND INSTITUTIONAL COOKING

Special attention must be given to preparation of plant material for consumption at home or in mass catering establishments

Problems in this area have either been overlooked in the past or arise from new methods of cooking and catering. (Steaming, simmering or deep fat frying are examples of techniques which have widely differing effects on final product quality).

The effect of holding foods at warm temperatures is also worthy of study.

FREEZING

Freezing, which enables products which have been largely unchanged to be stored for long periods, is an area requiring extensive research.

Research is needed to understand fully the basic phenomena accompanying freezing (eg ice formation, cell damage caused by changes in concentration and mechanical stresses, etc.) The change in the mechanical properties of the structure of the cell should also be examined.

The reaction kinetics at different temperatures during storage and the resulting changes in sensory and nutritional properties (particularly texture and flavour) are other fields of interest. Research work in this field will be of value in the formulation of guidelines for legal requirements concerning production, storage and distribution of deep frozen food products.

QUALITY EVALUATION

In addition to research concerning the interaction of food components with each other and with the environment and the study of reaction kinetics, the Panel recognised the need for research into the interaction of foodstuffs with human sensory and digestive organs.

The significance of quality changes can only be assessed when adequate information on individual nutrients is available.

There is also an urgent need for the development and standardisation of more reliable and more accurate sensory evaluation techniques which would lead to a better understanding of regional differences.

Similarly, methods for the rapid and accurate assessment of nutrients are desirable.

The development of a vocabulary of quality-terms which would facilitate communication between research workers in different centres is also required.

NUTRITION CO-ORDINATION GROUP FINAL REPORT

Chairman:
G. Varela

Co-Rapporteurs:
J.P. Kevany and F. Cremin

INTRODUCTION

The basic function of food is to provide energy and nutrients to meet human population requirements for health and well being. A range of factors affect the nutrient content of foods from the farm to metabolic utilisation. For the purposes of this report, this sequence of events can be divided into two major sub-divisions: from production to consumption and from consumption to metabolism. It is the first of these sub-divisions that is the concern of the food technology sector and includes such factors as choice of cultivars and stock, harvesting, transport, storage, processing, distribution and ultimately domestic preparation. The object of these activities is to obtain a product that is nutritionally valuable while meeting requirements of quality and price.

This situation demands the effective co-ordination of technological and nutritional interests and resources in determining appropriate conditions to guarantee an optimal end product. Such co-ordination should not be limited to the consideration of nutrient losses and changes but should also include the use of specific foods as vehicles for nutrient fortification as a mechanism for improving the quality of the diet as a whole.

Ideally, it would be desirable to plan the nutritional aspects of food technology research and development on the basis of identified population dietary requirements. In turn, such requirements would be related to specific food sources on the basis of dietary intake studies. At this point, the processing techniques applied to these key food sources, in particular thermal processing, should be examined to determine their effects on nutritional value. In this context, it is important to consider the availability of reliable and practical methods for determining nutritional properties and the use of standardised procedures for evaluation at all stages of production and processing.

It is evident that this ideal approach is rarely followed in practice for lack of precise information on such factors as population requirements, dietary patterns and food composition. Furthermore, there is an absence of appropriate methods for determining in a simple manner the full range of nutrients requiring evaluation.

It is the purpose of this report to identify nutritional R and D needs and priorities within this conceptual framework as a basis for long-term programme planning. It was noted that the areas of food additives and toxicology, though important, were outside the terms of reference of this COST Food Technology Seminar. Furthermore, the question of the nutritional value of live/dead yoghurt was discussed and it was concluded that this was a matter for clinical research at this time.

The following research priorities represent a consensus of opinion among the participants of the Nutrition Group drawn from the various commodity panels and representing institutions and services in thirteen of the nineteen countries participating in the Seminar. These priorities reflect current research activities in the majority of COST countries.

1. Development of methods and procedures for monitoring new products and new processes in regard to protein quality and vitamin content.

2. Development of rapid, reliable laboratory tests for the assessment of protein quality, in both high and low content foods, with reference to enzymatic, dye-binding, microbiologically available amino acid assays, FDNB lysine, MMR and other physical methods. The results of such tests

should be subsequently validated in animals and human subjects, where feasible.

3. Effect of thermal processing on the amount present, the chemical form and the efficiency of absorption of minerals and trace elements with particular reference to iron, iodine and zinc.

4. The effect of thermal processing, at both the industrial and domestic level, on the form and content of B_6, folates and B_{12}. This is to be considered as complimentary to existing studies on B_1 and other heat-labile micronutrients.

5. Effects of thermal processing on the properties of dietary fibre components in cereals and other plant materials. Such effects cannot yet be specified in terms of human metabolism and must await the outcome of current research projects. In addition, the effects of processing on polysaccharide structure and physical properties should be determined with particular reference to starch as a source of carbohydrate; a matter of special importance in relation to baby foods.

6. The effect of polysaccharides on viscosity in the rheology of foods and the diet as a whole may have important nutritional consequences. Collaboration between rheologists and nutritionists involved in this area is required to study the interaction of these factors.

7. The effects of thermal processing and storage on the chemical and physical characteristics of unsaturated fatty acids, including the production of unusual substances of unknown nutritional significance, requires further study. This may be of importance in relation to high-fat fish, particularly if these become a significant source of EFA in the future.

In addition to these proposals, the group offered the following recommendations for consideration:

1. The growing importance of processed foods in the diet means that a knowledge of their nutritional composition is essential for those concerned with monitoring nutrient intakes of populations and in studies of nutritional status. All studies of processing should therefore, include nutritional assessment of the value of the product. Such an assessment <u>should not</u> be limited to protein and the major vitamins. Fat, carbohydrates, vitamins such as B_6, B_{12}, and folate, and trace elements should also be examined.

2. A limited amount of information is available about nutritional changes in laboratory experiments and in manufactured foods as purchased but almost nothing is known of the nutritional content of foods <u>as eaten</u>. Foods may be stored for long periods, cooked well or badly and kept hot for varying periods before being eaten in both institutional and domestic situations. Until we know more about this we cannot assess the relative importance of any processing losses.

3. The role of dietary fibre and its consumption in relation to the incidence of disease suggests that studies whereby higher extraction cereal products can be produced and which are acceptable to the general public, should be encouraged.

LIST OF PARTICIPANTS

Prof. Agregado — Universidad Complutense
Madrid
Spain

Mr. D. Aitkenhead — Milk Marketing Board for Northern Ireland
Belfast
Northern Ireland

Dr. P.E. Andersen — Engineering Academy of Denmark
Laboratory of Biotechnology
Lyngby
Denmark

Dr. J. Anusic — "Pliva"
Zagreb
Yugoslavia

Mr. C. Bailey — Agricultural Research Council
Meat Research Institute
Bristol
England

Mr. F. Bannister — Department of Economic Planning and Development
Dublin
Ireland

Dr. M.J. Beaufrand — Laboratoire de Nutrition et des Maladies Métaboliques
Nancy
France

Dr. G. Bellucci — Stazione Sperimentale
Per l'industria dell Conserve Alimentari,
Parma
Italy

Prof. A.E. Bender — Nutrition Department
Queen Elizabeth College
University of London
London
England

Dr. J. Bernstein — World Food and Nutrition Study
National Academy of Sciences
National Research Council
Washington D.C.
United States of America

Prof. Dr. B. Blanc — Federal Research Station
Berne
Switzerland

Dr. F. Bramsnaes	Food Technology Laboratory Bygning Lyngby Denmark
Ms. R. Brew	National Science Council Dublin Ireland
Prof. G. Brubacher	F. Hoffmann-La Roche AG Basle Switzerland
Dr. J. Buckley	Department of Dairy and Food Technology University College Cork Ireland
Dr. H. Burton	Process Engineering Department National Institute for Research in Dairying Reading England
Dr. D.H. Buss	Nutrition Section Ministry of Agriculture, Fisheries and Food London England
Prof. R. Buzina	Institute of Public Health of S.R. Croatia Zagreb Yugoslavia
Prof. C. Cantarelli	University of Milan Italy
Dr. J. Carballo	Instituto Nacional de Investigationes Sgaarias Madrid Spain
Mr. E. Carden	Department of Agriculture Dublin Ireland
Dr. M. Caric	Faculty of Dairy Technology University at Novi Sad Novi Sad Yugoslavia
Dr. G.C. Cheeseman	National Institute for Research in Dairying Reading England

Dr. J.J. Connell	Ministry of Agriculture, Fisheries and Food Torry Research Station Aberdeen Scotland
Mr. A. Cotter	National Science Council Dublin Ireland
Mr. M.J. Cranley	Institute for Industrial Research and Standards Dublin Ireland
Dr. F.M. Cremin	University College Cork Cork Ireland
Prof. G. Crivelli	IVTPA Milan Italy
Mr. J. Cronin	City Analyst's Laboratory Eastern Health Board Dublin Ireland
Dr. R.F. Curtis	ARC Food Research Institute Norwich England
Mr. H.K. Dahle	Department of Food Hygiene Veterinary College of Norway Oslo Norway
Prof. A. Dahlqvist	Department of Nutrition Afdelmgen Fur Industriel Naringslara Kemicentrum Lund Sweden
Mr. S. Devine	Department of Industry, Commerce and Energy Dublin Ireland
Mr. B. Devlin	Department of Agriculture Dublin Ireland
Mr. M.B. Dorgan	Department of Fisheries Dublin Ireland

Mr. P. Dowling	Department of Agriculture Dublin Ireland
Dr. W.K. Downey	National Science Council Dublin Ireland
Dr. F.E. Escher	Department of Food Science Swiss Federal Institute of Technology Zurich Switzerland
Prof. G. Fabriani	International Association for Cereal Chemistry Rome Italy
Mr. J. Feehily	Institute of Food Science & Technology of Ireland Dublin Ireland
Prof. F. Fidanza	Universita Degli Studi di Perugia Perugia Italy
Prof. J. Foley	Department of Dairy and Food Technology University College Cork Cork Ireland
Prof. P.F. Fottrel	Department of Biochemistry University College Galway Galway Ireland
Prof. P.F. Fox	Department of Dairy and Food Chemistry University College Cork Cork Ireland
Prof. R. Garcia-Olmedo	Universidad Complutense Madrid Spain
Prof. M.Sp. Georgakis	Aristotelian University of Salonika Greece
Prof. Dr. Ing,K. Gierschner	Universität Hohenheim (LH) Institut für Spezielle Lebensmitteltechnologie Stuttgart W. Germany

Mr. T. Godfrey	Department of Industry, Commerce and Energy Dublin Ireland
Mr. R. Griffith	Technical Information Division Institute for Industrial Research and Standards Dublin Ireland
Dr. A. Hansen	Institute of Fishery Technology and Research Tromsoe Norway
Dr. F. Hill	City Analyst's Laboratory Eastern Health Board Dublin Ireland
Mr. G.O. Hogan	Department of Agriculture Dublin Ireland
Dr. N. Healy	Irish Sugar Company Limited Dublin Ireland
Dr. K.M. Henry	Arkendale Road Dublin Ireland
Dr. S.D. Holdsworth	Campden Food Preservation and Research Association Gloucestershire England
Mr. J.C. Holloway	Department of Industry, Commerce and Energy Dublin Ireland
Dr. A.W. Holmes	The British Food Manufacturing Industries Research Association Leatherhead Surrey England
Mr. A.T. Hunter	Food Technology Department Institute for Industrial Research and Standards Dublin Ireland
Mr. W. Ituk	Ministry of Industry and Technology Ankara Turkey

Dr. H. Johannsmann   Bundesministerium für Ernährung
Landwirtschaft und Forsten
Bonn
W. Germany

Dr. R.H.W. Johnston   Applied Research and Consultancy Group
Trinity College Dublin
Ireland

Prof. M. Jul   Danish Meat Products Laboratory
Copenhagen
Denmark

Mr. L. Kearney   Department of Food Technology
Institute for Industrial Research
and Standards
Dublin
Ireland

Prof. W.R. Kelly   Veterinary College of Ireland
Department of Veterinary Medicine
Dublin
Ireland

Prof. Dr.-Ing, H.G. Kessler   Institut für Milchw. Machinenwesen
Technische Universität München
W. Germany

Dr. J.P. Kevany   Department of Community Health
Trinity College
Dublin
Ireland

Mr. A.A. Kinch   EEC Commission
Directorate General for the Internal
Market and Industrial Affairs
Brussels
Belgium

Dr. A.G. Kitchell   Ministry of Agriculture, Fisheries and
Food
London
England

Prof. B. Krol   TNO
Central Institute for Nutrition and
Food Research
Zeist
The Netherlands

Dr. O. Kvaale   Norwegian Food Research Institute
As-NLH
Norway

Mr. J. Langan	Department of Food Technology Institute for Industrial Research and Standards Dublin Ireland
Prof. L. Leistner	Bundesanstalt für Fleischforschung Kulmbach W. Germany
Prof. H.A. Leniger	Department of Food Science Agricultural University Wageningen The Netherlands
Dr. J. L'Estrange	Department of Agricultural Chemistry University College Dublin Dublin Ireland
Prof. Dr. P. Linko	University of Technology Helsinki Finland
Dr. G. Londahl	Frigoscandia AB Fack Helsingborg Sweden
Prof. T. Lovric	Faculty of Technology Zagreb Yugoslavia
Prof. D.M. McAleese	Agricultural Chemistry Department University College Dublin Dublin Ireland
Mr. P. McAllister	National Science Council Dublin Ireland
Mr. P.J. McArdle	Department of Agriculture Dublin Ireland
Dr. R. McCarrick	Irish Meat Producers Limited Leixlip Co. Kildare Ireland
Mr. J. McCarthy	Mitchelstown Creameries Co. Cork Ireland

Mr. P. McGovern	Industrial Development Authority Dublin Ireland
Mr. B. McKenna	Agricultural Engineering Department University College Dublin Dublin Ireland
Mr. N. MacLiam	Department of Industry, Commerce and Energy Dublin Ireland
Prof. J.V. McLoughlin	Physiology Department Trinity College Dublin Ireland
Mr. M.C. Manahan	Department of Industry, Commerce and Energy Dublin Ireland
Dr. E. Markham	Malt and Hops Department A.J. Guinness Son & Co.Ltd. Dublin Ireland
Mr. M. Madden	Department of Agriculture Dublin Ireland
Mr. E. Magner	Department of Agriculture Dublin Ireland
Prof. Dr. D. Martin	Institute of Food Science and Technology Edif. del Instituto del Frio Cuidad Universitaria Madrid Spain
Dr. C. Mercier	INRA Centre de Recherches de Nantes Nantes France
Prof. Dr. P. Mildner	University of Zagreb Laboratory of Biochemistry Zagreb Yugoslavia
Dr. J.H. Moore	The Hannah Research Institute Ayr Scotland

Dr. F. Moran	Science Division Institute for Industrial Research and Standards Dublin Ireland
Mr. K. Moyles	City Analyst's Laboratory Eastern Health Board Dublin Ireland
Prof. J.A. Munoz-Delgado	Instituto del Frio Cuidad Universitaria Madrid Spain
Dr. D. Murphy	National Science Council Dublin Ireland
Mr. M.F. Murphy	Department of Dairy and Food Technology University College Cork Cork Ireland
Prof. Ir.M. Naudts	Government Station for Research in Dairying Melle Belgium
Prof. Dr.R. Negri	Instituto Superiore di Sanita Rome Italy
Dr. R.J. Nichol	Institute for Industrial Research and Standards Dublin Ireland
Dr. S. Nielsen	National Science Council Dublin Ireland
Mr. M. O'Connell	Department of Economic Planning and Development Dublin Ireland
Mr. S. O'Connor	Department of Agriculture Dublin Ireland
Mr. J.C. O'Connor	Institute for Industrial Research and Standards Dublin Ireland

Mr. D. O'Doherty	National Science Council Dublin Ireland
Mr. M. O'Doherty	Delap & Waller Consulting Engineers Dublin Ireland
Ms. J. O'Donovan	Department of Food Technology Institute for Industrial Research and Standards Dublin Ireland
Dr. K. Oestlund	Swedish Meat Research Centre Kavlinge Sweden
Prof. C. O'hEocha	University College Galway Galway Ireland
Mr. O'Kelly	An Bord Bainne Dublin Ireland
Dr. J. Olkku	Technical Research Centre of Finland Espoo Finland
Mr. D. O'Malley, T.D.	Minister for Industry, Commerce and Energy Ireland
Mr. L. O'Toole	Department of Agriculture Dublin Ireland
Dr. T. O'Toole	Department of Agriculture Dublin Ireland
Mr. F. Ozgun	Ministry of Industry and Technology Ankara Turkey
Prof. M.E. Paneras	Faculty of Agriculture Aristotelian University of Salonika Salonika Greece
Dr. K. Partridge	National Science Council Dublin Ireland

Dr. S.M. Passmore	University of Bristol Long Ashton Research Station Bristol England
Dr. J.W.G. Porter	National Institute for Research and Dairying Reading Berks. England
Mrs. M. Proctor	The Insitute of Food Science and Technology of Ireland Dublin College of Catering Dublin Ireland
Dr. N. Rajcan	EEC COST Brussels Belgium
Prof. J.A.F. Rook	The Hannah Research Institute Ayr Scotland
Ir. D. de Ruiter	Institute for Cereals, Flour and Bread TNO Wageningen The Netherlands
Mrs. J. Ryley	Catering Research Unit University of Leeds Leeds England
Dr. D. Schlettwein-Gsell	Institute für Experimentelle Gerontologie Basle Switzerland
Dr. G. Schuster	EEC Commission Research Science and Education Directorate Brussels Belgium
Prof. Dr. W. Seibel	Institute of Baking Technology Federal Research Center of Cereal and Potato Processing Detmold W. Germany
Mr. B. Sheehan	BEMRA Dublin Ireland
Prof. Dr. J. Solms	Swiss Federal Institute of Technology Department of Food Science Zurich Switzerland

Mr. J. Somers	Food Technology Section Irish Sea Fisheries Board (BIM) Dublin Ireland
Dr. D.A.T. Southgate	Medical Research Council Dunn Nutrition Unit University of Cambridge Cambridge England
Dr. B. Spencer	Flour Milling and Baking Research Association Hertfordshire England
Prof. W.E.L. Spiess	Federal Institute for Nutritional Research Karlsruhe W. Germany
Mr. B. Sucur	UPI Sarajevo R & D Institute Ilidza Yugoslavia
Prof. E.C. Synnott	Dairy and Food Engineering Department University College Cork Cork Ireland
Prof. E.L. Thomas	National Institute for Research in Dairying Reading England
Dr. R. Tiebach	Max Von Pettenkofer-Institut Des Bundesgesundheitsamtes Berlin W. Germany
Dr. Miroslava Todorovic	University of Novi Sad Novi Sad Yugoslavia
Mr. O. Tolboe	Jydsk Teknologisk Institute Arhus C Denmark
Prof. G. Tomassi	Instituto Nazionale Della Nutrizione Rome Italy
Mr. D. Twomey	Department of Food Technology Institute for Industrial Research and Standards Dublin Ireland

Ms. G. Ni Uid	National Science Council Dublin Ireland
Mr. P.K. Upton	University College Dublin Veterinary College Dublin Ireland
Dr. K. Van Derkooi	NRCO TNO The Hague The Netherlands
Prof. G. Varela	Instituto de Nutricion (CSIC) Facultad de Farmacia Cuidad Universitaria Madrid Spain
Prof. V. Veinoglou	Higher Agricultural School of Athens Greece
Prof. E. von Sydow	The Swedish Food Institute Fack Goteborg Sweden
Dr. M. Upton	University College Dublin Ireland
Prof. G.F. Vujicic	Department of Dairy Science and Technology Novi Sad Yugoslavia
Dr. W. Vyncke	Fisheries Research Station Ministry of Fisheries Ostende Belgium
Mr. M.P. Walsh	CBF - Irish Livestock and Meat Board Dublin Ireland
Mr. T. Walsh	Department of Economic Planning and Development Dublin Ireland
Dr. V. Wenner	Nestle Products Technical Assistance Co.Ltd. La Tour de Peilz Switzerland

Prof. Dr. F. Wirth   Bundesanstalt für Fleischforschung
                          Kulmbach
                          W. Germany

Dr. M.J. Woods   Associated Producer Groups Limited
                          Dublin
                          Ireland

Dr. R. Zacharias   Institute of Domestic Economics
                          Stuttgart
                          W. Germany

Proceedings prepared by:
JANSSEN SERVICES, 14 The Quay, Lower Thames Street, London, EC3, UK